The Story of Garum

W0018757

The Story of Garum recounts the convoluted journey of that notorious Roman fish sauce, known as *garum,* from a smelly Greek fish paste to an expensive luxury at the heart of Roman cuisine and back to obscurity as the Roman empire declines.

This book is a unique attempt to meld the very disparate disciplines of ancient history, classical literature, archaeology, zooarchaeology, experimental archaeology, ethnographic studies and modern sciences to illuminate this little understood commodity. Currently Roman fish sauce has many identities depending on which discipline engages with it, in what era and at what level. These identities are often contradictory and confused and as yet no one has attempted a holistic approach where fish sauce has been given centre stage. Roman fish sauce, along with oil and wine, formed a triad of commodities which dominated Mediterranean trade and while oil and wine can be understood, fish sauce was until now a mystery.

Students and specialists in the archaeology of ancient Mediterranean trade whether through amphora studies, shipwrecks or zooarchaeology will find this invaluable. Scholars of ancient history and classics wishing to understand the nuances of Roman dining literature and the wider food history discipline will also benefit from this volume.

Sally Grainger is an independent scholar with degrees in ancient history and archaeology. She is a food historian, chef and experimental archaeologist. She has worked with many university institutions and museums helping to interpret the foodways of ancient societies. She has published widely in food history, and jointly with Andrew Dalby she wrote the acclaimed *Classical Cook Book*, and with her husband Dr Christopher Grocock she edited and translated the recipe text known as *Apicius*. She continues to collaborate with archaeologist in research into the various ways in which ancient fish sauces were made, traded and consumed.

The Story of Garum

Fermented Fish Sauce and Salted Fish
in the Ancient World

Sally Grainger

Routledge
Taylor & Francis Group

LONDON AND NEW YORK

First published 2021
by Routledge
2 Park Square, Milton Park, Abingdon, Oxon OX14 4RN

and by Routledge
52 Vanderbilt Avenue, New York, NY 10017

Routledge is an imprint of the Taylor & Francis Group, an informa business

© 2021 Sally Grainger

The right of Sally Grainger to be identified as author of this work has been asserted by them in accordance with sections 77 and 78 of the Copyright, Designs and Patents Act 1988.

All rights reserved. No part of this book may be reprinted or reproduced or utilised in any form or by any electronic, mechanical, or other means, now known or hereafter invented, including photocopying and recording, or in any information storage or retrieval system, without permission in writing from the publishers.

Trademark notice: Product or corporate names may be trademarks or registered trademarks, and are used only for identification and explanation without intent to infringe.

British Library Cataloguing-in-Publication Data
A catalogue record for this book is available from the British Library

Library of Congress Cataloging-in-Publication Data
A catalog record has been requested for this book

ISBN 13: 978-1-138-28407-4 (hbk)
ISBN 13: 978-0-367-68312-2 (pbk)
ISBN 13: 978-1-315-26982-5 (ebk)

Typeset in Sabon
by KnowledgeWorks Global Ltd.

Marito caro et mihi et huic operi devotissimo

Contents

Figures

Tables

Preface and acknowledgments

This book has been in part a labour of love, and in part an obsession with this obscure and little-understood commodity of the Roman world. The research came out of a need to understand fish sauce myself but also to allow others to see and taste and understand ancient fish sauce in all its manifestations. It was just a jumble of disparate forms of evidence, from bones, amphora and empty vats, and it now has a visceral reality – pun intended!

Many people have contributed and encouraged me over the years in this journey. First and foremost, my husband Christopher Grocock for his support and devotion, Andrew Dalby who took me under his wing, when the journey into ancient food began, and especially Tom Jaine at Prospect Books. Thanks are due to my MA adviser Gundula Muldner at Reading University, the library staff at the Hellenic and Roman Library and at the Sackler at Oxford. I am extremely grateful to luigia Melillo and Grete Stefani for allowing me permission to visit the Pompeii archaeological store and to the Museo Archeologico Nazionale di Napoli to view the collections related to fish sauce. I could not have begun without the many conversations that I have had with like-minded food historians and archaeologists. I am grateful for their advice and encouragement: Dimitra Mylona, Rebecca Nicholson, Angela Trentacoste, Robert Curtis, Edward Biddulph, Paul Roberts, Alessandra Pecci, John Wilkins, Lee Grana Nicolaou, Sheila Hamilton-Dyer, Alison Locker, Andrew Jones, Ilias Anagnostakis, Alfredo Carannante, Dario Bernal-Casasola, Emmanuel Botte, Cuong Pham (Red Boat fish sauce). Nevertheless, all errors are mine own.

Sally Grainger
June 2020

Introduction

General introduction

From a purely literary perspective, the elite Roman was a passionate consumer of the rare, novel and therefore luxurious. The layman will have in mind such things as eating dormice and larks' tongues, along with many other amorphous yet disturbing forms of obscure offal. Over-indulgence and gluttony are concepts which we can understand, and prized offal can be delicious, but many people are disturbed and confused by the idea that the process of fermenting fish offal could produce something desirable. The scholar and layman alike have always had their doubts about this particular obsession. The myth of the 'rotten Romans' consuming their 'rotten fish sauce' is sadly all too often found in food history, while the lay public interested in food have moved on. The ubiquitous fish sauce of ancient cuisine and trade is now recognised as essentially the same as the South East Asian sauces that are valued for their transformative qualities in culinary circles. However, even a recent publication on food studies maintains that the very idea of fish sauce, which is defined as smelling of rotten urine, was invented as a covert strategy to maintain class boundaries. Effectively, this argument is that Roman food was made to be purposefully 'bad' so as to exclude those who would naturally reject such flavours (Feldman 2005; 2014: 416). Feldman maintains that those who did accept fish sauce in their diet did so *despite* its unpleasant taste, in order to be accepted in an elite culinary club, a kind of inner circle of Roman gastronomy. It is little disturbing that such an idea can still be found in print, and the very idea of a kind of exclusive Roman 'bad food club' seems far-fetched. Feldman only understands *garum* in terms of the luxury end of the market and has no understanding that fish sauce was also cheap and ubiquitous.

Ancient and modern fish sauces are magical commodities that, when used with experience, have the potential to transform everyday foodstuffs. In ancient Rome they were used extensively across all social classes and did not taste bad in any definition of that term: they were certainly 'different', 'striking', 'piquant', full of 'umami', 'strong' and 'savoury' of course compared with modern Northern European and North American cuisine but not 'bad'. The idea of 'bad' taste' is subjective, up to a point: an uneducated palate raised on the bland and feeble flavours of the West may reject strong and pungent foods, while someone raised on salted and fermented flavours will undoubtedly do the opposite and reject the blandness of Western food. It is all about personal taste and a naïve palate will often reject what is simply unfamiliar. Some types of fish sauce were clearly very popular, and along with oil and wine were consumed widely across the entire Roman world; their popularity is attested by the presence of abundant amphora finds where the Romans colonised and settled. Fish sauce amphorae are often found alongside wine amphorae (and before oil vessels arrived) at Roman colonial settlements of military origin (Martin-Kilcher 2003:79). These commodities must have represented an essential food resource to the military and from amphora distribution we have to assume that these products rapidly become popular among natives as well. Crucially, we cannot distinguish between a solid salted fish and a sauce in terms of amphora shape, and this is a fundamental issue as salted fish is clearly a more straightforward commodity. The concept of *garum*, however, was very complex: on the one hand, the term was used for something both elite expensive and valued and at the same time was rejected and viewed with suspicion by many writers, who appear entirely conflicted by the very idea of it. On the other hand, *garum* also apparently designated a product that was commonplace. In fact, ancient fish sauce did have a dual identity, and this is because it came in two forms. There was the original, basic idea of fermenting small fish with salt into a liquor that resembled Southeast Asian sauces, known as *garos* in Greek and *liquamen* in Latin. This was so ordinary as to be barely acknowledged by the elite. But there was also something quite different: *garum* proper, the fermentation of fish blood and viscera, the very nature of which was and is inherently disturbing, though its resulting product was viewed as exquisite. Hence, it was both valued by the elite and regarded with great suspicion. As we see from the archaeological evidence, it was the simple *liquamen* that was traded widely in the empire and became essential to the Roman way of life, while there is some doubt regarding how widespread the other elite blood viscera *garum* actually was, and is very much more difficult to comprehend in literary and archaeological sources.

The idea that *liquamen* and *garum* were 'rotten' to the taste, as summarized by Feldman (2005), discussed earlier, is largely based on the idea that all fish sauce was as bad as the blood viscera *garum* as perceived by ancient and modern commentators. This is still very problematic for archaeological and historical approaches to ancient fish sauce, as the negative image of *garum* colours how all fish sauce is studied. It is my contention that making a distinction between *garum* and *liquamen* is fundamental and essential to any discussion of the nature of the fish sauce trade. This distinction, however, that I have set out here, is not universally accepted; many scholars do not consider that there were two distinct types of sauce. If anything, it is this failure to distinguish the various types of sauce that has been the main motivation for my research and this book. The arguments are immensely complex, and the story of *garum* needs unpicking very carefully to reveal its dual nature in the ancient sources. It is fair to say that the use of some form of fermented fish sauce was a defining characteristic of the ancient Mediterranean cuisine that we think of as Roman. These sauces were used across the social classes: rich and poor consumed their daily food seasoned with a form of salted fish liquor. A poor student living on an entirely vegetarian diet consumed his lupins and beans with *garos* ('of course') and with oil according to Galen (*On the Properties of Foods* 1.25.2). *Garos* was so commonplace that Galen emphasises that even the poorest has access to it. The use of this product in cooking seems to have begun in earlier Greek culinary practices, where originally a salted fish brine called a *halmer* was utilised. This was the brine generated from the salting of cleaned fish to create a solid product. *Halmer*, later *muria* in Latin, was a sauce mild but salty in taste as it contained no digestive enzymes and was thus not fermented, a necessary stage in the process of making a dissolved fish sauce. At the same time as this sauce was on the tables of the elite, a commonplace *garos* made from tiny fish was consumed by the poor in Greek society. The origins of *garos* are discussed in detail in Chapters 1 and 2 and *passim*. We often think of *garum* as Roman because it was under Roman influence that these fish sauces were manufactured on a huge scale, and it was the Romans who almost certainly introduced the new and more fashionable types of sauce made with species-specific fish and brought sophistication to their consumption, but *garos* was essentially Greek in origin and also quite cheap and unsophisticated in its first manifestation. It is also apparent that it was through Greek cookery books that fish-sauce use in an emerging 'cuisine' spread across the Mediterranean.

The question that seems to dominate in archaeological circles and which is at the heart of the dilemma we have in understanding ancient fish sauce is this: should we consider 'fish sauce' in the same way as the other amphora-borne commodities? Wine and oil are simple and uncomplicated products: an 'essential substance', if you will. We know that wine is fermented grape juice and oil comes from fruits, seeds and nuts ground to extract a liquid fat. We are comfortable with the idea of oil and wine as they correspond to the knowledge and experiences we have in the modern world. Each of these commodities can have a multitude of varieties and qualities within the basic definition, but nevertheless we understand them as 'essential substances'. Did the ancients have a similar approach to fish sauce? We are not so confident about fish sauces because in the West such things have never been part of our culinary repertoire. We cannot comprehend them, as we do not know what they look like in production and in use, despite a more informed view from modern South East Asian sauces, modern culinary uses and recent experimental archaeology. Should we consider that the differences in variety and quality were more fundamental? This question seems to be at the heart of the dispute in academia over the very nature of *garum*.

Rotten fish viscera sauce

It is not surprising that doubt persists among lay and academic circles, as the ancients themselves were rather conflicted and confused about the identity and processes involved in making *garum*. Seneca is often quoted: 'do you not realize that *garum sociorum*, that expensive bloody mass of decayed fish, consumes the stomach with its salted putrefaction?' (*Ep. 95*). Seneca is almost certainly talking about the fermented blood and viscera sauce here, and he clearly doubts its value. Many have assumed that this negative attitude from Seneca also applied to the ordinary fermented fish sauce, that is *liquamem,* but I think this would be an error. Seneca is passing comment on the over-indulgence and gluttony of elites in demanding a fish-blood sauce. The amount of blood that one can harvest from mackerel or mullet is tiny and measurable in ml, while one needed immense quantities of mackerel to harvest enough viscera to generate a true *garum*. *Liquamen* on the other hand utilised small, otherwise undervalued, but not worthless, fish that actually represented a very cheap and commonplace commodity in terms of fish consumption and in stark contrast to *garum*. In archaeological circles, the 'single sauce hypothesis' has been formulated to understand fish sauce and it has led to the belief that the cheap and commonplace, that is, little fish, were blended with what was perceived as the rare and exotic but actually rubbish, that is, blood and viscera, to make an all-encompassing *garum*. None of the ancient didactic recipes or detailed literary sources indicate that this happens, but it is nevertheless what is believed (Bernal Casasola et al. 2014, 2016; Palacios 2016:92; Rodríguez-Alcántara et al. 2018:150). This belief in a single sauce has led to a reduction in the significance of the blood viscera *garum*, in some cases denying its very existence altogether.

We must dispel the myth of putrefaction at the outset. The salt levels and pH of fish sauce manufacture contribute to a bacterially safe environment. The bacteria that are found in these conditions thrive in salt and produce lactic acid, which ensures that any pathogenic bacteria are destroyed. Additional acid in the form of wine in one recipe also contributes to a higher pH. The process of fermenting fish to make *liquamen* is allied to the processes involved in fermenting all kinds of foods such as sourdough bread, beer and fermented cabbage such as sauerkraut and kimchi. These commodities exploit the benefits of friendly lactobacillus to break foods down into their component parts and are valued today for their health-giving properties. However, it appears that is was primarily enzyme hydrolysation rather than bacterial fermentation that resulted in the dissolved fish flesh yielding its nutrients into the liquid brine. It is rare for fish sauces to go off, as the salt levels are too high, but they do oxidise and become darker and more pungent. A modern gourmet fish sauce such as Red Boat (see page 134) is naturally quite sweet smelling, while an old and stinky bottle of squid brand left in the back of the cupboard for a year is pungent and verging on the unpleasant, though not in any way dangerous to health and still usable. Ancient fish sauces utilised fish viscera extensively, both as an addition to a whole-fish sauce to aid the enzyme digestion process and as a standalone product in its own right with the addition of fish blood.

The Seneca quote above is specifically referring to this special and often expensive form of fish sauce, known by satirists as *garum sociorum*. The definition of *sociorum* is disputed (Leon (2001:176; see the discussion on page 28), and seems to have evolved over time. Its initial meaning seems to have been a blood viscera sauce made from mullet; later mackerel was incorporated in the definition. Pliny recounts that one had to pay 1,000 SH for 6 pints of this *garum sociorum* and that no other product apart from unguents had come to be more valued (*HN.* xxxi.93). These prices are in stark contrast to the more modest prices for fish sauce that we find in Diocletian's price edict, which demonstrates that the ordinary *liquamen* was within the means of most ancient consumers. The more expensive and selective *garum* sauces were described as 'black' and 'bloody' by later medicinal writers and Pliny describes the fish sauce he understands as being like an aged honey wine, that is, a pale amber yellow or light brown. The Greeks also use terms like '*loukos*' meaning white but also bright and light to describe *garos* and this is occasionally contrasted with *melanos* meaning black (page 95). This colour distinction would seem to be crucial to understanding fish sauce and was also important in allowing the ancient purchaser to identify products. Modern fish sauces can be many shades of pale yellow to dark brown depending on which species is used, the processes involved and how long they are stored and aged.

What's in a name?

The leading scholar in fish sauce, Robert Curtis, once said that understanding fish sauce terminology was like pinning jelly to the wall: you think you have understood the basic terms and what they seem to be and it does initially seem simple enough, but the peripheral evidence from satire, and the late Roman and Byzantine sources provide a very different picture. As the sauce made of whole small fish resembled South East Asian fish sauces such as *nam pla* and *nuc nam*, and the more elite blood viscera sauces were seemingly distinctive by colour and expense, it should be easy to separate these ancient sauces out by colour and quality and attribute them to the fish sauce terminology that survives, *garum* and *liquamen*, but for reasons that will become clear this has proved very difficult. We do not even have consensus as to whether the terminology for fish sauce should have a universal meaning that is consistent across time and space. Should we assume that the labels on amphorae, known as *tituli picti*, had the same meaning at the beginning of the trade in fish sauce as at the end? Those ancient people who needed to use fish sauces in their daily work, principally manufacturers, traders and the people who bought these products, namely cooks, veterinarians and doctors, would have needed to be able to trust the terminology to know what they had purchased. The terms simply cannot be as fluid as some scholars appear to want them to be. Currently, the general consensus among archaeologists and historians is that there was a single substance which was a fish sauce named *garos/garum*, and this term is used to name that substance in the early Greek/Roman period. This theory determines that at a later date the use of the term *garum* went out of fashion for some reason and was replaced by *liquamen*, the universal term for fish sauce in the later Roman period. This 'single-sauce hypothesis' is still commonplace in scholarship, but it is unsatisfactory as a way of understanding the nature of these products, as it is possible to see at least three types of sauce named as *garum* by production methods alone, and by species utilised there are were even more; these distinctions needed to be identifiable to the trader, merchant and purchaser. This chronological separation of the terms into 'early' and 'late' usage in literature is simply not accurate, as we can find plenty of evidence for the continued use of *garum* in late Latin didactic sources. It is true that the 'later' term for fish sauce, *liquamen*, is absent in an early Roman context and is always considered obscure by modern scholars (Curtis 2016:175). Crucially, for the 'single-sauce hypothesis' to work effectively, *liquamen* would have to have functioned as a generic term in the later empire, in same way that *garum* appears to do in the early period, and it clearly did not: the blood viscera sauce is nowhere described as a *liquamen*. The answer to this dichotomy is relatively simple and concerned largely with the invention and short-lived popularity of the blood-viscera *garum*. The elite expensive blood-viscera *garum* was immensely popular in the 1st century AD but thereafter becomes less valued and much less visible in both texts and archaeology. The obvious decline in the blood viscera sauce in the late period may ultimately be connected to the blood prohibition found in Rabbinic sources, the early church and the Greek Orthodox church (page 108). One simple explanation for the confusion in later texts might be that, if the expensive *garum sociorum* was not longer in regular use, people began to revert to using *garum* for the original product, that is, the Greek idea of *garos/garum*.

The multiple definitions of the key terminology for fish sauce remains has led to such an intricately entangled quagmire of debate that it would seem almost impossible to disentangle; it has become a Gordian knot! This author believes that the knot is largely of our own making and that a radical rethink is necessary. I want to go back and start again if you will to a basic review of all the ancient literature and recognise where the confusion lies. I have tried to do this with an entirely open mind, setting aside most of the long-held preconceptions as to what these sauces were and re-engaging with them afresh. I have found that with this approach, it is possible for the knot to disentangle itself.

Fish sauce in the kitchen

The modern culinary world has embraced umami-rich products like soy sauce, kimchi and the Thai and Vietnamese *nam pla* and *nuc nam*. What is not fully understood is precisely why these similar products were so desirable and successful in the ancient world. We can be fairly certain that in the case of the Northern provinces, Gaul, Germania and Britannia, fish sauces were unknown before Roman contact and it is precisely these provinces that appear to have embraced fish sauce use enthusiastically. The distribution of Spanish fish sauce amphorae in the northern provinces speaks of a huge trading infrastructure set in place to service a great economic and social need for these products. It is generally understood that this enthusiasm for fish sauce was initially reflecting the invaders consumption, and this was followed by native elites aping Roman food-ways. It was only later that the practice of using fish sauce filter down to the less wealthy natives. There is a possibility that the huge increase in trade in these products that occurred from the late 1st century BC through to the early 2nd century AD was prompted not by an increase in elite use but a commercial pull from beneath: fish sauce had a transformative effect on the diet of ordinary people in the provinces. Adding cheap and readily available *liquamen* sauce and even the less nutritious *muria* to their daily boiled meal made from pulses and vegetables with minimal meat content that served as the staple for most people would have transformed the taste of these foods. It is likely that eating food with just salt and without these magical ingredients would very quickly be unthinkable.

Among chefs and cooks today, it has long been understood that fish sauces are magical ingredients that can transform bland foods. The magic is known as umami, and it gives an enhanced taste perception in the mouth. Monosodium glutamate is a naturally occurring substance in lots of foods, including mushrooms, tomatoes, dry-aged meat, soy sauce, parmesan and of cause anchovies. The compound glutamic acid also functions as an excitatory neurotransmitter meaning at a basic level our brains are excited by the taste of food that has glutamates added or in other ways included or generated. In general, the more umami present, the more our perceptions of flavour are heightened, making food more satisfying. Umami also alters the perception of other tastes: salt is enhanced, sugar is sweeter, and sour and bitter flavours are perceived as more pleasant. The effect is one of balancing the numerous flavours that are found in ancient recipes. Roman and Greek recipes are notorious for what has been perceived as the over-use of spices. These multiple flavourings have been misunderstood by modern commentators who make assumptions about the quantity of spices used in recipes that have no quantities indicated. Some of the spices used can be sharp and acrid, particularly if used without care and can nevertheless create tastes that are perceived to be bitter and certainly out of balance and 'discordant'. Fish sauce brings these diverse flavours into a surprising harmony that can only be really understood if a 'before' and 'after' taste test is undertaken. Fish sauce is even utilised in dishes that we might consider a dessert, where it is also remarkably successful. A popular Roman dish in *Apicius* is a *patina* (frittata), a recipe for a mixed nut omelette of eggs, almonds, hazelnuts, pine nuts, pepper, sweet wine, honey, milk and fish sauce, and which formed the basis of a simple 'taste/preference' experiment which I conducted at an English Heritage site. This was conducted with a group of young volunteers whose palates had not been too conditioned and whose judgement was readily offered! The simple procedure of splitting the mixture into three dishes and seasoning one with fish sauce, the other with a little salt and the final one with nothing was quite a revelation. All the children who took part in the tasting selected the one made with fish sauce as its simply tasted better. The flavours had become more complex, the dominant flavour of eggy sweetness had been pushed to the back of the flavour spectrum and one could taste the individual nuts, pepper and wine first. It was simply a better experience and the children knew it. Some of the children found the knowledge that the fish sauce version was better rather difficult to process. It was clear that even with the logic of a better taste experience their negative preconceptions coloured how the food was experienced. In the West there is always going to be something wrong with there being so much fish in ancient food. There is therefore a stark contrast between the perception of fish sauces as something bad and the reality that they were (and are) in fact, products with unique qualities to transform everyday foods.

The structure of the book

Archaeological scholars studying the ancient trade in fish products could be forgiven for being daunted by the enormity of the task, as Dario Bernal Casasola has pointed out (2016:187). There are 10 interconnected but largely isolated disciplines that must be brought together and studied holistically in order to bring any kind of fully integrated picture of ancient fish sauce to life.

1 Zooarchaeology: the identification of fish bone residues and classifying them as sauce or *salsamenta*.
2 Ceramic archaeology: the production, typology and distribution of fish amphorae across the empire and the data concerning 'consumption' of the products they carry.
3 Archaeology of production sites: the structure and function of the fish processing factories.
4 Marine archaeology: the study of shipwreck data where fish products are found in amphorae at the bottom of the sea.
5 Archaeology of consumption sites: the study of ports, trade routes and markets in the Roman empire linked to the disposal of residues.
6 Epigraphy: the study and translation of *tituli picti* for fish products.
7 Classical elite literature: consumption evidence from Roman and Greek drama, satire, letters and natural histories.
8 Ancient didactic literature: the recipe literature from veterinarians, doctors and cooks.

9 Modern fish sauce: scientific research into fish sauce production in South East Asia.
10 Experimental archaeology: duplicating ancient recipes to reproduce authentic sauces.

It goes without saying that no one person could possibly possess sufficient expertise in all these disciplines to deal with all the material at an expert level. The task of bringing them all together would seem almost impossible in terms of the logistics alone. It is also challenging in the sense that there are good reasons why some of these separate approaches to this study into ancient fish sauce are not often brought together. For example, that pure archaeology does not fit well with the study of Roman satire is not surprising. What I am planning is immensely challenging because in order to achieve my goal I must extract only that which is essential from each discipline to carry an understanding of fish sauce forward through the book. My contribution to this topic is proudly unconventional in that it tries not to use the inaccessible conventions of much of academia. My approach is also unique in that wherever possible the practical and empirical perspectives of these products are the primary focus. I have degrees in ancient history and archaeology, but I began this journey as a chef and a sense of the look, taste and smell of these products are at the forefront of my research and is hugely valuable to a holistic approach. I firmly believe that it is advantageous to view fish sauce holistically, rather than, as is often the case, as separate entities depending on who is commentating and from within which discipline. Each individual approach conjures up a subtly different image of what ancient fish sauces were like so that it is often impossible to argue across disciplines or move forward with new ideas. I am fully aware that leaves me open to the criticism that I am claiming that no one but I can understand fish sauce, but nevertheless – while I remain open to such criticism and may have to adapt my conclusions with fresh evidence or argument – the attempt made here to comprehend ancient fish sauce from this holistic perspective is a unique one and in my view the only way to solve the problems that we face.

It has proved extremely useful to have that empirical perspective, developed through experimentation, as often complex questions concerning the characterisation of the products, how they inhabit the various vessels and how they appear at point of consumption, can only be understood with this perspective. There is a legitimate argument to be made that, in looking at the bigger picture from too far away, one is liable to miss the fine detail. That should not deter one from approaching the subject holistically, but it is a disadvantage. I shall no doubt be challenged and/or disregarded for missing some fine detail or being perceived as failing to approach these topics with the required academic rigour. That will have to be accepted. When I first broached the idea of this project, a colleague made the disturbing suggestion that it would be impossible to unite all the material. Much of the minutiae of detail resulting from the skills and expertise within each discipline, which scholars within their own field value and respect, may have to be recast and viewed with a fresh angle to achieve the current objective. This book may therefore disappoint specialists in their respective fields. Including all the possible details from every quarter would have resulted in a concoction which was indigestible in a way its subject-matter was never intended to be. There were areas in which a collaborative approach was absolutely necessary and I have developed connections with leading specialists in zooarchaeology to provide advice, and special thanks are due to Dimitra Mylona and Rebecca Nicholson in this regard. That the separate disciplines are isolated and inaccessible to each other is not surprising. Fish bone specialists deal in the literal minutia of species and size identification and cover many eras and cultures and inevitably have a limited concept of literary context to the product they are attempting to identify. There are some valuable exceptions such as Mylona (2008). The ceramicist's primary purpose is to identify the complex typology of amphora shards and to map their production and distribution while the wider issues of the shape and corresponding function of amphora has only recently become an area of study. We are still compelled to talk in terms of 'fish amphora consumption' without any real comprehension of what product was being traded in them and 'consumed' (Martin-Kilcher 2003). The processing sites are being excavated with increasing precision and though we are able to map the factories and calculate capacity of the tanks, there is still great doubt as to how they were precisely used. Shipwrecks also provide enormous amounts of macro-scale data on amphora typology and trade routes for a wider understanding of the Roman economy (Scheidel 2012; Wilson 2007). The problem is fundamentally one of too much information leading to more and more complex discussions about the minutiae of the issues, while the bigger picture of what these products looked like and how they were distinguished, the 'whole', remains a blur. Dario Bernal describes the situation we face as one of 'hyper specialisation' within each discipline and 'excessive generalisation' outside of each discipline (Bernal 2016:187), which is very astute.

It is almost impossible to find a specialist willing to step outside of their comfort zone and fully engage in a multidisciplinary research, which often means they are unaware of the debates and disputes among the other scholars in other areas. There is every possibility of errors and misunderstanding going unchallenged and becoming accepted fact, which in turn becomes more and more difficult to challenge. In what follows I will challenge many long -held beliefs about the nature and origin of fish sauce and many will find these ideas difficult to accept and prefer not to abandon the received tradition. This is a radical re-think in which none of the facts are off the table. Any radical challenge to received tradition runs the risk of being charged with a circular argument: one's premise is often formed within one discipline, the proof of which has to be chased across numerous boundaries in research, and fellow researchers are often unwilling to follow an argument that leads them into other areas where they lack the confidence (or foolhardiness?) to follow. In answer to the charge of a circular argument, I advocate looking at all the evidence using the only permissible circular argument in logic, known as the 'hermeneutic circle' as expounded by Friedrich Schleiermacher. Dilthey says of this theory that 'meaningfulness fundamentally grows out of a relation of part to whole that is grounded in the nature of living experience' (Palmer 1969:120). An approach based on this model, following Schleiermacher, to the interpretation of a text, focuses on the importance of the interpreter understanding the text/object as a necessary stage to interpreting it. Understanding involves repeated circular movements between (all) the parts and the whole, and the experience is fundamentally empirical in nature. The whole is the reality of the solid multi-faceted existence of fish sauces in the ancient world, the parts are the study of the various forms of residual evidence of that existence that survive today, and each and every one has to be fully understood empirically before any real understanding of the whole is possible. The main aim of the book and the long-term research project has been to incorporate amphora studies, epigraphy, fish bones, ancient texts and experimental archaeology to create a theoretical framework, a model of the fish sauce trade in the Roman Mediterranean. I will begin from the premise that a rational system of nomenclature was set in place by the manufacturers when these sauces became popular, but it may have been poorly understood right from the beginning by the ancient consumer as it was ultimately designed by those who managed the trade. The alternative approach is to assume a chaotic system right from the start, where each region applied the terminology for salted fish products indiscriminately according to its own rules and a rational and coherent system of nomenclature was inherently impossible. It goes without say that the latter is unacceptable as a means of understanding these issues. Given the scope of this book, it would be impossible to do justice to the vast amount of research that has previous been published on the ten separate disciplines that inform its subject-matter. A brief survey of the material will have to suffice. Equally, within this volume, some aspect of the traditional approach to fish sauce related studies will have to be dealt with fairly briefly, in order to find room for the necessary debate. The minutiae of amphora typology and, sadly, much of the extensive distribution and consumption data cannot be incorporated into this volume. Amphora will be dealt with only in terms of their functionality. It is hoped that others better equipped can manipulate such data into the theoretical model of fish sauce production that will be presented here.

Chapters 1–5 deal with the ancient literature for fish sauce, while Chapters 6–12 approach the topic from an archaeological perspective. It was my original intention to try and combine all the relevant evidence from each of the 10 disciplines in discussion defined by the practical issues such as nomenclature, consumption, disposal, trade, etc., but this has proved almost impossible. There are contentious issues everywhere you look within a discipline and between disciplines, which include disputed terminology, controversial interpretations and inaccurate translations which need to be clarified before any degree of clarity is possible. With this in mind, I would recommend that readers do not dip into this book but begin at the beginning and take the journey through the evidence. If you do dip, be aware that you may miss key arguments that are necessary to grasp before you reach the issues you are seeking and also there is the possibility that you will lose the thread of the argument, and give up, before you get to the end! For reasons that will become clear I have postponed the traditional literature review where previous scholarship is discussed until the archaeological section begins at Chapter 6. The book begins with Chapter 1 and a detailed re-evaluation of the key ancient didactic texts, which discuss the recipes themselves and the fundamental nomenclature issue is outlined. Here I review the issues of handling ancient literary sources for Roman foodways and consider to what extent elite sources can and should be utilised to illustrate actual practices. Chapter 2 is a diachronic account of fish sauce from 5th-century Athens through to the late Roman Empire in the 5th century AD. This is divided into sections and at each stage the basic archaeological accounts that correspond to or contradict the literary evidence are outlined. Chapter 3 gives an account of the practical uses of fish sauce in recipes and remedies. Chapter 4 examines the hugely valuable and unique source of information from Greek and Egyptian papyri on the everyday consumption of these products within middle-rank households. Chapter 5 provides an account of how fish sauces evolved and changed in late Roman, Byzantine and early modern world. Chapter 6 comprises the postponed literature review in which the scholarship from each of the main European countries that either produced or consumed fish sauces is described from the traditional archaeological perspective. Chapter 7 deals with modern fish sauce in South East Asia and the different methods employed to make the numerous forms of fermented fish products. Chapter 8 includes modern fish sauce experiments conducted by the author and others. Chapter 9 is an overview of fishing in the Mediterranean in relation to the species associated with fish sauce and salted products. Chapter 10 deals with the infrastructure of fish sauce manufacture, based on evidence for the large-scale production process in *cetariae* and small-scale production process from dolia and bottles and jars. In Chapter 11, I examine the numerous studies of fish sauce residues found associated with amphorae, wrecks and processing sites. Chapter 12 deals with the amphora typology and the ways in which amphorae may have been used to accommodate fish sauce and here the inscriptional evidence from amphorae *tituli picti* is incorporated. The 'Afterword' attempts to round up the ideas and conclusions found in the previous chapters.

1 Fish sauce in classical literature

Dealing with the ancient literature: methodology

There is considerable information about consumption practices contained within the world created by poets and prose writers. The question as to whether the imaginary scenario found in fictional accounts of ancient life is useful historically is rather more difficult as there is a natural doubt that fiction can reflect actual practice. Classicists such as Gowers in her innovative work *The Loaded Table* (1993) reject the idea that actual practice could be found in these ancient forms of fiction. She seeks for the hidden meanings, which are by their very nature subjective. It is of note that historians of Roman social history and classicists have not shown a great deal of interest in the minutiae of consumption practices. I suspect the current perception of *garum* as a bit odd, certainly rotten and stinky is one that scholars have no desire to change. It seems to fit the rather distorted image of Roman decadence and gluttony that is the standard view, yet there is also considerable confusion as to why *garum* was even desirable. Rarely has anyone attempted to engage with the elite consumption literature, largely satire, from a practical perspective and discussed the wider context of the practicalities of cooking and dining. The valued and much-admired work by Robert Curtis, *Garum and Salsamenta in Materia Medica* (1991) did recognize the value of fish sauce and also demonstrated a multidisciplinary classical and archaeological approach but still managed to relegate the majority of elite textual evidence to the (albeit extensive and comprehensive) notes, where some of the anomalies were noted but not addressed. For Curtis' work and the other key commentators see Chapter 6. The fields of Roman archaeology and classical studies are worlds apart in terms of methodology and empathy. I have heard archaeologists and historians argue that the poet is compelled to write something other than what he wants to write in order to comply with what is called "stylistic and metric constraints" (Marzano 2013:70). These rules are understood to impede the flow of accurate information. Meter might limit a poet's choice of words, but not his aim at conveying meaning: these two operations are not necessarily mutually exclusive and the tendency to doubt the veracity of the poet's words is not sound thinking. At the heart of this rejection there is a simple matter of choice. Many Northern European archaeologists cannot and do not want to deal with text in the detail that is required and find excuses to remove, discard or reject the information contained within the textual form.

Classicists view the world in which the poem exists as a literary construct, in which things of fantasy happen, and one cannot determine reality from what was an entirely imaginary scene, where exaggeration, farce, chaos and the unreal are all that is presented. Emily Gowers, writing of the food in Roman satire, suggests that "as the literary sources represent a tiny fraction of the elite, they ought to make us suspicious of their historical value" and she concludes that it is a "mistake to use literary sources as evidence of what the Roman ate" (1993:2). I would fundamentally disagree. As we will see, it would be wrong to suggest that the dining practices associated with fish sauce consumption, which we can see in Horace and Martial and also the recipe texts were unique to the elite. From papyri we see that the same basic fish sauce consumption practices were utilized by all classes, and this is mirrored in ancient Greek texts and also in the Latin and Greek phrase book (see page 102). These practices involve the composite sauces, found in satire, which blend oil, vinegar and wine with fish sauce. There is evidence to suggest that these simple practices were mirrored across the social classes and across the empire in what was undoubtedly a form of "Romanization": the act of willingly copying Roman cultural practices (which the Romans leant from the Greeks) for purposes of advancement or prestige but also because these practices were appealing and desirable. In current archaeological and historical fields this term is considered unsuitable to convey the transfer of cultural practices, as its use is seen to convey colonial preconceptions of a superior culture, but I use it nevertheless, as it is entirely apposite. It conveys, without judgement, the phenomenon of behaviors identified at the center of the Roman world being practiced at the periphery of that world.

For many the information contained within a constructed reality cannot be taken as evidence of actual practice but must be engaged with using only the tools available to a literary critic. Food and dining is a loaded subject replete with metaphor to suggest political and social messages way beyond the reality of the food. What one eats is able, through culinary language, to convey identity and social status, and to contribute to the maintenance of social hierarchies. Emily Gowers has written the only entirely food-related approach to Roman satire but she has over-loaded the genre with the weight of its metaphor and rejected all the physical attributes of the food found within the poetry. The total denial of the physical world in this approach is self-defeating, as so much of the meaning that scholars seek is to be found in the empirical and physical world. Dunbabin has noted "how little is known about the normal family meal in antiquity" and stresses that the literary world is "filled with satirical exaggeration.... archaizing and idealized references" (Dunbabin 2003:3). This may be true to some extent, but I would argue that this "constructed world" is not made up only of invented exaggeration. This was always a small part of the imaginary world, but much of the content has to be real and familiar for the reader to engage with the work at all. Ancient writers did not invent a new world with imaginary practices, they quite naturally used the one that they and their readership and audience knew and understood and could relate to. This "background reality" had to be truthful in order to supply a backdrop for the satire or exaggeration. It is therefore theoretically possible to distinguish the actual practice from the exaggeration and spin, so long as the former can be identified. If sufficient knowledge of what constitutes this "background reality" in literature which stresses consumption can be established, then the anomalies, the exaggeration, can be seen in sharper focus. The flip in reality, the humor that is intended by the poet to appeal to his audience, is more effective because it is "other" than expected. If you can work out what is expected you can see what is out of place. The background reality in relation to fish sauce is the culinary world of ancient Greece and Rome, embraces many facets concerned with culinary practices, dining etiquette and how the diner engages with food and the size and nature of the serving vessels. By accumulating sufficient empirical knowledge of these areas than it is often possible to see what is normal in satire. The "normal family meal" may be largely absent from ancient literature but it is possible to see through the fog at what was happening on a daily basis beneath.

The early Greek dramatic evidence, from Tragedy and Old, Middle and New comedy, is replete with culinary references, and fish sauces figure widely in these texts. Old Comedy is dominated by Aristophanes, who melds farcical unrealities peopled by gods and heroes with the political and philosophical actuality faced by 5th century Athens, while Middle Comedy (which developed in Athens from about 400 BC to about 320 BC) represents the transition toward New Comedy, which was characterized by gentler domestic themes with characters resembling normal Athenian people (323–260 BC). Aristophanes will figure a little in what follows. In the play "Peace," first performed in 421 BC, the Gods of War have a giant mortar in which the ingredients, leek, garlic, cheese, and honey, are placed, ready to be pounded up. The ingredients represent the communities of Prasiae, Megara, Sicily, and Athens, either by similarity of name or because these were their principal traded goods; as an image of the impending damage to be done (by the pestle) to Greece from yet more conflict, it is a fantastical scenario, made comic when War's accomplice, is unable to find a pestle either from Athens or Sparta; here there is a clear reference to the hawkish leaders Cleon and Brasidas, who both perished in the fighting just a year before ("Peace", 242–288). However, at a very basic level we can also see from other sources that these and similar ingredients could actually be placed in a mortar and ground into a paste or *myttōtos* (Dalby 1996:107, 244), an *opson* or relish for bread (Columella, xii. 49;

Pseudo-Virgil, *Moretum*; *Apicius* 1. 35). For all the chaos, the comic world is very much grounded in reality. In discussions of the historical value of Archestratus and his hunt for fish in 4th century Sicily, John Wilkins has a less extreme approach. Archestratus is still a "work of instruction for the acquisition of food" and at the same time an "epic parody to entertain and inform" (Wilkins 2011:13). Wilkins has no problem in allowing it to be both. The many works of Andrew Dalby (1996, 2000, 2003, 2013) are conceived around the idea that drama poetry and prose writing in the ancient world reflected different levels of "actual practice", Why would they not? Numerous Middle and New comedies were written, but almost all are lost, and much of our knowledge about their contents comes from fragments, largely preserved, in Athenaeus' work the *Deipnosophists* (*The Learned Banqueters*; Olson 2006). Many of the fragments appear without context and they are often preserved for obscure reasons of grammar. The evidence for fish sauce in early Greek society is scarce, and largely dependent on many of these fragmentary sources; consequently, the picture of fish sauces in 5th and 4th centuries Athens is not complete, and there are many difficulties to extracting any useful evidence. But it is possible to extract something and to suggest scenarios that may or may not be corroborated from the Hellenistic/early Roman sources. According to Wilkins "things of the ordinary world were incorporated into comic texts with profusion" (2000:xx), and he asserts that "Athens was a culture which gave a prominent place to food in its literature"; and "The comic world reflects the real world of the polis but at a remove" (2000:xviii).

A greater understanding of the culture of 4th century Athens and fish sauce particularly can be found in works written in c. 2nd century AD. The letters of Alciphron are responsible for a style of literary imitation known as the "second sophistic." These writers compose literature in imitation of the Attic comedy world of 4th century Athens. Alciphron's fictional letters are written between stock comedy characters such as farmers, fishermen, courtesans and parasites: "We are dealing with a kind of literature that is based on literature" (Benner and Fobes 1959:6). However, we need not be so skeptical, as it is clear that many now-lost Middle and New comedies were almost certainly extant and available when Alciphron composed his letters, and it can also be demonstrated that some passages were lifted verbatim or were simply rewrites of portions of the lost plays (Benner and Fobes 1959:12). The question then remains, can we learn about actual practice in terms of fish sauce from Attic comedy? The answer is a guarded yes providing care is taken to acknowledge the difficulties.

Greek and Latin didactic sources and the prejudice toward technē

The vast majority of ancient medicinal veterinary and culinary literature and particularly that which discusses fish sauce is Greek in origin or is direct translation from Greek sources that do not survive, and there were few original Latin works to consult on these topics. Greek practitioners of the "compounding arts" were not necessarily slaves, their skills were valued within Magna Graecia and they had time and leisure to publish. Many elite Latin writers, such as Columella, Pliny, Vegetius, and Pelagonius make valuable contributions to the genre by including veterinary formulas, recipes, and compounds but their source material was essentially Greek (Adams 1995:1–65; Vegetius 1. *Prol.* 2). The same situation applies to medicinal works where Galen and Soranus dominate. The genre of culinary recipes is apparently linked to the famous Late Republican/early Imperial Roman gourmet Marcus Gavius Apicius but the recipe collections associated with his name have internal evidence to suggest an original Greek source for many of the recipes which were subsequently translated into Latin (Grocock and Grainger 2006:17). Not only did they invent the cuisine that used fish sauce but it is through the spread of Greek literary texts that the skills required to make these sauces and use them in recipes was promulgated. By contrast, there was a prejudice in Rome toward *technē* or manual, physical work involved in a practical skill, and this is very apparent in the nature of the literary output of the Roman elite. The skills (or at least the theoretical knowledge) directly associated with farming were acceptable topics to write about, but the labor and practical experience involved in the preparation of medicinal, veterinary and culinary compounds fell to the slave practitioner and within a Roman context these people did not have the leisure to publish. Elite writers collected compounds, remedies and recipes and put them in their writings but they did not make them up or devise new ones. The practical knowledge is separated from the theoretical.

Ancient fish sauce: the basic recipes and production method

In our hunt for the essential substance that is fish sauce we must start with the didactic literary evidence for its production through basic recipes and the few ancient authors who make attempts to explain what these sauces were. Each of these texts has its particular problems of interpretation. We are dealing with texts stretching over 500 years of culinary and cultural history and we are dealing with authors who were almost certainly consumers of fish sauce in some form or other rather than producers. We therefore have a dilemma when interpreting the literary accounts of ancient fish sauce. The authors are detached from the production process and their accounts reveal huge gaps in their understanding. In ancient societies this ignorance might well be considered desirable as the process of fish liquefaction was then and is now viewed with considerable apprehension and distaste.

Our initial problem concerns whether the four names associated with fish sauces and their residues (*garum, liquamen, muria,* and *allec*) remained essentially the same products (though varying in what kind of fish was used) throughout the Roman period, or did they fundamentally evolve and change to such an extent that we cannot be certain from one period to the next what is being discussed? The order in which these key literary sources are viewed and the value placed on each source is very important; they are not, as one might expect, ideally viewed chronologically. Beginning with what can be perceived as the "wrong" text can form a distorted view of the products we are trying to understand, and each subsequent interpretation of the remaining source material is then subtly distorted by the initial impression that has been formed. Previous studies have always begun with the account of *garum* from Pliny the Elder's *Natural History* in the belief that he was able to convey the essentials of what was always believed to be a *Roman* culinary ingredient. This premise itself is false, as *garos* was essentially Greek and remained so, to the extent that, after the Latin west had discarded *garum* it remained ubiquitous in Byzantine Greek cuisine into the 14th century. Two texts preserve the principal fish sauce recipes. One is in Greek and one is in Latin, but both are apparently Byzantine in date; there are no detailed recipes from an earlier period. The late date of these texts has led to them being regarded as untypical of earlier periods, especially of the 1st century AD, when fish sauce usage appears to have become most prevalent, but as we shall see, they are valuable witnesses to ancient fish sauce.

The Geoponica

The *Geoponica* is a Byzantine Greek farming manual of the 10th century AD which contains material from a long tradition of preserving and copying agricultural knowledge including veterinary and medicinal recipes from the wider Roman Empire and over many centuries. Within the text there are four separate fish sauce recipes, three fermented, one cooked. They are straightforward and uncomplicated, and use fish and salt with few additional ingredients (Dalby 2011).

> 46. Making *gara*.
>
> 1. The so-called *liquamen* is made thus. Fish entrails are put in a container and salted; and little fish, especially sand-smelt (*atherina*) or small red mullet or mendole (picarel = *spicara maena*) or anchovy, or any small enough, are all similarly salted; and left to pickle in the sun, stirring frequently. When the heat has pickled them, the *garos* is got from them thus: a deep close-woven basket is inserted into the centre of the vessel containing these fish, and the *garos* flows into the basket. This, then, is how the *liquamen* is obtained by filtering through the basket; the residue makes *alix*.
>
> 3. The Bithynians make it thus. Take preferably small or large picarel, or, if none, anchovy or scad or mackerel, or also *alix*, and a mixture of all these, and put them into a baker's bowl of the kind in which dough is kneaded; to one *modios* of fish knead in 6 Italian pints of salt so

that it is well mixed with the fish, and leaving it overnight put it in an earthenware vessel and leave it uncovered in the sun for 2 or 3 months, occasionally stirring with a stick, then take [the fluid?], cover and store. Some add 2 pints of old wine to each pint of fish.

5. If you want to use the *garon* at once, that is, not by ageing in the sun but by cooking, make it thus. Into pure brine, which you have tested by floating an egg in it (if it sinks, the brine is not salty enough) in a new bowl, put the fish; add oregano; place over a sufficient fire, until it boils, that is, until it begins to reduce a little. Some also add grape syrup. Then cool and filter it; filter a second and a third time until it runs clear; cover and store.

6. A rather high quality *garos,* called *haimation*, is made thus. Take tunny entrails with the gills, fluid and blood, sprinkle with sufficient salt, leave in a vessel for two months at the most; then pierce the jar, and the *garos* called *haimation* flows out.

The first recipe begins its list of ingredients with fish viscera, from an unde-signated species. The fact that viscera are added first does not in itself make this ingredient the principal one, or more important to the recipe than the solid fish that are added next. The whole fish are undoubtedly the principal ingredient (see the discussion on Pliny the Elder, below). That the viscera are added first is deceptive and has led some to confuse this sauce with that described in Pliny and later in this text, where we find that viscera alone are used. The most striking feature of the text is the way in which the two terms *garon* and *liquamen* are used. *Gara* in the plural implies that in Greek, *garos* functioned as a generic term to convey various kinds of fish sauce.[1] The absence of any mention of *garum* in its Latin sense here is note-worthy (see below and page 94). We do not have a firm date for this text though it is likely to be relatively late, that is, 3rd/4th century see below. At this time, as we will see, the Latin term *garum* is a difficult term to pin down and rarely seen. The use of *garum* to indicate the other sauce, the blood viscera sauce was relegated to use by specialists, namely cooks, doc-tors and vets. *Liquamen* was the only common term for the essential fish sauce hence its presence here. The interchange in the text between the "so called *liquamen*" and the use of *garos* twice, before returning to *liquamen,* provides clear signals that identification of these products was a major issue for people and the recipe attempts to clarify precisely what was being made and what was generally understood by these terms because the general populace were not entirely sure. The first and second recipes make similar fermented sauces, though it is noteworthy that the second sauce makes no mention of extra viscera and this is also true of the third sauce which was a quick-cooked liquefied fish sauce but is essentially the same product. The additional *allec* in the second recipe has been interpreted by Curtis as the addition of many other kinds of small fry, that is, *cleupids* and *sparids,* but I consider that as anchovy is listed individually and the other fish is also specified in larger forms (mackerel and horse mackerel) it is not likely that a generic term such as this would be used here. The use of the Latin term *allec* to mean many kinds of "small fry" is not well attested until the 6th century and largely in Northern Europe (Curtis 1984:147). The use of *alix* in this context where the term is defined within the same text as a fish sauce residue makes it clear that some of the left-over residue from a previous production of fish sauce was left in the tank to seed the next batch. It is very important and in fact crucial that it is precisely this recipe that does not include any extra viscera as the *alix* provides the very necessary digestive enzyme that aid liquefaction of the fish when they are too small to be cut open. If this is correct this actually means that the ordinary small/ medium fish *garos/garum/liquamen* did not necessarily require viscera to be added at all, rather only that there was some means of ensuring that there were sufficient digestive enzymes to dissolve what was present in the tank. This could be provided by what was naturally present within each fish or added in the form of viscera or residual *alix*. These issues are discussed more fully in the Chapter 8 on experimental archaeology. The final phrase of the first *Geoponica* recipe in the Greek is critical to understanding the nature of the fish sauce residues and the nature of *allec*. It states τὸ δὲ λοιπὸν πάτημα γίνεται ἅλιξ, "the residue *makes or becomes allec.*" The Greek is very precise, it does not state that the residue is *alix,* only that *alix* is created from the residue. The residue, rendered in the Greek as "λοιπον" = that which remains (after the sauce is taken) and as πατημα = "refuse" from πατεω = that which is trodden on. The sentence is governed by γινεται, from the verb "to become." The residue as a whole, rather than its component *alix,* will no doubt be a mixture of semi-dissolved fish flesh, the flesh as a paste and the bones, in various stages of degradation. Pliny is far less precise and simply says that *"allec* is the sediment of it" (*garum,* but essentially *liquamen*). Pliny's definition dictates how *allec* is currently understood: the bones and fish paste are believed to be a fundamental part of the product and in this state it is believed to be commercially viable and fit for consumption by slaves and the poor. This is highly problematic, for various reasons discussed on pages 229. It is true that a substance which appears to be a bony fish paste which scholars assume was an *allec* was apparently traded widely, and it is difficult to understand why this should be (Van Neer and Ervynck 2002; the relevant sites are discussed in Chapter 11). A nuanced analysis of the Greek has revealed what was being said in this simple sentence.[2] It will become clear with reference to experimentation that *allec* could exist in two forms: a smooth paste and a fish sauce mash: and the liquid sauce, the bones and the paste can be efficiently separated very early on in the production process, discussed in Chapter 8.

The final sauce in the *Geoponica* is very different, principally because there is no fish meat of any kind in it. Fish viscera and blood are essentially rubbish that would normally be thrown away and it has always been difficult to comprehend this sauce as a separate commodity to *garos/liquamen.*

The absence of the Latin term *garum* in this recipe text is difficult to understand. If the Latin can be given for the ordinary *garon* why is it not also given for the special blood/viscera *garon?* The tunny viscera, fluid, gills and blood are specified, and only these things. It is only with this recipe that we can affirm that a sauce existed derived from blood, and viscera harvested from the inside of a fish along with salt and specifically not the meat from that fish. The unusual nature of these ingredients means that the processes involved may have been very different from a standard *liquamen*. There is no muscle tissue to dissolve into the liquid and the very fact that the blood is, or ought to be, still a liquid when it is harvested, means that it does not necessarily need the enzyme activity to render the protein soluble. It is however essential that the blood is harvested fresh, as once it has coagulated inside the head and body of the fish it becomes inaccessible. It is also of importance to note that the *Geoponica* makes reference to an *alix/ allec* only in reference to the manufacture of a *liquamen/garos*. There is no indication *whatsoever* that the blood/viscera sauce could generate a residue or paste of any value. Whatever residue remains after the sauce is taken is left inside the production vessel (see page 200).

From the *Geoponica* text, then, we can identify various methods for making fish sauce, as follows:

1 A unspecified amount of viscera from other unspecified fish and sufficiently small whole fish of various kinds such as smelt, anchovy and mullet. Salt is added and it is allowed to liquefy open to the sun until pickled and then taken when the sauce flows through a basket and can be ladled out.
2 A mixture of somewhat larger fish such as horse mackerel, mackerel and anchovy which are kneaded with salt at a ratio of 7/1 (15%). The kneading may imply that the larger fish are cut up. An addition of the residue from a previous fish sauce production known as *alix* is added and no additional viscera is needed. Extra liquid can be used (wine or brine and vinegar is also listed[3]). This is pickled for 2 to 3 months.
3 A quick and clearly domestic method where whole fish are boiled in brine until all flavor and nutrients are transferred to the liquid and then it is fully strained (we note that just one herb is added to this recipe).
4 A luxury sauce made with viscera and blood from tuna – but no fish meat - and salt that is allowed to ferment for two months and then taken by piercing the vessel.

Pseudo-Gargilius Martialis

We now turn to the other comprehensive recipe for fish sauce, that is now understood to be a later chapter appended to the 3rd century AD text of Gargilius Martialis. The recipe survives in a single 9th/10th century manuscript of the Latin work *Medicina ex holeribus et pomis* ("Medicines from fruits and vegetables") which is purported to be written in the 3rd century. The recipe clearly does not fit well with the title of the work and it is believed to have been added at a later date as a medieval gloss (Curtis 1984:148). The most recent *Belles Lettres* edition of Gargilius does not include the recipe and makes no reference to it (Maire 2002). The technique of the recipe for the fish sauce is relatively problem-free, but the additional ingredients and subsequent *confectio* are problematic as they include large quantities of herbs and spices, evidence for which is lacking in the archaeological record and in the technique of residue sampling currently available (see Chapter 6). The spices added to the *confectio* are also anachronistic for the Roman period and this certainly adds to the belief that the recipe is medieval in date.

> 62 A confection of *liquamen* which is called *oenogarum*
>
> Naturally oily fishes are caught/taken, such as are salmon and eels and shad and sardines or herrings, and an arrangement of the following kind is made of them along with dried fragrant herbs with salt/they are put together with fragrant dried herbs and salt in this way: a good, sturdy vessel. Well-pitched, with a capacity of three or four modii, is got ready, and dried herbs with a good fragrance are taken – these can be garden or field herbs – namely dill, coriander, fennel, celery, sicareia, sclareia?, rue, mint, sisymbrium (?wild thyme), lovage, pennyroyal, oregano, bettony, argemonia, and the first layer is spread out at the bottom of the vessel using these. Then the second layer is laid down using fish – whole if they are small, cut in pieces if they are larger – over this is added the third layer of salt two fingers deep, and the vessel is to be filled right to the top in this, with successive triple layers of herbs, fish and salt. It should then be closed up with a lid fitted and put aside as it is for seven days.
>
> When the seven days are over, the mixture should be stirred right to the bottom, using a wooden paddle shaped like an oar, twice or three times every day. When this process is complete, the liquor which flows out of this mixture is collected. And in this way liquamen or oenogarum is made from it. Two *sextarii* of this liquor are taken and are mixed with half a *sextarius* of wine, then single bundles of (each of) four herbs – viz. dill, coriander, savoury and *sclareia*. A (one) little handful of fenugreek seed is also thrown in, and of the aromatics thirty or forty grains of pepper, three pennies of *costum* by weight, the same of cinnamon, the same of clove, and when pounded up finely these are mixed with the same liquor.
>
> Then this mixture should be cooked in an iron or a bronze pan until it reduced to a *sextarius* in volume. But before it is cooked half a pound of purified honey ought to be added to it. When it has been cooked it ought to be strained through a bag like a medicine until it is clear – it needs to be boiling when it is poured into the bag. When clarified

and cooled it is kept in a well-pitched vessel in order to give flavor to *opsonia*.[4]

The title of the recipe is principally concerned with the final '*confectio*' or compound sauce which is being made. This sauce requires a two-stage technique in which a fermented fish sauce is made and then this *liquamen* is used as the basis of the *confectio/oenogarum*. There are other views which consider the technique employed here are a single process and that the cooking stage could potentially be an integral part of the manufacture of *liquamen* generally much earlier when sauces were made in Spain (Varges *et al.* 2014:70) This issue is discussed more fully in Chapter 4. The differences between this sauce and the one in the Geoponica are quite important: the fish are of larger species and are cut open to expose the viscera to the salt; there is no extra viscera; the salt to fish ratio is very high compared to that stipulated in the *Geoponica*; the vessel is sealed and intermittently airtight between episodes of stirring; it is not apparently put in the sun; and finally, large quantities of herbs are used.

Why should we choose the Geoponica recipe over Gargilius"?

This question has proved to be quite a controversial one. The first consideration is, what kind of fish sauce are we trying to replicate? The recipe sources advocate two quite distinct techniques: one open to the sun, while the other is enclosed. This mirrors exactly the two techniques utilized in South East Asia, which are described in Chapter 8. The *Geoponica* recipes appear on the surface to be the more complex one for experimental purposes, while the Gargilius recipe seems ideally suited to a small-scale experiment.[5] However, a detailed analysis of the modern techniques and experimental approach makes it clear that the sauces resulting from the Gargilius recipe could potentially be of low yield and of low-quality in comparison to one made with the Geoponica, as discussed in Chapters 7 and 8. This is due to the high salt and large amounts of additional herbs which reduce the yield. There are also linguistic, and culinary anomalies in the Gargilius recipe that make it a problematic text. The medieval spices used in the *confectio* recipe are anachronistic for the imperial period, as clove and cinnamon were not used in Greek and Roman recipes of that time, and can be considered an indicator of the transition into Byzantine early medieval culinary spicing. True cinnamon (*cinnamomum zeylanicum*) is not attested, according to Dalby, until found in an 11th century Byzantine food source by Simeon Seth (Dalby 2003:27; Langkavel 1868:55). The Romans used cinnamon cassia as incense for sacrifice and at funerals and it also had a medicinal role in antidotes, but it was not found in any culinary recipes considered Roman.[6] Clove (*caryophyllon*) was also known in Rome but was recognized for its medicinal qualities according to a number of later physicians and as an aromatic (Miller 1969:49). Clove makes its first appearance as a culinary spice in the separate short recipe collection attributed to *Apicius* but made by the *Vir inlustris Vinidarius* known as the *Excerpta Vinidarii* (1a), probably in the early 5th AD (Grocock and Grainger 2006:32).

There are other ingredients that are anachronistic in Gargilius' recipe. The use of salmon, eel and herring are not typical of the Mediterranean fish species used to make fish sauce, though recent residues do attest to the occasional random catch of herring (Curtis 1984).[7] The typical species that are found in the Mediterranean are the sparids and clupeids along with occasional scombrids and these species are mentioned in the *Geoponica* and are also typical of species found in residues associated with fish sauce at processing sites in the late empire.[8] Fish bone evidence suggestive of a fish sauce residue containing herring and sprat have been found in a refuse dump in Tienen in Belgium (Van Neer *et al.* 2005; Lentaker *et al.* 2004) and also on the Thames in Roman Britain (Bateman and Locker 1982), which suggests that it was used to make fish sauce in northern Europe. The herbs that are used in the initial fermentation stage of the Gargilius recipe are also somewhat problematic. The *Geoponica* suggests that a single herb, oregano, is suitable, while direct evidence for herbs in residue sampling is not extensive. A herb signature has been seen in amphora and *doliola* from Portugal and Galicia (Morais *et al.* 2016:107) and the pollen from the *Asteraceae* family, which is largely associated with wild flower meadow, has been found in residues from a single amphora but there is every possibility that this is a reflection of the surrounding vegetation (see page 204 note 4; Bernal *et al.* 2016:746).

In support of the Geoponica recipe, Andrew Dalby"s translation of the *Geoponica,* the first English version since the 17th century, notes that the text is remarkably similar to the works of Columella and Palladius (Dalby 2011). There are 47 Greek and Latin writers cited in the *Geoponica*, and though many are obscure and unknown other than in this work, others are familiar sources preserved elsewhere. Of the known writers, eleven of them were writing in the Hellenistic period and even earlier.[9] A further eleven were writing between the 1st and 4th centuries AD and the latest source cited is Cassianus Bassus, who wrote in the 6th century AD.[10] The *Geoponica* manuscript may well be dated to the 11th century, but the works cited are essentially Roman and Greek sources contemporary with the period when fish sauce production was at its height. The *Geoponica* is a unique Romano-Greek agricultural and medicinal text, and though we have no attribution for the fish sauce recipes themselves, they are likely, given the early and also wide scope of the sources used, to be a reliable indicator of the nature of the principal fish sauces in Greece as well as Rome in the early, middle and late Empire.

There are therefore several issues which argue against using the Gargilius recipe to identify fish sauce from the imperial period. Against all this, however, what seems most pertinent in querying the value of this recipe in assisting us to understand the Mediterranean fish sauce trade in the early imperial period is the fact that when this recipe was written, it appears that a regular supply of a predictable bulk fish sauce does not seem to have been available.

In the context of this recipe the fish sauce itself has to be made domestically before it is turned into a *confectio/oenogarum*. Given the archaeological evidence it seems more than likely that the vast majority of the urban populace of the 1st century AD Roman world would have had regular access to fish sauce from the market and would not need to make their own. What characterizes the late empire is the increasing need to make it at home, as the supply of a regular product had effectively stopped. Many fish processing sites close by the end of the 2nd century and while others begin to make sauce in the 4th and 5th century the supplies are greatly reduced; wreck evidence for fish sauce is in permanent decline and many of the typical amphora also cease to be made, suggesting considerable reduction in the trade in fish sauce (Wilson 2007; Bernal and Sáez Romero 2008:69).

Pliny the Elder on garum

Pliny the Elder's account of the different kinds of fish sauces available in his world is fraught with problems. Pliny has an economy of expression that can lead to ambiguity. He tends to omit what would normally be obligatory elements in a sentence because they can or should be understood from the preceding context. This causes few problems so long as the subsequent discussion is unambiguous, but there is often confusion over precisely what he is talking about when he introduces multiple subjects, and also some doubt that he is not always sufficiently familiar with his material (Pinkster 2005:239), and this may explain the considerable interpretative problems which arise from his text on fish sauce. In particular, Pliny contradicts what is found in the *Geoponica* and consequently seems confused about the different types of fish sauce, and what they are derived from. The sauce that he initially calls *garum* is an expensive "choice" product which is ironically made from fish rubbish and it is this factor that he stresses.

> "There is yet another kind of choice liquor called *garum*, consisting of the viscera of fish and other things that would normally be thrown away, soaked with salt so the *garum* is really a putrid exudation. Once this used to be made from a fish that the Greeks call *garos*."[11]

Having read the *Geoponica* we can immediately recognize that he is appears to be talking about the special blood viscera sauce here. Pliny describes a fish sauce made from fish viscera and "other things" that would otherwise be thrown away, mixed with salt, but there is no specific mention of fish blood or the addition of little fish, which I would argue should never be considered rubbish. Pliny's sauce appears to be made principally from viscera as he stresses this as the primary ingredient. I maintain that he is referring to the concept of the blood viscera sauce here and is not referring to the basic *garos* of Greek culinary practice. Pliny refers to the sauce he is describing as being the result of *putrescentium sanies* (putrescent exudate) or that which oozes from wounds, which certainly has an indirect association with blood. If he did understand that another liquid sauce existed, made from small fish, he does not make this clear at this point. The reference to a fish of Greek origin called *garos* is no indication that he does understand this. Many commentators have attempted to reconcile what Pliny says with the knowledge we have from archaeological residues, namely, that the vast majority of traded fish sauce was made from insignificant small-fish and considered a *garum* too. This has led them to interpreted the "other things that would normally be thrown away" as little fish, and in so doing they have brought the two types of sauce together into one generic type of *garum*. That little fish could be considered as rubbish and otherwise thrown away is unclear, and in fact highly unlikely. It seems more likely that Pliny is interpreting these products from the perspective of an elite diner, a consumer who has little understanding of what he is being offered at table. And more over, a Roman whose interest in gourmet food was at best noncommittal: his nephew Pliny the Younger talks about his daily eating habits as being traditionally very modest and unremarkable (Ep. 3.5). Pliny the Elder's social circle was nevertheless one of the highest in terms of status and wealth, and it is more than likely that the only form of *garum* that he ever encountered or heard other people talk about was the very expensive and desirable form, that is, the blood-viscera *garum*. It is I think likely that Pliny's experience of the everyday whole-small-fish *garos*/*liquamen* was limited if not nonexistent. He will no doubt have uncounted elite *garum* at the table of his contemporaries and simply not recognized that it could be either a black bloody *garum* or a good quality *liquamen*. We will discuss the various ways in which the diner engaged with fish sauce at table in more detail in Chapter 3 but for now it is important to point out that diners may not have been able to distinguish the two principal types of fish sauce visually from their appearance at table. The difference may have been more readily apparent from taste. The color of a neat sauce is crucial to the identification of the product, as we shall see, that sauces could be black or brown and also light or bright when neat, and when blended into a compound sauce: *oenogarum* with wine oil and vinegar indistinguishable. Here is our essential confusion right at the point of consumption. One may legitimately ask, if the ancients cannot see the difference, how can we? We will discover that the other sauce, *liquamen*, was mainly used in the kitchen and would not be visible as a separate sauce to a guest. A guest such as Pliny would not necessarily have understood that two sauces existed, or cared to inquire as to which one had been used. Use of the small fish *liquamen* would have been hidden from the diners' view during the food preparation process or in a premixed *oenogarum*. Pliny's suggestion that the sauce he understands as *garum* "used to be made from a fish that the Greeks call *garos*" is difficult to understand as we do not know whether Pliny is referring to an elite expensive sauce or is he, (or the source he is using) actually referring to the original Greek sauce called *garos* which becomes *liquamen*. The latter explanation is more likely and would rather indicate that he has begun to talk about the small-fish sauce though he clearly does not realize there is a difference at this point. The "fish that the Greeks call *garos*" is unknown, but we can say with some certainty that *garos* corresponds to the modern Greek term γαύρος which is still used for anchovies today (Carannante 2019:21; Demir 2007).

There then follows a very obscure and indecipherable statement in Pliny's text: *capite eius usto suffitu extrahi secundas monstrantes*. The Loeb attempts a translation as follows "noting that by burning and fumigating the head, the afterbirth is extracted." It is obviously entirely unconnected phrase inserted from a different part of the book. We note at *HN.* 32.130.2 a similar line referring to an afterbirth being fumigated to extract it and with a reference in the previous line to the head of a fish being burnt which is surely a scribal error.

Pliny continues:

> "Today the most popular *garum* is made from mackerel in the fisher-ies of *Carthago Spartaria*. It is called *sociorum,* 1,000 sesterces being exchanged for about two *congii* of the sauce. Scarcely any other liquid except unguents has come to be more highly valued, bringing fame even to the nations that make it. The mackerel is caught also in Mauretania and at *Carteia* in *Baetica*; the mackerel enters the Mediterranean from the Atlantic but it is only used for making garum. *Clazomenae* too is famous for *garum* and so are Pompeii and *Lepsis* just as Antipolis and Thurii are for *muria* and indeed also Delmatia."[12]

Pliny understands that in his day the most popular and the most expensive form of *garum* fish sauce was called *sociorum* and that it was made in New Carthage from mackerel, but the Latin does not exactly specify that the term *sociorum* was only applied to a sauce from Carthargo Spartiae. It only states that the best sauce was known as a *sociorum*, and later sources would seem to corroborate this, discussed below, as the original defining characteristic of this *garum* is the presence of blood, from mullet, not nec-essarily blood from Spanish fish = mackerel. Pliny later gives more infor-mation about *garum sociorum* elsewhere: "Marcus Apicius... thought it especially desirable for mullets to be killed in a "*garum sociorum* (a *garum* of its companions/allies) – for this thing also has procured a designation (*nam ea quoque res cognomen invenit*" (Pliny, *HN.* 9.66.4)." This passage requires delicate and detailed unpicking as it is the most important piece of evidence we have on the origin of this elite Roman sauce. When he says that even "this thing" as got a name, it surely cannot be the original Greek *garos* itself which, as we will see in the chronology in Chapter 2, had been around in Rome for decades and centuries as a Greek sauce. The sauce in ques-tion must refer to something new that was being used to flavor the mullet dish. The background to this gourmet behavior sheds more light on Pliny's remark. Seneca provides a description of mullet being cooked/asphyxiated in front of the guests; he writes that they are in a glass vessel and change color as they do so, for the guest's entertainment. He then says *alios necant in garo et condiunt vivos*, "they kill others (mullet) in *garum*, and season them while alive (*Quaest. Nat.* 3.17.2.9)." This process is also described in the feast of Trimalchio (see page 73). A little later Seneca remarks "How inconceivable it would sound to them to hear that a fish swam in *garum* and was killed during dinner" (*Quaest. Nat.* 3.17.3.4). There is probably some exaggeration here, and Seneca is nothing if not contemptuous of gourmet behavior, but the fact that fish were cooked at table in front of the guests is entirely feasible. The fact that they were alive requires that they are uneviscerated, which is difficult to equate with an actual cooking process; nevertheless, they would have been poached in water flavored with *garum* rather than entirely in *garum*. This cooking liquor has a name, *hydrogarum*. In *Apicius* (2.2.5) a sauce called *hydrogarum* has a ratio of 1part *liquamen* to 7 parts water, plus a little pepper. Each mention of this liquor shows it being used as a cooking medium not unlike a court bouillon to poach meat and fish (*Apicius* 2.2.1; 2.2.2; 2.2.5). This is the context of the Seneca, Petronius and Pliny passages, and it demonstrates that it was not out of the ordinary to use fish sauce to flavor water in this way.[13] Apicius the gourmet – rather than the named author of the recipes – has apparently created a different kind of *garum* to flavor the *hydrogarum*– poaching liquid.

Initially, the idea that the *socii*, "allies," were some sort of Spanish trading consortium making this high-quality sauce was taken for granted, especially as *garum* was seen as a single commodity of varying qualities (Curtis 1991:63; Etienne 1970; Leon 2001:175). It is surely the case that the "allies or friends" in the original formation of the concept were not trading "allies" but the more general idea of companions or friends, which Leon also fully acknowledges (2001:176). That *societas* = groups collaborating to do business, existed in some form seems to be clear from other sources but the link between *sociorum* with this particular Cartagenan one seems to be unproven. Etienne in 1970 also concluded that the *socii* and *sociorum* were connected in a general idea of fine fish-based sauces from Spain, rather than a specific one, clearly the best when from Spain but still potentially made anywhere (Etienne 1970:313). It would seem probable that *sociorum* was initially conceived as a very specific name for a very specific sauce, that is, one made with blood and viscera of mullet, used in the process of cooking/poaching/killing "their fellows or allies," in other words more mullet, and it continued to have that meaning, albeit that it evolved into the name for the sauce as "fish blood essence," rather than the dish as conceived by Apicius. The idea that the sauce in question was simply dissolved mullet as in a *garos/liquamen* seems utterly prosaic and could not have elicited the same response in Pliny or anyone else. Leon also acknowledges that the connection to Apicius the man is strong and that he almost certainly conceived of the idea of this sauce, but nonetheless maintains that the term eventually came to signify Spanish fish sauces generally as made by a Spanish consortium (Leon 2001:177). This is a theory that only works if, once again, fish sauce is a single substance, with *sociorum* coming to mean a trading group and or geographical origin not a specific type of sauce. I am not convinced by his argument and we need to unpick them carefully. Leon seeks corroboration of this idea from a reference in the *De Materia Medica* of Dioscorides, who was a contemporary of Pliny (AD 40–90) concerning the Romans and *garos*: γάρος ὁ σπανός ὅς σοκιῶρουμ λέγεται, "and Spanish *garos* which is called *sociorum*."[14] Leon takes this to mean that *sociorum* is only applied to *garos* from Spain but he is assuming a single substance of cause, which is disputable. Regardless of its meaning, the reference in Dioscorides cannot be securely associated with the original text. The passage cited by Leon does not appear in the most recent Spanish edition of Dioscorides, either where the reference is cited or where *garos* is discussed (*Dios. De Materia Medica* 2.34; Valdes 1998). Valdes notes that the original text of Dioscorides' *De Materia Medica* is extremely difficult to extract from the numerous additions made to it throughout the late Roman and medieval transmission of the text.[15] It is very likely that the information contained within this disputed passage cannot be dated to the 1st century AD; it may well be a later gloss only preserved on one manuscript. Leon then links this phrase from Dioscorides to the only other reference in Greek to a sauce with the designation *sociorum*. This is in a text attributed to Galen which mentions Spanish products and also describes a sauce as black.

> "For the stench of wounds that (remedy) which is called 'of the Spanish.' Take: black *garos,* called *oxyporum* by the Romans, 1 *sextarius*, squill vinegar, 1 *sextarius,* Attic honey, 1½; boil until it binds, and put it away in a glass vessel and use."[16]

This is an undated Medieval Latin translation of Galen and it is included, in the Kuhn edition from 1965 which is derived from the 1823 Olms edition. There are extra lines not found in the Greek: "black garum which the Romans call *sociorum*," *gari nigri quod Romani sociorum appellant*. Though undated this reflects what I believe to be the correct way to identify *sociorum*. The terms of the original passage may have been jumbled up and in fact reversed during transmission. The remedy is clearly called an *oxyporum* and this term should logically belong where "of the Spanish" is found in the line. In this case the "black *garos*" might logically be termed "of the Spanish" clearly reaffirming what is clear that the best black *garum* was a Spanish product but it seems entirely probable that black *garum* could be made anywhere fish could be bled in sufficient quantity. Clearly this reference in Galen is also not secure, and for similar reasons. Cribiore (2019) considers that didactic remedy collections were being amended even in the lifetime of the author and in the decades after publication, so that a secure, datable text is almost impossible to pin down. This has long been the case with the books in the Galen corpus where we find that later additions, not made in his lifetime, but added too throughout the medieval period using Latin and Arabic translation, and were all attributed to Galen (Scarborough 1981:24). Scarborough has unfortunately affirmed that the Galen work known as *De compositione medicamentorum secundum locos* where we find the reference to a black sauce is one of these. These links found in medicinal text between *sociorum* and Spanish products are insecure in themselves because undatable but it has never actually been possible to determine that only Spanish fish sauces were called *sociorum*. In Pliny the Elder's text we find the same implication concerning *sociorum*: the term (*sociorum*) and the geographical origin (Spain) are not mutually exclusive; the sauce is not defined by the origins of the fish but by the method and recipe. Leon sees *garum* and *garos* as the same thing, yet if they are separated it becomes clear and patently demonstrable from the archaeological residues and *tituli* labels that *both* varieties of fish sauce were made in Spain and therefore *sociorum* cannot signify both and must refer to one of them. Pliny does not state that Apicius invented the concept, though it is certainly implied. If he created the dish, he necessarily also conceived of the sauce as mullet essence, while at some point in the future the term ceases to be applied to the dish and was applied to the sauce alone. The date of the inception of this idea is unknown but is nonetheless crucial for an understanding of the nature of *garum*. There is every expectation that the Greek term *garos* was initially transliterated into *garum* and undoubtedly represented a 'single' dissolved small-whole-fish sauce and it is this single sauce that was incorporated into Roman cuisine from the late Republic (see page 52). That a blood viscera *garum* was very different from a *garos* is clear: the sauce has complex umami flavors, but they are not fishy, as iron seems to dominate on the tongue and its black bloody color is very different from a *liquamen* which is amber in color. Pliny wrote his Natural History in c. AD 50–60, and certainly before AD 79, while Apicius seems to have had his culinary heyday during the reign of Tiberius, before AD 37, and evidence suggest an even earlier date for the beginning of his influence at court. Pliny himself tells us that Tiberius was annoyed that his son Drusus was encouraged to reject cabbage by listening to Apicius (*HN.* 19.137). We cannot know how old he was when this all too familiar rejection took place, but it is likely to have been in his youth, in which case the influence of Apicius at court may have begun as early as Augustus' early reign; Drusus was dead by AD 23 aged 36. If we have a putative age of 16 for the time when Drusus was rejecting cabbage, it puts Apicius' influence at court as early as AD 3. We are also told that Maecenas and Apicius dined together (Martial 10.73.1–4), and Maecenas was dead by 8 BC. This clearly cannot tell us when *Apicius* invented his new sauce but it gives a much wider perspective on his influence. For the purposes of convenience, I place the invention of *garum* at about I BC–AD 1.

Returning to Pliny's original quote, he states that mackerel was not used for anything else (*HN*. 31.94), a statement which we may doubt, as mackerel was evidently a staple solid salted product across the ancient world for a time, and therefore may have been a commonplace food beyond his understanding. Pliny names a small number of very specific geographical places in Spain and Italy where this kind of high quality *garum* is made; they correspond to places where we know mackerel was known to be caught in bulk according to the seasons (Morales-Muñiz and Roselló-Izquierdo 2016) We must ask, why does he not mention any of the other places across Southern Spain and North Africa and Italy where we know, from archaeology and fish residues, that the basic whole-fish sauce was made? I would argue that even if he is himself unaware of the differences between the sauces, the information he is quoting is not discussing *garos/liquamen* here but specifically the luxury black sauce. The vast majority of fish sauce of the *liquamen* type was made either in industrial quantities from undervalued, cheap small fish of multiple species, or on a small scale and in a local way *everywhere*, and as such it was not considered valued, special or luxurious. The specific places named, Mauretania and at Carteia in Baetica, Clazomenae, Pompeii and Lepsis, very likely made a valued and recognized blood viscera *garum*, that is, a potential *sociorum,* which if accurate is clearly being made outside of Spain.

That Pompeii is noted to have made a good *garum* is hard for archaeologists to comprehend as they expect to find the same scale of fish processing factories designed for a bulk *liquamen* with numerous large scale *cetariae* as those on the Baetican coast, and such structures are apparently absent from the bay of Naples region.[17] However, true *garum* manufacturing does not require huge infrastructure in terms of *cetariae* processing basins, as it is likely that the process of macerating the viscera and blood, which, as we will see (page 163) was likely harvested in relatively small quantities, in all but the most industrial processes and probably took place in very much smaller vessels such as dolia and very smaller tanks, which would remain unidentifiable to archaeologists looking for the manufacturing infrastructure of *liquamen* production.

When Pliny says that Antipolis, Thurii "and indeed also Delmatia" are known for *muria*, he is surely stating that these places concentrated on or had a reputation for first producing *salsamenta*. There is little doubt that *muria* was initially and basically a salt water brine but it also clearly functioned as a culinary sauce too and a by-product of the processing and storage of whole cleaned solid salted fish products.[18] Isidore of Saville (below) distinguishes *salsugo* as pure brine with that called *muria* and Pliny also makes reference to this prior to his discussion on *garum* (Pliny *HN* 31.92): It is brine of a higher salinity than sea water. Pliny elsewhere refers to a *muria salsamentorum* "brine of salted fish" (Pliny, *HN*. 31.83) but this secondary term is rarely included in many texts and one has to establish some kind of context to understand which *muria* is meant in a given context.[19] However, the precise definition of *muria* is still very much disputed. The one made by Isidore of Seville (see below) has been mistranslated and mis-interpreted, so that current thinking is that *muria* was a by-product of *liquamen* but only when *liquamen* is different from *garum!* (Corcoran 1963:205). Such is the confusion. The vast bulk of *muria* must have been taken from cleaned solid salted fish though others disagree (Vargas *et al.* 2014:66; Palacios *et al.* 2016). References to *muria* in Roman satire are also taken as an indication that *muria* could sometimes even refer to different kinds of blood viscera *garum* but they can be understood in different ways (see page 104).

Pliny goes on to say that "*Allec* is the sediment of it (the *garum* Pliny understands), the residue neither completely 'used up' or strained."[20] He appears to consider that the valued and expensive blood viscera *garum* would generate a fish paste residue yet, when this is made without whole fish it can be demonstrated that this particular type of sauce was unlikely to have been able to generate a paste suitable for consumption, discussed in Chapter 9. Pliny continues "It (i.e., *allec*) has, however, also begun to be made separately from a tiny fish otherwise of no use. We call it *apua*, the Greeks *aphye*, because this tiny fish (seems to be) born out of rain." It is important to note here that Pliny does not make the link between his "other things that would normally be thrown away" in the previous passage, that go into his *garum*, and these unspecified worthless (*inutile*) little fish (*pisciculo*) that make a "separate" fish paste. The term *aphye* literarily means "unborn." It probably refers to very young fish which are tiny and transparent and it may refer to a transparent goby which has a very thin pale and foamlike body and big eyes (Dalby 2003:15). The term 'separate' here implies an unfermented fish paste which has probably not generated a clear sauce first. It is very likely that this separate product resembles pissalat, the purpose made bone-free fish paste made in the Nice region of France which is made from anchovy but also with poutine: the tiny anchovy hatchlings which resemble foam. (Sternberg and Delaval 2007, 80–81; see page 229). Pliny goes on to say *Foroiulienses piscem, ex quo faciunt, lupum appellant*, "the people of Forum Julia call *lupus* ('wolf') the fish from which they make (it). It seems logical to regard this as another statement about *allec*, as the next sentence resumes the topic, with the statement "Then (*allec*) became a luxury and its various kinds have come to be innumerable," though the translation in Jones' Loeb edition of Pliny unhelpfully adds *garum* at this point.[21] It seems far simpler to take this as referring to *allec*, and the "innumerable" ways of making it may also refer to *allec* not *garum/liquamen*. Pliny does revert to the subject of *garum*, but only after the connective *sicuti*, when he says directly "for instance *garum* has been diluted to the color of aged honey wine and to taste so sweet as to be drunk." In order to highlight this apparent increase in luxuriousness he then seemingly returns to *allec* with the next sentence which has a possible *lacuna* which makes it difficult to be sure he is still on topic: "but another kind... (lacuna) is devoted to superstitious sex abstinence and Jewish rites and is made with fish without scales."[22] This line is generally assumed to be connected to the idea of *garum* or *liquamen*, but I suspect it refers only to *allec* and the purposes of *allec*. Susan Weingarten has addressed this issue recently and highlighted that rabbis rejected some forms of *allec* but not others, and the distinction seems to be that between a residue still containing identifiable fish parts and one that was fully sieved and a smooth paste (see page 233; Weingarten 2018). Pliny reverts back immediately to *allec* in the next sentence, despite the grammatical error and the error in understanding Jewish dietary law; the seafoods that he lists are precisely those that would make a scale-less and boneless *allec*: "thus *allec* has come to be made from oysters, sea urchins and sea anemones and mullet livers, and salt to be corrupted in numberless ways as to suit all palates."[23] Pliny appears to imply that the luxurious *allec* made with scale-less seafood was introduced to accommodate Jewish dietary law, which is both entirely wrong as Rabbinic law precludes the consumption of precisely this kind of seafood, and accommodating any such laws are hardly likely to have been the motivation for its introduction. The culinary appeal of this kind of product may well have been its smooth, bone and scale-less texture. Pliny throws up far more questions than he answers as he is relying on external sources and he does not understand them. It seems we should be wary of relying on him as our primary source, but can see that the source he was using was more informed.

Isidore of Seville on garum and liquamen

The *Etymologiae* of Isidore of Seville is a vast encyclopaedia of ancient knowledge concerned with the origin and meaning of ancient words compiled in the 7th century. Isidore is writing in a Christian world using ancient pagan sources. According to Barney, "Isidore preserved the accumulated learning of the classical world" but has determined that Isidore had limited access to original sources and his work is considered derivative and is often rejected as an unreliable source (Barney 2006:11) Andre's (1986:13–22) analysis of the book 12 demonstrates that there are altogether nearly 600 citations and 293 uncited borrowings and most importantly there are 45 from Pliny the Elder.[24] The section of Isidore's work on fish sauces in *Etymologiae* 20. III. 19–20 appears authoritative, but there are signals that he is insecure particularly about *garum*:

> *Garum est liquor piscium salsus, qui olim conficiebatur ex pisce quem Graeci γάρον vocabant; et quamvis nunc ex infinito genere piscium fiat, nomen tamen pristinum retinet a quo initium sumpsit.*
>
> *Garum* is a salted liquid of fish, which formerly was made from the fish that the Greeks called *garos*, and although it is now made from an infinite variety of fish, it retains the original name that it had at its beginning.

The phrase "A salted liquid of fish" is very vague, and suggests that Isidore did not actually know what precisely *garum* was. We will see in Chapter 5 that the concept of the blood viscera sauce was largely unknown in the 6th century and his vagueness therefore is understandable.[25] The reference to the original fish which the Greeks call *garos* is a direct citation from Pliny (see above) and the use of *infinito genere piscium* is also a near quote from Pliny – *innumerisque generibus* which, I have argued above, was not referring to *garum* per se at that point but to an infinite variety of fish being used to make a purpose made *allec*. The fact that he uses Pliny for his passage about *garum* and is markedly silent about the viscera and *sanies* (exudate or blood) that Pliny provides, makes me suspicious that he is not sure what is distinct about *garum* and his vagueness is deliberate. The next section on *liquamen* has more information and Isidore definitely appears more confident.

> 20 *Liquamen dictum eo quod soluti in salsamento pisciculi eundem humorem liquant. Cuius liquor appellatur salsugo vel muria. Proprie autem muria dicitur aqua sale commixta, effectaque gustu in modum maris.*
>
> "Liquamen is so called because little fish, dissolved in (the process of) salting fish (*salsamentum*), liquefy (*liquare*) as a liquor(*humorem*) of the same type, the liquid of which (= *salsamentum*) is called *salsugo* or *muria*, but properly *muria* (i.e., 'brine') is the name for water mixed with salt, to produce the taste of the sea."

Isidore reaffirms the idea found in the *Geoponica* that *liquamen* is a product derived predominantly from dissolved small fish. The liquefied little fish forms a *humorem*, an "essential liquid" of the same "type" as the garum in the line above, that is, a salted fish liquid but clearly not exactly the same. The phrase *cuius liquor*, "the liquid of which," has been taken to mean that the liquid derived from the making of *liquamen* was called either *salsugo* or *muria*. This is simply inaccurate. Grammatically, there is no doubt that *cuius* goes with *salsamentum*, not *liquamen,* and this is confirmed in numerous references to the idea of a *muria salsamentorum*, "brine of salted fish" (Pliny, HN. 31.83). There is ambiguity about how to render *salsamentum* in this passage. The translation by Barney (*et al.* 2006:399) gives "pickling brine," which is not technically accurate as *salsamentum* is generally understood to be a solid product (Curtis 1991:8), not the medium of that process, which is in fact *muria*.[26] Corcoran (1963:205) was the first to mistranslate and misinterpret Isidore's words. He switches it round and regards *salsamentum* (= solid product) in this sentence to mean that *liquamen* was derived from a process of salting solid fish, but clearly the fish will not be solid, but liquid. The use of *salsamentum* here is confusing, but was I think used for convenience only. Technically they were *salsamentum* at the beginning of the process, but the little fish were whole and uneviscerated, and the only thing that would have happened in those circumstances was that they would dissolve. True salted fish is by its very nature eviscerated of its digestive enzymes and therefore it cannot dissolve and remains a solid product. Corcoran assumes that *liquamen* is somehow a by-product of both the liquid and the solid product (1963:205). A number of scholars continue to make the assumption that *muria* was a by-product of *liquamen* because of this misreading of Isidore. The Barney translation of Isidore associates the *cuius* with *liquamen*, (2006:399) and the Corcoran interpretation (1963:205) is still retained by many. Curtis (1991:8) does not acknowledge the anomaly, while Botte (2009:21) is quite clear that *muria*, when it was a culinary sauce, was largely a by-product of solid salted fish. It is because of these mis-readings that Enriques Vargas *et al.* (2014:66) consider that this so called *muria* was potentially taken from the *liquamen* in the early stages as a separate product. These issues are discussed more fully on pages 104; 119 and 157. *Muria* was of course in its most basic form salt and water, as Isidore notes. The multiple meanings of these terms give the impression of confusion if not chaos, but context and an empirical perspective can help disentangle their meanings.

Fish sauce in Manilius' Astronomica

The *Astronomica* of Marcus Manilius is a didactic poem on the science of astrology which dates from some time in the early 1st century AD, around the end of Augustus's reign and the beginning of Tiberius; we have no information about him apart from what can be gleaned from the poem. In book 5 of the work, he describes the journey of the constellations across the sky, and at line 656 fishing becomes a suitable topic. This text is important because after some preliminaries Manilius describes in detail the process of harvesting fish and the content of fish to make what I would identify as the three main ancient fish sauces, that is, *garum, liquamen,* and *allec*. The detail is exceptionally rich and really quite visceral, even deliberately so, and this gives an impression that his knowledge is not second-hand: it is possible to imagine that the author has witnessed the processes he describes and has some, albeit vague, understanding of the species used, the various components utilized and how the sauces are used in the kitchen and dining room. In most accounts of fish sauce this description in Manilius is rarely given a sufficient airing, largely because as a poem it is inherently unreliable, as we have seen from the discussion of modern approaches to it at the beginning of this chapter. It is self-evident that it does not provide a straightforward description of fish sauces with nice neat labels! It is a difficult piece of Latin containing a number of ambiguities, and the text must be approached very carefully to extract useful meaning. The information it contains is open to different interpretations: there are questions as to what the author intended to convey, and how what he says may (or may not) be equated with the other forms of evidence. My approach to this poem is dictated by my original idea that two sauces existed and I have seen strong evidence of these two sauces in this passage but other do not. The section in question in Manilius begins at 5. 661 with reference to fisherman in general and those that catch seals in ocean prisons; then we are told how tuna are drawn along a network of meshed areas where they are killed in a mass slaughter, "and their capture is not the end, the fish struggle against their bonds, await new assaults, and suffer death by the knife; and the sea is dyed mixed with blood" (5.664–666).[27] At this point we must engage closely with the Latin to get a sense of what is really going on.

> *Tum quoque cum toto iacuerunt litore praedae,* 5.667
>   *altera fit caedis caedes: scinduntur in artus,*
>   "Then again when the (a) catch lies all along the shore, a second slaughter is done to the slain: they are cut into pieces ..."

The usual interpretation of this passage is to assume that the tuna in the previous section are still the topic when the line starts at "*tum quoque.*" Botte (2009:19) and Low (2018:468; Curtis 2001:404) all consider that the next passage concerning fish sauces continues to use the same tuna already mentioned. I consider that "then again" is more emphatic and takes you to another scene entirely and a different catch of fish altogether. The reason for the switch is complex but will become clear as we move forward. The Latin continues

> *corpore et ex uno varius discribitur usus.* 5.669
>   *Illa datis melior, sucis pars illa retentis.*
>   "and from the one body different purposes are allotted. One part is better with its juices drained, another with them retained."

These lines refer distinctly to two separate products derived from the preparation of this particular catch. Some scholars have interpreted the product with its "juices drained" as a reference to a cleaned, solid, salted tuna-fish product (Botte 2009:19) and the second product the sauce that is assumed to be the single all-purpose *garum*. However, from what follows it is very clear that the two products being discussed are in fact both sauces. It is the internal juices (i.e., the viscera) that are utilized in the first product not the empty carcass; they are almost incidental and not referred to in the proceeding passage. The meaning is somewhat obscure, but there is a clear reference to the making of a blood/viscera *garum* or *haimation* in the next lines in the poem, which refer to the gore flowing out of the fish to make one of the products (see below; Goold 1977:352; Botte 2009:19). If tuna is still the fish being processed here, then we have to envisage the large tuna fish being cut up but with its intestines left in place so that it could turn into a *liquamen*-type fish sauce. There is no corroborating evidence for a *liquamen* like this from residues and epigraphy. The evidence for the *liquamen* type sauces is dominated by *clupeids* and *sparids* and by mackerel from epigraphy, and the size range of the fish used is medium to small, that is, 5 to 20 cm. We do not find tuna bones in a form that could be identified as a fish sauce residue and it is not feasible to imagine the wasted labor of removing the meat in order for it to dissolve in its own viscera. This is also contradicted by the terminology here, which implies the viscera are actually retained *inside* the fish, and this is simply not feasible if it is tuna, because tuna caught for consumption are almost always too large a fish for that to happen safely as the fish are expected to burst open and dissolve. It is true that tuna bones are very scarce everywhere in the Roman world (Morales Muniz & Rosello-Izquierdo 2016:38; see page 174). The absence of bone evidence for a tuna *liquamen* is not in and of itself proof that tuna was *not* used in the production of a *liquamen,* but if it was, I do think we might expect the literary and epigraphic sources to say so as such a sauce would be prestigious and noteworthy. Others disagree, see page 176 (Djaoui 2016). A blood viscera *garum* made with tuna viscera is first attested in the late 1st century AD, from residues (Van Neer and Parker 2008) and subsequently in the Geoponica. The original form of black bloody *garum* was initially made with mullet and later in Manilius' time it was clearly from other sources, mackerel. It is also of note that in Manilius' description, the tuna in the previous section has already bled into the sea, and so the desirable juices (which from the text below definitely include blood) have already been lost. The lines continue:

> *Hinc sanies pretiosa fluit floremque cruoris* 5.671
>   *evomit et mixto gustum sale temperat oris:*
>   "Here a precious exudate flows out, which vomits up the flower of the gore, and mixed with salt balances taste in the mouth."

The construction using *hinc* and *illa* (here and there) contrasts the two products again. 'Here' we find a straightforward description of the blood/viscera *garum* in which the blood is allowed to flow out. The *sanies* is a term for fluid that exudes from wounds! while *cruor*, bloody gore, is equally evocative. Pliny the Elder and Seneca both use *sanies* for fish sauces that seem to be expensive and exclusive, and blood is a common descriptor. This bloody viscera sauce is said to "balance taste in the mouth." This implies that the sauce in question is experienced at the point of eating. This requires a discussion of table etiquette and dining practices which are explored in Chapter 3. Manilius continues:

> *illa putris turbae strages confunditur omnis* 5.673
>   *permiscetque suas alterna in damna figuras,*
>   "there that whole mess the of decaying host sinks to the bottom and mingles its shape in a second demise ..."

This surely indicates the second separate sauce referred to above being formed from the "decaying hosts" of the first sauce, that is, the empty mackerel, but this surely also (I would argue) included some uneviscerated mackerel referred to above, where the viscera is retained either inside the body or alongside the body. What is important is that the viscera remain with the fish in order to dissolve the flesh. It is my belief that this product is a good quality mackerel *liquamen*, as opposed to the less desirable form that we will see shortly made from very small fish. A precise picture of what is actually happening at this point, with the mackerel sinking and dissolving into a mass, will have to be delayed till we can talk about the experiments. It should be noted however that the proposed mackerel or even tuna that were emptied of their viscera to make the *garum* are not otherwise mentioned as *salsamenta*. The essential point of these passages is that the one species of fish, mackerel, is used in two different ways to make two types of fish sauce, one with the internal viscera left inside or beside (*liquamen*) and one where the juices are taken out and used on their own (*garum*). Manilius rounds off his depiction of this process with

> *communemque cibis usum sucumque ministrat.* 5.675

The mass of decaying hosts forms itself into two further products; however, the passage is somewhat unclear grammatically. In this line, do we take *usum* and *sucum* as (1) referring to two separate things, both quantified by *communem* (useful), or (2) two separate things, with *communemque cibis* referring only to *usum*, or (3) a hendiadys of *usum* and *sucum* – a "useful juice?"[28] Thus three interpretations are possible:

1   And it serves up a common condiment (*allec*) and a useful juice for foodstuffs.
2   And it serves up a multipurpose ingredient for foodstuffs and a juice
3   And it serves up a handy multipurpose liquor for food

The line is definitely referring to the fish paste residue of fish sauce known as *allec* and so (3) is the least likely, but are there clearly two products being proposed and do they have separate roles? It has been possible to observe the residue formation and understand that there will indeed be two separate products from this mass of decaying flesh, one of which is an *allec* "the common condiment" derived from the mackerel flesh and the second a juice, that is, another fish sauce, a *liquamen*. It is important to note that in this description it is this second fish sauce product that generates the *allec*. The "useful juice" is to be contrasted with the earlier blood viscera sauce which "balances taste in the mouth." There is a precise contrast between the direct taste sensation of the consumer at table reflected in the first and the commonplace usefulness of the second which speaks of a separate role for these commodities. The many uses for fish sauce are dealt with in detail in Chapter 3 but, for now it sufficient to point out that *liquamen* had a general purpose to flavor food in the kitchen and as such the reading (1) a "useful juice for food" would seem particularly apposite. That mackerel is the intended species in this passage would seem to be clear from the many sources of evidence for this species in residues (Chapter 11), the evidence from amphora *tituli picti* (Chapter 12), and from literary descriptions of fish sauce (Chapter 2). The next lines in the *Astronomica* take us to another place and time on the beach, where a third sauce, which we have learnt from Isidore of Seville was a more ordinary *liquamen*, is described.

> "*Aut cum caeruleo stetit ipsa simillima ponto squamigerum nubes turbaque immobilis haeret, excipitur vasta circum vallata sagena ingentisque lacus et Bacchi dolia complet umorisque vomit socias per mutua dotes et fluit in liquidam tabem resoluta medullas.*"
> 5. 676–681,
> "Or when the cloud of scaly creatures stands still just like the blue green sea clinging together in a stationary mass, it is hauled out, besieged all around by a vast net, fills huge pools and wine jars, exuding its wealth of fluid upon each other, their innards' melt and issue forth as decomposition."

This passage is less problematic. We have the basic small-fish *liquamen* identified by Isidore of Seville, described in the *Geoponica* and corroborated by zooarchaeology, giving credence to the idea that this kind of sauce represents the every-day essential substance that we are looking for. That it is another form of *liquamen*, rather than a purpose made *allec* is clear from the lines 'wealth of fluid' and 'their innards melt and issue forth' are a precise description of what happened to each small fish in the presence of salt and warmth as we can see in Chapter 8. We should note that this sauce would also generate an *allec*, though Manilius considered it unnecessary to mention it here, possibly because it was considered so cheap. It would seem logical to assume that the *allec* from the larger species would be more desirable as there would be a richer and more nutritious paste derived from fish with a greater volume of flesh, though others disagree (Bernal *et al.* 2016:196; Bernal *et al.* 2013:2529).

Later Latin recipes for "quick-cooked" sauces

There are a further three fairly obscure texts which provide recipes for types of fish sauce that are cooked. These quick-cooked sauces are then blended with other ingredients to make other kinds of cooked (reduced) *confectio*, compound table sauces using spices, wine and honey. As we will see these compound sauces resemble the original *oenogarum* dressing found in *Apicius* and Galen though they do seem to have evolved from their original form (see pages 85–87).

Pseudo-Rufius Festus' *Breviarium rerum gestarum populi Romani* ("Summary of the History of Rome") contains a medieval gloss added to the late 4th century work of Rufius Festus which is a quick-cooked recipe for a *confectio gari*. Small fish, salt to make a brine and wine are boiled together and reduced and then filtered and bottled (Curtis 1991:192).

Sextus Julius Africanus (160–240 AD) composed an encyclopaedic work in Greek entitled *Kestoi* (Κέστοι means "Embroidered"), which has been called as "a work of agricultural and natural history" (Thee 1984:145). This includes a recipe for a *garum* sauce called *sokkios* (from Latin *sociorum?*) "which was praised more than all other sauces." It is boiled with spices and honey and made with one-part pure brine and two parts sweet wine and, apparently, tuna meat from the belly rather than blood and viscera which is normally associated with the original concept of a *garum sociorum* (Varges *et al.* 2014:68). This is a long way from the idea of a blood viscera sauce and we should not be tempted to see the inclusion of tuna meat in this recipe as an indication that a true blood viscera *garum* of the *sociorum* type originally contained fish flesh. The fact that this is a boiled and cooked sauce places it in the category of a late development where cooking had apparently become the normal way to make quick sauces and at a time when *garum sociorum* had lost its original identity and meaning (see pages 76 and 88).

A Latin medicinal and culinary text from the 11th century (Paris, Bibliothèque Nationale, manuscrit Latin 11219) includes a recipe for *confectio ad garum faciendum*. In the recipe, two parts fish and one-part salt are blended and stirred together and then boiled with numerous spices (more than 25) and herbs, wine and honey; it is then reduced, sieved and bottled (Curtis 1991:193).

All of these three texts are written in Latin and demonstrate that in the late and medieval world the Latin term *garum* had to a large extent become synonymous with the general idea of a single fish sauce, with no distinction of quality or characteristic. As we shall see this clearly does not mean that it always had that general meaning. The sauces and the names evolved over time as we shall see and using this kind of late material to support the idea of a single sauce in the early empire will always be problematic as the evidence for two sauces is so strong. These sources provide little useful information about fermented fish sauce, as they describe a separate technique.[29] However, they do shed useful light on the later development of the cooked *liquamen/confectio*, and in that context we will return to them in Chapter 3.

In conclusion

We have reviewed the four key pieces of ancient literary evidence for fermented fish sauces. Manilius and Pliny the Elder provide a view from the early 1st century AD, while the *Geoponica* and Isidore of Seville demonstrate a perspective from the late empire. The time scale between the two perspectives is immense, but there is a substantial amount of conformity and agreement. If these four texts were all we had to work with in defining ancient fish sauces, we would not have much to argue about. There are four varieties of sauce;

1 *Garos/garum?/liquamen:* a liquefied sauce of small and medium sized fish and salt, sometimes with extra viscera, sometimes the fish are cut open and sometimes additional *allec* residue is added.
2 This generates a residue called *allec*, composed of the undissolved fish flesh as a paste, with or without bones.
3 *Garum/garum sociorum/haimation* are terms which indicate a sauce derived from the fermentation with salt of internal organs and blood of many different species (specifically mackerel, but also tuna) and potentially any other species that can be bled and eviscerated.
4 *Muria:* a sauce derived from the long-term storage of solid salted fish.

Notes

1 We note that the sauce known as *halmer* (ἁλμυρ) and *muria* in Latin, when this is a brine associated with salted fish, does not seem to be considered a fish sauce under the heading of a *garon*.
2 There are numerous different spellings of this term but they almost certainly indicate the same product, that is, a fish paste derived from the manufacture of a *liquamen : allec, alec, hallec, hallex allex*. Van Neer (Van Neer *et al.* 2010:163) notes the sources for these spellings but follows the normal line of assuming that the residue is the *allec*.
3 Athenaeus suggest that a sauce could be made or brewed together with vinegar as opposed to wine at this early stage (*Ath.* IX 366C). Vinegar was also added to the finished sauce when blended as an *oenogarum*.
4 Translation by Christopher Grocock.
5 The archaeologists at Cadiz used the Gargilius technique when attempting to replicate the fish sauce that was found in the dolia from *Garum* Shop in Pompeii. The numerous problems with this recipe were not taken into account when choosing to make fish sauce this way (Varges *et al.* 2014).
6 Many medieval spices were originally considered aromatics and desired as incense for their perfume, and the transition from experiencing and desiring their smell to desiring their taste is one shrouded in mystery (Miller 1969:1–26).
7 A rare sample of herring has been found in residues at Herculaneum from the Cardo V sewer (Nicholson *et al.* 2018:276).
8 Quinta do marin Alhoa (Algarve) 3rd century AD, sardine and anchovy; La Trevessa de Frei Gaspar (Setubal) late 3rd century AD, sardine and bream; Troia (Setubal) 5th century AD, sardine and bream (Desse-Berset and Desse 2000:91).
9 Aratos, 3rd c. BC; Aristotle, 4th c. BC; Demokritos (Belos of Mendes), Hellenistic; Cassius Dionysius, 2nd c. BC, citing Mago, the Carthaginian agricultural source; Diophanes, 1st c. BC; Paxamos, c. 150 BC; Plato, 4th c. BC; Theophrastos Nicopolis, 4th c. BC; Varro, 1st c. BC; Xenophon, 5th/4tc c. BC (Dalby 2003:36–49.)
10 Africanus Sextus Julius, 3rd c. AD; Cassius Bassus, 6th c. AD; Florentinus, 3rd c. AD; Nestor of Laranda, 3rd c. AD; Oppianos, 2nd c. AD; Pamphilos, 1st c. AD; Pelagonius, 4th c. AD; Plutarch, 2nd c. AD; Ptolemy, 2nd c. AD; Quintilius, 2nd c. AD; Theomnestos, 4th c. AD; Berytius 4th c. AD.

11   Aliud etiamnum liquoris exquisiti genus, quod garum vocavere, intestinis piscium ceterisque, quae abicienda essent, sale maceratis, ut sit illa putrescentium sanies. hoc olim conficiebatur ex pisce, quem Graeci garon vocant: *HN. 31.93.1.*

12   nunc e scombro pisce laudatissimum in Carthaginis spartariae cetariis sociorum id appellatur, singulis milibus nummum permutantibus congios fere binos. nec liquor ullus paene praeter unguenta maiore in pretio esse coepit, nobilitatis etiam gentibus. scombros et Mauretania Baeticaeque et iam Carteia ex oceano intrantes capiunt, ad nihil aliud utiles. laudantur et Clazomenae garo Pompeique et Leptis, sicut muria Antipolis ac Thurii, iam vero et Delmatia: *HN* 31.94. Adapted translation from the Loeb (Jones 1963)

13   The origins of *hydrogarum* are obscure, but it was originally a Greek concept (ὑδρογαρον) though its uses are obscure. Heliogabalus, emperor from AD 218 to 222, is said to have introduced *hydrogarum* to the table: 'he was the first Roman emperor to serve *hydrogarum,* which before was only a soldiers' dish, which habit Alexander promptly restored' (*Scriptores Historiae Augustae,* Aeli Heliogabalus 29.5.1). The implication is that before it was rather more commonplace, and certainly not served to consume, though it may be that the soldier drank it.

14   1.45 (3.p. 176,6) cited by Leon (2001:176).

15   Valdes, who bases his edition on the text edited by Max Wellman (Berlin 1958 = 1907–14), notes that the textual transmission of Dioscorides is fraught with problems, as editors and copyists added and removed contributions, while there were numerous versions in Latin and Arabic displaying many different readings. (Valdes 1998: V Textual transmission).

16   Galen, *Opera Omnia*, ed. C.G. Kuhn (1965 reprint of 1823 ed., Georg Olms, Hildesheim) 12.637. (comp. med. sec. loc).

17   The evidence for a similar scale of production in the bay of Naples may simply have not been found yet. To calculate the volume of sauce that could have been generated by dolia compared to *cetariae* is problematic. Re-brining of concentrated sauces allows for a considerable increase in the volume of potential sauce from a batch of fish which could indicate that the production capacity of the smaller vessels was far greater than appears. See page 163, but See Pena 2007, Wilson 2007 and Ellis and Devore 2010 for a different view.

18   Brine from designated fish is also a regular occurrence in didactic recipes. Dioscorides 2.34 says that a *muria* of sprat was efficacious to the ears.

19   In Apicius an instruction to wash a stomach with *muria* clearly refers to simple brine while references in Horace and Ausonius, discussed page 104, refer to a culinary sauce. Apicius 7.1.6;

20   Vitium huius est allex atque inperfecta nec colata faex. coepit tamen et privatim ex inutili pisciculo minimoque confici. apuam nostri, aphye Graeci vocant, quoniam is pisciculus e pluvia nascatur (Pliny, *HN.* 95.1).

21   transiit deinde in luxuriam, creveruntque genera ad infinitum, sicuti garum ad colorem mulsi veteris deoque suavitatem dilutum, ut bibi possit (*HN.* 31.95).

22   aliud vero <est> castimoniarum superstitioni etiam sacrisque Iudaeis dicatum, quod fit e piscibus squama carentibus. Jones (1963:436 note *b*) says "since *allex* is feminine and *aliud* neuter, it seems best to suppose that there is a lacuna here, but Pliny may be thinking of *garum*, to which he has just reverted."

23   sic allex pervenit ad ostreas, echinos, urticas maris, mullorum iocinera, innumerisque generibus ad sapores gulae coepit sal tabescere.

24   The most cited are Vergil (over 190 citations), 39 Cicero (over 50), and Lucan (some 45). Other much-cited writers are Plautus, Terence, Lucretius, Ovid, Horace, Juvenal, Martial, Ennius, Sallust, and Persius (Barney *et al.* 2006:15).

25 The Islamic tradition in Spain was an inland culture, and what fish was con-
    sumed was smoked and dried (Bernardes 2015:64). A small-scale and local
    production of similar products may have continued on the coast in small
    pockets, escaping archaeological and historical notice.
26 There are complexities here. If the little fish were eviscerated as a cleaned
    salted fish product as in the *Colatura di alici* (see page 231) then they would
    not dissolve easily. Cleaned salted fish get harder and harder in brine.
27 This is a new translation by Christopher Grocock of the Latin text by Gould
    (1977).
28 Hendiadys: in Latin, a figure of speech where two nouns are used in a phrase
    connected by 'and', where one of them functions as a descriptor.
29 Other interpretations seek to join the fermented technique with the cooked in
    one process discussed in Chapter 3 (Varges *et al.* 2014:70).

2 Fish sauce in the consumption literature:

a literary and archaeological chronology

It has been noted by Wilkins that the incidence of references to *garos* in Greek sources is remarkably scarce (Wilkins 2005:28). Despite its apparent popularity among elite Roman diners, Athenaeus also refers to it only rarely in the dining culture of Rome in the 2nd century AD. It appears in the numerous quotations that he takes from his sources, namely Old, Middle, and New Comedy from Athens in the 4th and 3rd centuries BC. It does not figure at all in Archestratus, which is remarkably odd given that this writer has fish as his main theme. These absences are strange indeed, particularly as many didactic writers such as Galen, Dioscorides, and Columella make reference to it in relation to everyday culinary medicinal and veterinary writings, and it also figures in *Apicius*. These anomalies are indicative of a situation in which *garos/garum* is perceived very differently by different social groups in ancient Greece and later in Rome. It can appear to be valued but it is also looked on with mistrust while it is also apparently both insignificant and largely ignored by many. We can see that fish sauce did have a dual identity: *garum* was valued and yet feared, while *liquamen* was so ordinary as to be barely acknowledged.

Fish sauce from 5th to 3rd century Athens

The earliest reference to a fish sauce is in Greek drama. A satyr play by Sophocles called *Triptolemus*, dated to the 5th century BC, contains the first reference to *garos*. Satyr plays generally occurred at the Athenian Dionysia as comic relief to break the oppression of hours of gloomy and fatalistic tragedy, and were typically farcical with sexual inuendo and slapstick. There is no context of any kind for the line in question (it is cited by the later writer Athenaeus because of its connection with food) and it is very intriguing. A "wretched women" (τάλαινα δοῦσα) has brought a ταριχηροῦ γαροῦ, pickled fish *garos* (*Triptolemus*, fr. 606); τάριχος just means preserved or pickled fish, the implication being that *garos* was made in the process of preserving fish. This may be a reference to the whole fish or a sauce it is not clear. The term τάλαινα has a strong association with poverty and misery. Can we make the assumption that, as the women was

poor, the product she brought was also of low value? This is the implication, though where she was and why she brought what she brought is lost:
it is likely to be associated with some sort of religious offering. A fragment
from a satyr play by Aeschylus called Proteus simply refers to ἰχθύων γάρῳ,
"and the fish *garos*" (frag. 211).

References in 5th century Old Comedy are equally ambiguous. We have
an unknown play by Pherecrates (frag. 188, Athenaeus ii. 67c) which suggests that someone could get their beard dirty with *garos*. *Garos* in later
Roman sources is a crystal clear, limpid liquid, but one might think that a
product that was thicker, more like a paste and somewhat more clinging,
would be more visible on a beard. Does this mean that this early *garos* was
a thicker paste? A fragment from Cratinus, a 5th-century comic writer
contemporary with Aristophanes, is intriguing as it says "your basket will
be full of *garos*" (Frag. 312, Athenaeus ii. 67b–c). The basket (τάλαρος) is
associated with cheese making and was surely full of holes, in which case,
we are entitled to wonder how such a basket could be full of a liquid or even
a sauce as paste. We may have a reference to the whole small fish freshly
caught. If sauce it may have been made by local people and on a small scale
by characters in the play, possibly fishermen, and there is a direct conversation about this process in the play. It may also refer to a pile of small fish in
the basket still in the early stages of fermentation and intact. The final 5th
century comic fragment is from Plato: "they are going to choke me to death
by dipping me in rotten *garos*" (ἐν σαπρῷ γάρῳ, Athenaeus ii. 67c, frag. 215).
The use of σαπρός ("rotten," "putrid") is typical from later sources and
tells us that this product right from the beginning was viewed with distaste, but it also implies, I think, that *garos* was a semi–liquid product
and being made locally in quite large amounts, given that the vessel was
in principle large enough to get a person inside! The characters are talking
about *garos* as though the production process was sufficiently familiar to
a drama audience to engender humor. The audience may well have been
familiar with the sight of vessels of dissolving fish on quaysides in coastal
communities.

From Alciphron's letters, written in the 2nd century AD but potentially
using extant plays from this earlier period, discussed on page 16, we learn
that fishermen supposedly from the 4th/3rd centuries BC made *garos*, but it
seems they made it by boiling the fish. A fisherman called Sosias is known
for "boiling that tasty and useful *garos* from the tiniest fish that he catches
in his net" (Alciphron 18.2). It is quite possible, therefore, that the earliest
garos were actually boiled, that is, made quickly with the simplest technology, rather than fermented. It is not that clear as yet that fermenting *per se*
is actually happening. In these circumstances, the resulting mixture could
be more paste-like if a minimum amount of liquid was used, or clear if fully
filtered; we cannot be sure. The *garos* may also be just cooked and eaten.
The context to this 2nd century AD description of fish sauce production is
key. The supply of fish sauce in the Roman world from the mid-2nd century

AD onwards does become more unpredictable, as many factories ceased to trade and fish sauce amphorae become less visible in the archaeology. It is therefore possible that supplies of *garos/liquamen* were limited in some places during this time. We are certainly told that the incidence of cooking a *garos* sauce became more common in the late empire (see page 39). Alciphron is perhaps describing fish sauce as he knew it, that is, cooked. We cannot be sure that cooking was the original technique, though as cooking permeates the evidence, it must be given consideration (see page 88).

The earliest information we have about *garos* does not indicate that it was particularly desirable. In fact it is associated with poverty, and fishermen seem to make it from their haul of the tiniest fish, which are always viewed as low status in ancient texts. We have reference to fishermen being condemned for taking fish too small, that is, not letting them come to full size, in a play by Alexis (*Odysseus Weaving*, frag. 159; Athenaeus. vii 303a) and this may be connected to the fact that fishermen were largely poor men who were exploited by the middle men selling fresh fish in the markets and were reduced to making this kind of *opson* or "relish" with the tiniest fish for themselves (Mylona 2008:67–74). The tiniest fish do not need to be eviscerated, just as whitebait do not, but as they get larger and reach 5 to 10 cm. then evisceration may have occurred if the product was cooked. It is from surviving Middle and New comedy that we know about the use of another fish sauce which seemed to be much more popular and desirable in the elite culinary world of Athens in the 4th and 3rd centuries. This is a brine known as ἅλμη (*halmē*) and ἁλμυρίς meaning "saltiness," from which the Latin term *muria* was derived. There are numerous references to what was called "Thasian *halmē*," which was almost certainly a pickled fish brine from the island of Thasos. Thasos is in the northern Aegean, close to the mainland, and this sea is richer in nutrients than that further the south (see page 172). Thasos nowadays generally supplies sardine and anchovy to its own fish markets from local waters, and while there is a belief that the seas in ancient times were more productive, it is possible that the smaller species (clupeids: sardine and anchovy) were the main source of this ancient *halmē*. Tatiana Theodoropoulou has recently highlighted that fact that faunal evidence from Neolithic sites on Thasos demonstrate that small tuna, mackerel and mullet were processed at the site and this may suggest that larger species were potentially processed in later periods (Theodoropoulou 2018:395). Thasos is known to be on the route that many migrating Scombridae species take to and from the Black Sea, and it is also very likely therefore that a brine from a tuna product was available (Theodoropoulou 2018:397; Mylona 2008:50). A later 3rd century AD papyrus refers to a Thasian ceramic jar containing *kubia* and *horaia*, the former being small pieces of tuna or possible bonito, which is believed to have been transported in reused jars and would naturally produce a brine for culinary use (Lytle 2018:411). However, a fragment from Aristophanes refers to the "unfortunate one who was first to be immersed in pilchard-brine (ἐν ἅλμῃ τριχίδων;

Aristophanes, *Merchantships* frag. 426; Athenaeus vii. 329b), which was apparently a brine made from small, particularly bony, insignificant anchovy, τριχίδιον. Lytle actually doubts the validity of the *Thasia halmê* as a sauce derived from a commercially traded fish, and suggests that it was a type of locally made sauce, largely made from small fry which he suggests were "reduced" in brine, though what he means by this is not totally clear as it has to be somehow different from *garos*? I suspect he sees this as a kind of *garos* with little culinary value (Lytle 2018:414). This I do not find convincing. *Halmê/muria* has substantial culinary capital in later sources, and a tuna *halmê* made in an experiment (page 63) certainly resulted in a product which tasted quite exceptionally good, and thus may have had the culinary value assigned to it in the sources, discussed below. Brine-salted tuna and mackerel in some form of unidentified ceramic jars or reused amphorae would seem to have been a regular commodity throughout the classical period, and the difficulty with the archaeology – lack of processing sites and ceramics – would seem not to be sufficient justification to doubt the evidence of the literature. (Theodoropoulou 2018:401). Species-specific *muria* are rare on later Roman amphora *tituli picti*, but medicinal sources from the later empire do stipulate particular fish brines such as a brine from a catfish (*siluris glanis*; Dioscorides, *De Materia Medica* ii. 29), and tuna and mackerel brine are found in Martial *xenia* gift poems (*Ep.* xiii. 103). A salt fish brine, as opposed to a fermented sauce, is delicate and subtlety fishy in flavor, and of course dominated by salt. A much later description of *muria* by Horace suggests that it has the essence of the sea about it, and that the juice from a sea urchin could be superior to *muria* (Horace, *Sat.* ii. 4. 63–71). Greek salt fish brine is offered as a dip that is served with roasted or barbecued fish and is eaten with the hands. Archestratus serves tuna steaks hot from the coals, dipping them in pungent brine; Thasian brine is sharp (Athenaeus iv 164e), and it could also be used as a cooking medium to stew fish. Saltfish brine could also be mixed with other ingredients such as oil, and this is quite surprising. Fragments from a comedy by Sotades, *Captive Women*, has a slave serve red mullet ἄλμη τε λιπαρᾷ, "with brine and oil," "placed beside the fish" (Athenaeus vii. 293c). In Aristophanes *Acharnians* 671, the chorus recites a list of repetitive physical activities that happen in the kitchen, fanning the fire, kneading bread and beating the Θασίαν λιπαράμπυκα. It is not immediately clear what is happening: we have a Thasian brine, clearly, and with oil from the λιπα. An ἄμπυξ is a lady's headdress, so an oily band of oil on the surface would seem to be the best guess; the activity is the beating of the mixture to create an emulsion. One immediately thinks of a vinaigrette. This is very reminiscent of the later use of *garos* to make oeno*garum* sauces blended to make a similar type of dip. This implies that the very ideas of a blend of oil and fish sauce was developed using fish brine not *garos*.

A late 4th century comedy by Archedicus, *The Treasure*, has a cook who stews his fish and serves it with a "perfect brine which any free man could dip his food into" (Athenaeus vii. 292). This has a hint of exclusivity: only free men of status dip their food in brine. Is this in contrast to that "other" sauce so little mentioned in our sources that poorer people dipped their food into, that is, *garos*? In 4th and 3rd century Athens this salt fish brine seems to have culinary cultural value among elite diners, while γάρος, is both rare, and when it does appear it has a low social standing. One way to approach these two images of fish sauce is to consider who is making it and in what quantity. If this sauce was made locally, from small catches of little fish which were left to dissolve into a paste, it would be pretty unappealing in both a visual and olfactory sense, and particularly in relation to the clean and bright appearance of *murialhalmē*. Did the first *garos* appear as a paste? We have seen that a man could get his beard dirty with *garos*. What does this imply? A fine liquid would not be visible on a beard, but a paste would. There is an expectation that *liquamen* was a clear liquid when in use: that is certainly the impression one gets from the recipes (discussed page 126).

At some point we have to imagine that this embryonic *garos* changed from a thick fish paste eaten by peasants and poor fishermen into a crystal-clear, dark, honey-colored sauce that we would find more familiar. One suspects it did not become more widely used in Greek cooking until it had become more visually appealing and presentable, and even then, it seems to have been utilized predominantly in the kitchen rather than at table. Salt fish brine is fundamentally different from *garos* as the salting process is clean and devoid of the digestive enzymes that are present in the viscera. Viscera are fundamentally repugnant, but it is only when it is retained that the transformation takes place whereby solid protein is converted into a liquid form. Fish brine is less nutritious, which in turn makes it less tasty, certainly in relation to *garos*. One can definitely, from my own experience, say that it has limited *umami*, and *umami* is very desirable. A brine derived from freshly salted tuna has a delicate flavor, quite appealing in fact, with none of the "in your face" pungency of *garos*. A fish brine does not appear to be the result of decay and putrefaction, which was the image that clung to *garos*, and as a result it may have been more appealing to the elite. One can quite easily imagine someone attempting to unite the clean, crystal-clear image of άλμη with the umami rich taste of *garos* in a new processing technique that removed the fish paste and the bones and made a rich darker liquid that had the potential to add a remarkable transformation to everyday food. I fully admit this is no more than an imagined scenario. I have attempted to bring all the elements together and create a coherent picture of early fish sauce in Greece. We can see a different and possible new image of *garos* as a useful seasoning in the story in Athenaeus of Philoxenus, the 4th century food writer, who is said to have entered other people's homes to "season whatever was being cooked for everyone" with oil, *garos*, wine and vinegar, so that he could correct the seasoning of the household cook (Athenaeus i.6a). *Garos* here has become a regular and useful liquid, "tasty and useful" as Alciphron claims later, to season all manner of foods in the kitchen and this is how we find *garos* being used in the later Hellenistic and Roman periods.

Fish sauce in the archaeology of 5th to 3rd century Greece

The archaeological evidence in the form of fish bones that might be identifiable as sauce residues or evidence of salted fish are sadly lacking in this early period in Greek culinary history. Indeed, corroborating archaeological evidence of the act of salting fish from prehistory is scarce, and what is available to us is discussed Chapter 9. There are no processing sites in Greece comparable to those found in the Punic west that have identifiable *cetariae*, the absence of which, has led to the assumption that there could be no systematic commercial enterprise in fish processing in the Aegean, and this absence is true for the entire archaic and Classical period (Theodoropoulou 2018:393). There may be a problem with this assumption, as the facilities in the West should not necessarily be taken as a template for fish processing elsewhere and in all periods. Fish salting techniques are very varied and can be undertaken in *doliola,* and other multi-functional vessels and traded locally in reused vessels of all kinds and these techniques cannot be recognized or confirmed as fish processing. An estimate of the productivity of such small-scale processing techniques are difficult to determine and may be greater than expected, while the productivity of the vast *cetariae* in the West may in fact be on a scale even greater than currently envisaged, an issue explored in Chapter 10. Bones derived from a residue of fish sauce made from smaller fish are more fragile and less survivable in the archaeological record. However, if, as would seem probable, the early *garos* fish sauce was made in relatively small quantities as a paste with the smallest fish, then the bones would be particularly unlikely to survive in sufficient quantity to be recognized as a potential fish sauce residue and it is probable that the volume of fish sauce being made at this time was simply too small to make an impact on the environment. Alternatively, fish pastes were consumed bones and all and there would be no residue to find. There is however plenty of evidence for the consumption of the small fish associated with fish sauces. Mylona (2014) has indicated that many sites in the southern Aegean Seas in the Bronze Age have evidence of consumption of picarels (*Centracanthidae*), bogues (*Boops boops*), sea bream (*Diplodus annularis*), and pandora (*Pagellus erythrinus*). These are species typical of the southern Aegean, and are fished close to the shore with simple techniques. They are also typical of fish sauce residues found in later periods, such as those found at Saltsburg in an undated Roman amphora (Von Den Driesch 1980; Lepsiksaar 1986).

Evidence of a fish paste has been found in a storage vessel in Akrotiri. The product was preserved by the special conditions of the volcano which destroyed the site in 1650 BC. The product was a form of fish paste made from similar species, such as picarels or bogues, along with small sting rays, and cereal and a potential condiment was also present (Mylona 2014; 2016). Initially it was suggested that this might be evidence of an *allec*, that is, a fish sauce residue in the form of a paste, but as we will see this definition is in itself a distortion of the concept of *allec*, and the term is so contested that identifying this product as an *allec* might be unhelpful (see page 229). Subsequent analysis of the Akrotiri fish paste has redefined it as possibly a "cooked" and or "marinated" product (Mylona pers. comm.).

The unique Punic amphora building in Corinth dated to the mid-5th century represents the earliest evidence for a Mediterranean-wide trade in salted fish, albeit in what seems to be a modest volume. The species present were tuna vertebrae, and scales and sea bream, as well as numerous small *clupeids*. We know that the amphorae, on the site and probably the tuna fish, came from Punic Cadiz in Spain and/or North Africa and possibly Sicily (Zimmerman-Munn 2003; Theodoropoulou 2018:397). The most recent report on the site is Sáez Romero *et al* (2020). The small *clupeid* species at the Punic amphora building are in sufficient numbers that they may be indicative of a *garos* in the form of *allec*, which was probably a local catch (Theodoropoulou 2018:397). Literary evidence from Greek drama also confirms this early trade between Spain and Greece and a similar early trade between Greece and the Black Sea for a solid salted fish product, and as a result, a trade in the resulting fish brine as a condiment. In the 4th century BC Archestratus praises fish from around the entire Aegean in his fish lover's guild, and Aristotle recognizes Byzantium as a source of traded fish (Olson and Sens 2000; Mylona 2008:126). Antiphanes, a mid-4th century comic poet, praises tuna from Cadiz and Byzantium (frag. 78) and Cratinus, the 5th century comic poet, implies that Black Sea saltfish is available in Athens (frag 44; Athenaeus iii.118d/e).

From the current evidence we have for *garos*, it would seem unlikely that there would have been a market for this kind of fish sauce large enough to maintain a trade from outside of the Greek mainland. There is a marked absence of archaeological evidence for a commercial trade in any kind of salted fish or sauce throughout the classical and Roman periods in Greece which has proved difficult to understand. The implication for Lytle is that a local small-scale industry always existed but it was not of a scale to constitute a commercial endeavor (Lytle 2018:409). *Garos* was clearly at this time a locally produced low value and potentially poverty food with little opportunity for commercial exploitation and probably made in relatively small amounts in many households, and it does not appear that there was a viable market for a traded version of *garos* from elsewhere. One might expect that the demand for a foreign-made product would derive from the elite driving the market and, as it would appear that the consumption of *garos* in Greece was not sufficiently conspicuous to merit outside interest, it is unlikely that a trade existed in *garos* produced and traded from outside Greece in the 5th and early 4th centuries. Absence of evidence alone can never be sufficient to justify a point such as this and it is clear in fact that many others consider that a tuna *garum* was traded alongside *salsamenta* from Spain all over the Mediterranean at this time (see below). There was nevertheless a growing interest in the use of fish sauce in cooking, which we can see from Philoxenus in the 4th century, and we may reasonably conjecture that the demand for a consistent and regular supply of a good quality *garos* increased exponentially during the 3rd century. The first evidence suggesting a spread of *garos* outside of Greece comes from the late 4th/early 3rd century find at Olbia in Sardinia where small pickerel bones in a baggy jar are assumed to be a form of *allec* (Dellusi and Wilkens 2000:53).

Fish sauce from Hellenistic Greece

Greek literary evidence for fish sauce during the 3rd century BC through to the beginning of the conquest of Greece by Rome in 146 BC is very scarce. However, it is likely that there was a steadily burgeoning interest in fish sauces across the Mediterranean as the demand for them grew and their use became more widespread in the kitchen and among the slaves who worked with food, though we have a limited amount of concrete evidence for this. The sudden burst in productivity at the end of Republican period is rather indicative of what must have been a growing commercial trade. It has been suggested that there may not have been a commercial trade in fish sauce or salted fish within Greece. Lytle has suggested that the preference in Greece was always for fresh fish for the table and that the risks involved in a local commercial trade in wet salted products was too great. The presence of evidence of a salted product is assumed to be from external trade from either Spain or the Black sea. This does not accord well with the value Archestratus places on salted fish in his poem. He praises products from Sicily and the Black Sea, but he also has high praise for recently cured freshly salted three-day-old mackerel, even though this was never a valued table fish (Archestratus, *Frag.* 39). Much later, in the 2nd century AD, Galen says that fresh tuna is unpleasant and is best after pickling (Galen, *On the Properties of Foodstuffs* 3.25.6). It is not clear to what extent Archestratus' readers, who did not reside in Sicily, were expected to travel to get their Sicilian tuna from its source or to be able to acquire it closer to home. Wilkins suggests that all the salted fish dishes that Archestratus praises were part of a luxury cuisine, and that the only fish that the ordinary people of Greece could access were the small-fry, which were eaten rotten, either dried or salted down into *garos* (Wilkins 2018:231). I am always less convinced by arguments such as these which make such stark divisions in society in relation to access to commodities that were only defined as "luxury" because elites talk about them in their writing. The fact that elites discuss these commodities does not mean that they also make them inaccessible to anyone else. The problem here is the definition of luxury, ordinary and poor rather than any real evidence that salted fish was restricted to a very small percentage of the society.[1] Lytle argues that consideration such as profit and uncertainties in a commercial market would have overridden any rational desire on the part of the fishermen to preserve and sell excess catch, which denies the fishermen any intelligence or agency at all to act rationally. Such economic arguments as these surely go too far (Lytle 2018:410). A small-scale local trade in excess catch would almost certainly be invisible in the archaeological record, and absence of evidence is clearly never enough. Lytle points out that there is literary evidence for a cheap and commonplace dried salted fish from the Black Sea made with small species such as anchovy, sprat and horse mackerel (*saperdês*) that was traded in baskets called *sarganê* (Archestratus *Frag*.39. 3–4; Lytle 2018:410).[2] It is described as a "foul smelling, coarse common food" and just as foul as *garos* was perceived to be but clearly not quite as foul (Olson and Sens 2000:165). A dried product is much easier to trade and transport on land and leaves no trace, while there is an absence of amphorae either in Greece or from the Black Sea area to trade a wet salted product at this time. When wet salted products from the Black Sea are found in Greece in the 3rd century AD, they appear to have been traded in reused amphorae originally designed for wine and oil, which might suggest a low volume of trade but not necessarily one which was unsystematic. Reuse of amphorae is little understood and may have been far more normal than is often perceived (discussed in Chapter 12; Lytle 2018:411). Most importantly, the wet salted trade has to have been sufficiently organized to generate the desirable forms of brine that was utilized in the dipping sauces. The overriding impression is that in Greece in the Hellenistic period there was a widespread but relatively small scale and local production of all kinds of salted fish products: a dried small fry product, *garos* and *halmê* from the same small fry species, and a lightly salted small fry product, which may have been marketed as "fresh."[3] This is defined by Ørsted (1998:23) as a stage 1 subsistence or *oikos*-economy with minimal profit. Alongside this local product there was an indeterminate wider stage 2 trade which involved collaboration and infrastructure from Sicily and the Back Sea of wet salted tuna, mackerel, bonito and small fry species transported in purpose made and reused amphorae which is hard to see in the archaeology.

Fish sauce in the Roman Republic

The culinary Hellenization of the elite of the Roman republic is better illustrated in other aspects such as literature and the plastic arts (Dalby 1996; Dalby and Dalby 2000 Grocock and Grainger 2020), but there are many gaps in our knowledge in relation to how fish sauce was introduced to Rome and how it was perceived by the various social groups in this early period of Rome's gastronomic education. We are informed that Rome appears to fall under the spell of Greek gastronomy in the mid-2nd century BC. Prior to this "fall" the Romans considered themselves to be unspoilt "porridge-eating barbarians" and with simple tastes (Plautus *Mostellaria* 828; Leigh 2015:48; Gowers 1993:53). There is a little-known reference in Pliny the Elder which is rarely brought into the picture, the introduction of which has been postponed until now. When talking of the consumption of *caules*, spring cabbage shoots, Pliny says

> "Nor did the people approve very highly of *caules* as they do now, (mid-1st century AD) since they (looking back at the late Republic) condemned a *pulmentaria* (relish = that which is eaten with bread) which needed other *pulmentaria* to get them down. That meant sparing the oil (too) for the desire for *garum* was a matter of disapproval."
> (*HN*.19.58.1)

This needs unpicking carefully. *Caules* need cooking and it seems they need a sauce too, which the Republicans saw as extravagant. To have a sauce required not only oil but *garum*, but clearly *garos/liquamen* is meant here and in this context, they seem to go together: confirmation of this comes from the fact that fish sauce was always blended with other ingredients, principally oil or vinegar in our early Greek sources. Pliny appears to be saying that he believes the early use of *garum*, and in fact the use of dressings, was frowned on as extravagant, and a culinary idea from Greece that the "porridge-eating barbarians" disapproved of because it was foreign. This may potentially be read as indicating that fish sauces per se were not consumed widely in republican Rome. It is therefore difficult to say to what extent this passage is indicative of a widespread lack of "sauces" in the pre-Hellenistic Roman diet. The simple unpretentious Republican culinary culture may well have already included the consumption of various kinds of small-fish pastes, cooked or marinated, and particularly in coastal areas. The small fishermen may have developed their own version of *garos*, long before the elite were influenced by Greek culinary fashion, though there is no direct evidence of this. It nevertheless remains a strong probability. The Hellenization of Roman eating habits is documented by Livy.

> "the banquets themselves began to be prepared with greater care and expense. The cook whom the ancients regarded and treated as the lowest menial was rising in value, and what had been a servile office came to be looked upon as a fine art. Still what met the eye in those days was hardly the germ of the luxury that was coming."
> (Livy 39.6.9; Roberts 1905)

The Roman slave cook was not valued, it seems, but the Greek *mageiros* was a free man, and respected at least for his talent if not for his arrogance. Such men were able to write their own recipe books, though they do not survive beyond fragmentary references in Athenaeus (Grocock and Grainger 2020), and it is likely that Greek cooks considered Rome as a potential new market for their skills. Free-born cooks from the Greek colonies in Magna Graecia would appear to have arrived in Rome to offer their services as early as the mid-2nd century. This can be understood from the complaint from Cato the Elder that men "who happen to cook," but clearly cannot be slaves (though one had a Greek name) had statues erected in their honor in Rome (Cato, *Frag. 96*, Malcovati 1955). If such people could circulate and offer their services and gain a reputation for cooking, then Greek cooks and their cookery books must have circulated in high-status circles, and possibly further down the social scale as well. We have seen that *garos* has become a general seasoning along with wine oil and vinegar in food preparation as early as the mid-4th century BC in Greek comedy (see above). As we will see, this practice of using *garos* as a basic seasoning become very widespread in later Roman and Greek culinary sources. How long it took to be incorporated into the Roman way of life is hard to discern. That it was used by all social groups is evident from the fact that fish sauce amphora were found at small rural peasant communities such as Marzuolo in Tuscany in the early empire (Marzano 2018:441). The vessels were *Baetican* and we might conclude that the Spanish products were everywhere too, though there is evidence that fish sauce vessels were reused and often refilled with sauces, which may have been local products or otherwise cheaper. Reuse issues discussed Chapters 11 and 12.

Our first piece of evidence is a negative, as there is no *garos* or *garum* in either Cato's agricultural manual (c. 150 BC) or in any of the plays of Plautus (fl. 210–180 BC). However, in both of these sources we find evidence of something called *allec*, which we subsequently learn from Pliny and the *Geoponica* was the name applied to a product that was derived from the residue of *garos* production when it was made on an industrial scale, see Chapter 11 for a discussion on *allec*. In Plautus' plays, *allec* is a commonplace: a fragment from *Aulularia* has the line "those who offer me raw vegetables should add *allec*" and this is clearly a dip for the vegetables (frag v *Aulularia*; Wolfgang De Melo 2011). In the play *Persa*, which is peopled by characters from the world of slaves, prostitutes and unlucky parasites, *allec* is a suitable accompaniment to reheated ham (*Persa* i.III.107). The parasite is offered reheated leftovers, and shows a hint of disappointment when he says "*Ecquid hallecis,*" "Is there any *allec*?" The fact that the parasite asks for *allec* indicates that this combination, meat with a strongly flavored relish, was a recognizable everyday combination. He is expressing some distain for the plain leftovers, particularly as the slave indicates that other meats are being freshly cooked indoors which are clearly not being offered to the parasite. The fact that he asks for *allec* might suggest that he associates re-heated leftovers with this lower-value *opson* and the parasite is passing a sidelong comment on the stinginess of the slave by asking him for the *allec*. Two hundred years later, Martial can suggest that a poor miserly man who rejects luxury fare in public is apparently satisfied in private with "capers and onions floating in putrid *allec* and meat from a dubious ham" (Martial, *Epi.* 3.77.5). We may be dealing with a literary trope garnered from Plautus, or this may simply reflect what modest men could get to eat. The implication of these sources is that some sort of fish sauce was being made locally or traded from further away and the residue in the form of a fish paste was utilized as a food resource. The contexts to each of these references to *allec* imply it was commonplace and a lower value product. Cato offers *allec* to slaves as part of their ration when the figs have run out in c. 150 BC (Cato, *De Agricultura* 58), and this is normally viewed as evidence of Cato's stinginess, as it is assumed to have been a pretty unappealing food and even that it was all that they were given, which is unclear. It was also assumed to be full of bones, but we may be far too influenced by modern distaste for this product (and in any case it was almost certainly a bone-free paste; see page 229). We will hear shortly from Horace that in the 1st century BC some kinds of *allec* were considered elite fare. These incidences of *allec* consumption may therefore demonstrate that fish sauce residues were not necessarily low value or reject foods, but potentially desirable and appealing depending on species, freshness and recipe. The emphasis that Pliny places on *allec* is also indicative of a far more desirable product (see page 32).

The absence of any reference in Latin to the product that we know of as *garos* in Greek in this period is difficult to explain, and is problematic as it is the period of transition between Greek and Roman dominance in the Mediterranean. In order to acquire cultural capital, one had to emulate Greek culinary practices in Rome, and therefore conspicuous consumption of Greek cuisine including compound sauces and marinades. It is most likely that *garos* and *halmē* (*muria*) would form part of this cuisine, and emulating their use would therefore increase one's culinary cultural capital. As the Republic entered its last century, Greek cooks were passing on their skills, training slaves and selling them on (Athenaeus, xiv. 659d), exchanging recipes and developing techniques that would form the basis of an international Mediterranean cuisine which even at the height of the Roman empire we should not consider entirely "Roman." The situation might be equated with the culinary climate of Europe 50 years ago, when above a certain status, good food was based on the precepts of Classical French techniques. European chefs created a cuisine that was basically French in style. Ancient Greece might usefully be compared to France in this scenario while the rest of Europe maintains the same position in both eras (Grocock and Granger 2020). Greek recipe books were translated and widely disseminated in Rome, and we find numerous occasions where this Greek influence is to be found in Roman agricultural works such as Cato the Elder and Columella (Gowers 1993). It was one such Greek recipe book, that would form a substantial part of the later compilation of recipes known as *Apicius,* which took its final format in the 3rd/4th century AD. This is believed to epitomize Roman food yet the text is permeated with Greek culinary terminology, much of which is never translated as there was no equivalent. What happened is that many terms were simply transliterated into Latin forms, hence *garos* initially clearly became *garum*. Dalby has recognized that a particular coalescence and hybridization of the two languages was occurring.

> "To treat the phenomenon as one of borrowing from a foreign language is to oversimplify. Cultures were mingling and technical terms from one of the two languages might seem the only good way to denote a newly developed institution or concept."
>
> (Dalby 1996:183)

In the case of *garum* and *liquamen, muria* and *allec*, this coalescence of Greek and Latin in culinary language will prove very difficult to untangle. The demand for fish sauce is difficult to see in this period in the literature, and it is equally difficult to see in the archaeology. Nevertheless, in step with other phenomena of cultural assimilation, the increasing use of fish sauce was almost certainly infiltrating Roman society from the top down.

Fish sauce in the archaeology of the late 3rd and 2nd centuries in the Mediterranean

The evidence for the early trade in sauces is relatively scarce, and there are areas where commentators dispute the detail. At the height of the Roman fish salting industry the facts are relatively straightforward and I will do not more than set out the chronology so that the literary evidence can be seen in an archaeological context. Readers wishing for the finer detail should consult the Bibliography from the *Oxford Roman Economy* project[4] and the many papers of Dario Bernal Casasola (Bernal and Sáez Romero 2008; Bernal *et al.* 2018; Bernal 2018). A good basic outline can be found in Trakadas (2005) and Curtis (1991). The evidence for a trade in any kind of fish sauce at this time is that associated with solid salted fish, namely the *muria/halmē*, that is, the brine that these salted fish products were transported with. The amphorae used to transport solid salted fish products are well documented from the 3rd/4th century BC from Sicily (Botte 2009) and from the straits of Gibraltar at Malaga and Cadiz (Carrales *et al.* 2011; Bernal Casasola *et al.* 2008). The earliest vessels associated with salted tuna from Cadiz in the 5th century are categorized as Mana A4 and D type (T.11.2.1.3) and over the following centuries there are numerous Mana T variants with small changes in characteristics but they are all, from a practical perspective, designed to accommodate solid pieces of fish: the mouth is wide open to allow access, the neck is largely absent and the body is straight and/or much wider at the bottom and/or gently tapered (Bernal Casasola and Sáez Romero 2008:56). The overriding impression is that these vessels were designed to allow a maximum volume of solid pieces of fish to be contained within them and to allow easy access to remove these pieces, while the vessels were small enough to be handled easily (see Chapter 12). The distribution of the earliest solid-salted-fish vessels is relatively modest, apart from the finds in the amphora warehouse in Corinth, and largely within the Spanish mainland (Sáez Romero 2014:169). Over the period of the 4th to mid-2nd centuries BC these amphorae began to be distributed widely within the Mediterranean, indicating that the salted fish, assumed to be mainly Scombrid species such as tuna and mackerel, had reached Italy (Ostia), Gaul and north Africa, and Athens.

The question that concerns us here is how to pinpoint clear evidence of extensive trade in a product that we can define as "a liquified fish sauce" rather than solid fish in the archaeological record, and whether it was made within Italy for the Italian market or was imported from Spain. We have noted in Chapter 1 that it is currently taken for granted in many archaeological papers, that the manufacture of fish sauces, that is, *garos/liquamen*, was an integral part of the fish-salting process from the outset in the early salting factories in Spain (Corrales *et al.* 2011; Bernal *et al.* 2003). There is an assumption that if tuna and mackerel were salted whole then fish sauces, which are understood as a by-product utilising viscera substantially and also blood were an integral part of the same process, and they must also have occurred at the same time. In discussing the early site of la Redes in Baetica, the archaeologists assume that fish sauce was made on the site as early as the mid-5th century, a conclusion that cannot be corroborated. There is evidence of darkened soil at the factory which Vicente interprets as indicating decomposition of organic matter and probably associated with fish sauce production, but this is equally likely to be the disposal of organic matter without fermentation. The site was most active between 430 and 321 BC and was in fact in decline by 200BC and these dates simply don't accord with the idea of a growing fish sauce industry (Vicente *et al.* 2009:94). During the sites active period the amphora associated with the factory are characterized by the wide-open mouth and these amphorae are closely associated with *salsamenta,* particularly through faunal evidence of tuna scales and bones in Punic amphorae at Acinipo (Ronda, Malaga) dated in the 7th century BC and at Camposoto (San Fernando, Cádiz). In the late-Punic period (2nd century BC), the large scales of tuna measuring nearly 50 cm have also been found inside an amphora of the Ramon type T-7.4.3.2 at Baelo Claudia (García Vargas and Bernal, 2009:141; see also the discussion on amphora in Chapter 12). As noted, a trade in *garos* requires a commercial pull from a wider Mediterranean consumer and that cannot be identified with confidence in literary sources or archaeology of the Hellenistic period.

The archaeological report of the fishery at Cosa produced in 1987 by Anna McCann determined that a fish salting factory was present on the site alongside the fish ponds for breeding fish for the table from the late 3rd century through to the 1st century BC, though the evidence is not strong. The distinctive Dressel 1 amphorae generally assumed to be wine vessels were linked in their 1C variant to fish sauce products, though the evidence was conjectural (McCann 1987:33; Botte 2017:102). McCann suggests that when looking for a vessel to house his fish sauce, the trader Sestius who is linked to the site looked to a form that was actually a Spanish model, namely a form of Dressel 1C type 5 which were made from the end of the 2nd to mid-1st centuries BC in Cosa (Lyding Will 1987:201–203; Botte 2017:129). Botte has rejected this, countering that the Spanish versions of this form which Sestius is supposed to have copied in the late 2nd century BC were actually made at Baelo Claudia much later in the 1st century BC (Botte 2107:135). The port of Cosa was at its most active in the 3rd and 2nd centuries BC and ceased to operate by the 1st century BC, just when we expect to find evidence of an increase in trade in sauces. It has recently been outlined by Botte (2017) that the evidence for an early salting *cetariae* at Cosa is missing and in fact that all the subsequent salting workshops, 5 in total, in Etruria and Latium are imperial in date and correspond to the later thriving market in fish sauce from the beginning of the Augustan age. Evidence of fish sauce manufacture, that is, in small *cetariae* or dolia, and a wider trade in a sauce product, is not easy to see in Republican Italy.

There seems to be secure evidence of fish sauce production in Republican Pompeii at the end of the 2nd century BC which is associated with faunal remains. An industrial property along the via Consolare (VI. 1.2.5) has a series of tanks with water proof lining and in one evidence of articulated sparids (Jones and Robinson 2007, figs 25.1–6). Steven Ellis has put forward a comprehensive and complex theory that salted fish products (sauces and solid products) were being prepared in Pompeii from the 2nd century BC. This trade is identified from other evidence and the presence of what are understood to be small *cetariae* identified behind the threshold of a number of retail units within houses near the Porta Stabia, (two insulae at VIII.7.1–1S and I.1.1–10), in the Hellenistic town (Ellis 2011).[5] It is not universally accepted that these particular tanks were fish salting *cetariae* (Botte 2018:382).[6] They are subterranean and lined with a hydraulic lime render, which has been described as plaster by Ellis, which is potentially confusing. This form of render would have been temporarily impermeable but not as stable as the various forms of *opus signinum* with crushed tile and lime which is the normal waterproof lining for fish sauce tanks. The very existence of water proofing on a vat has a multitude of uses not necessarily connected to fish as noted by Bernal and Sáez Romero (2008:68). It is not impossible that these spaces were simply used for secure dry storage and were lined in order to keep moisture out. The placement of these tanks behind the doorways seems not to be ideally suited to the production fish sauce or solid products, especially as the smell alone would have been all pervasive, as Ellis notes.[7] Zooarchaeological evidence is rare in the case of these threshold tanks: there is just one tank with very small amount of fish bone evidence to corroborate the fish connection. Ellis accounts for the lack of zooarchaeology by suggesting that the process had to be a "clean" one, but this is not convincing (Ellis 2011:68). The tanks all ceased to be used at the same time in the early Augustan period, and each was filled in with rubble and leveled. The lack of residues such as staining that has been found associated with fish sauce tanks elsewhere in these circumstances would seem to be too significant, discussed on page 149 (Driard 2012). Ellis concludes that "given the relatively early, possibly pioneering, date for the Pompeian vats and the almost experimental nature of this kind of cottage industry, the Pompeian vats may belong to Republican prototypes for Roman fish-salting vats" presumably located elsewhere (Ellis 2011:67), which, given there is still doubt over the identification of these vats for salted products, stretches the evidence too far. For the later salted fish trade in Pompeii see Chapters 10, 11 and 12.

There are further three pieces of evidence that are cited suggesting an Italian trade in fish or fish sauce, though they are equally contested. The Grand Congloué wreck of 110 to 80 BC, was determined to be carrying fish products as wine residues were not found in the Dressel 1 variants on board, though fish bones are also absent (Botte (2017:149). A site at Populonia has produced a cut-down Dressel 1C amphora from a layer contaminated with medieval finds in which tuna bones have been found from small and young fish which we may call a *cordyla* (page 175; Botte 2017:139–144). As Botte has notes, there is nothing here that can indicate that Dressel 1 forms typically or regularly transported fish, whether a sauce or a solid product and their shape does appear to indicate this. Further residue evidence in the Mediterranean is dated to 50 BC from Olbia in Sardinia from a broken Dressel 2 to 4 amphorae with an accumulation of fish bones (80% pagellus of 11 cm and 20% horse mackerel of 20 cm) which we learn from later examples probably represent *allec* (Sternberg 2006:373). Sternberg suggests two possibilities: either the amphora was imported with its fish-based content or the amphora was imported with wine and reused for making a local preserve, both are possible. This evidence suggests many small-scale manufacturing facilities making a product that had not taken off commercially because it had not attracted the attention of the elite. One might consider that fish sauce was being made everywhere for the culinary markets and this simple sauce was distributed locally and is therefore largely archaeologically invisible. It does not appear to have been made in any substantial or industrial way or formed part of interregional/Mediterranean trade as the evidence is inconclusive.

Spain and fish salting in the 2nd and 1st centuries BC

From the earliest Punic occupation of Cadiz the fish resources of the area were exploited extensively. The region that Taradell entitled the "circle of the straits" (Taradell 1960) encompasses Cadiz, Algeciras and the opposite coast of Morocco, as they are believed to constitute one historical region in terms of organization and exploitation of fish resources. Baelo Claudia is the most extensively excavated fish processing site on the coast of Baetica, situated on the Atlantic coast near to the Straits of Gibraltar.

Fish processing has been documented at the ancient town from its foundation in the mid-2nd century BC. The foundation of the town and the fish processing facilities have been linked to Roman colonial influence from the colony of Cetaria founded for Roman soldiers in 171 BC in the Bay of Algeciras. Italian wine amphora and other Italian ceramics dominate the finds as do imitations of these amphora (Bernal *et al.* 2013:357). The small salting installations in the earliest Punic phases of the coast of Baetica such as la Redes and Puerto 19 are typically designed around two modest (1–2 m^3 capacity) rectangular tanks along with preparation and storage areas outside the residential areas (Bernal Casasola and Sáez 2008:56, 77). These facilities remained in place until a transformation in production facilities in the last decades of the 1st century BC, when capacity increased dramatically. We can only speculate about the products, as the faunal evidence is rare, discussed below. The size and capacity of the earlier tanks does not however imply a modest volume of salted fish, as we are informed that the initial salting process can take no more than three days in total. Archestratus, writing in the 4th century, tells us that the fish are put into the vessel with brine after just three days of dry salting (Athenaeus iii. 117a). Given such a time scale, the volume of product from these modest tanks would be considerable. One may imagine one tank being used to salt the fish, and one serving to hold the pre-prepared brine. On the other hand, fish sauce in its most developed form can take up to three months to make according to the *Geoponica*. This in theory would require other facilities altogether, either dolia or smaller vessels that would not be identified as being used for this purpose. However, we have to imagine that these twin tanks were also used to make fish sauce as the demand for it increased. There is evidence of many pieces of dolia (very large and stationary pot-bellied vessels) found at a later imperial factory at Baelo Claudia which has recently been excavated, which could suggest that dolia may have played a greater part in the early process of making fish sauce than has hitherto been thought (Bernal *et al.* 2018:79).

A new form of small ovoid amphora designed for liquids rather than solids appeared in the 90s BC; these are Dressel 9 and 10 types, and their association with a fish sauce is very strong later, but at this time there are no *tituli* or residues to indicate fish sauce. They have the distinctive narrow hour glass neck which precludes their use for a solid product (Martin-Kilcher 2003:73). We may conclude that these vessels were the new industrial form in use to ship the increasing volume of fish sauce around the Mediterranean. However, the small factories with twin tanks remained in place along the Baetican coast in numerous small units right up to the end of the 1st century BC, when a sudden and dramatic transformation in fish processing occurred. The factories increased in size and capacity dramatically under Roman control, and this transformation would seem to be linked to an apparently sudden and massive increase in the demand for fish sauce. Evidence for a *gradual* build-up in trade is lacking. We can be fairly sure that the new factories with their batteries of *cetariae*, giving them five times the capacity of the previous sites, were introduced to make a dissolved fish sauce. These factories were also linked to numerous new amphora shapes not seem before (Dressel 7–11), which we might conclude were specifically designed to accommodate fish sauces, given that solid salted products had been manufactured for a number of centuries before and their vessels are distinctly practical, being open necked and tub-like. The earliest evidence of a potential fish sauce residue has been found at Baelo Claudia (Bolognia). The remains of a processing tank, subsequently paved over and enclosing a well dated context of the second half of the 2nd century BC (phase C1 VI) was uncovered in 2004 and a number of amphorae, three of which contained fish bones, were found along with a large dump of tuna bones. The dominant species at this time was tuna, while when the site was at its most active during the early Empire, fish bones associated with the small fish-sauce species were also plentiful in *cetariae* (Bernal 2018:347). An amphora form known as Ramon T 7.4.3.2. (A9) was found to contain tuna scales, indicating a salted product. These are African amphora also known as Mana C2b[8] and were also made along the southern Spanish coast from Cadiz to Algeciras and Malaga, and are strongly associated with the fish-salting plants. These vessels display a distinct large open trumpet like mouth and a slightly narrower hour glass neck, long straight body and pointed spike. They were made from the 2nd century BC onwards but continued into the Julio-Claudian era alongside the recognized fish sauce vessels, that is, Dressel 7 to 14. Their content is assumed to have been fish related, based on a titulus which mentions *hal(ec)* from the Praetorian camp and the single example with tuna scales. There is nothing concrete to associate fish sauce with these vessels, though it is quite logical to infer it. Alongside this vessel were two imitation Greco/italic Dressel 1A wine amphorae found to contain a large number of disparate fishes, shell fish, snails and mammal bones, which have been interpreted as a kind of special mixed sauce (Bernal and Gonzáles 2008). These include peacock wrass (*symphodus tinca*), mullet (*liza romada*), bream (*diplodus vulgaris*), tuna scales from a large specimen (*Thunnus sp.*) as well as a small number of pig and some slivers of goat ribs bones. There is also a mole, mouse and hedgehog, which are naturally intrusive, as are four species of land snail, some of which are fragmentary while others are whole. The pig and goat have been taken to reflect the addition of special meat flavor to this "mixed sauce," which from our understanding of culinary fish sauces is entirely unrecorded. There are a number of issues with this interpretation. In the 2nd century BC there is no indication that special, elite and "different" sauces have become part of the culinary market. This so called "mixed" sauce is presented as an innovation, yet we are surely still only dealing with the most basic *garos*, that is, small fry sauce, and the very idea of innovation in fish sauces is not apparent until the last years of the Republic. The addition of meat to a fish sauce cannot be corroborated elsewhere in the immense body of fish sauce residue evidence that survives and would seem to be highly unlikely in this case. Animal bones are common at salting plants (Bernal 2016:201), but it is a leap too far to assume that meat and fish were blended in a sauce on the strength of this one example.[9] I am inclined to see this accumulation of faunal evidence as debris from a cleaning process.[10] How we interpret the content of an amphorae in a given context is complex. Of the many interpretations possible, one of the potential reasons for its disposal will often be a simple matter of removing rubbish and bad smells from the environment to detract vermin. As we have no direct evidence for a dedicated blood viscera *garum* at this stage we must call into question the idea that the production of *salsamenta* using Scombrids always resulted in a by-product of a fish sauce, that utilized the viscera and blood, which was first proposed by Curtis (2001:406) and now widely accepted (see Chapter 6 for references and discussion). I have argued that *garos/liquamen* from the small species was traditionally and regularly made separately and independently from the more sophisticated trade in Scombrid *salsamenta*.

Fish sauce in Rome from the Late Republic and early empire

During the first century BC, gastronomy took off in Rome. Cooks became de rigueur in the households of the powerful, and feasting in style became a necessity in the political climate of the time. The élite competed with more and more elaborate dinners, and numerous forms of sumptuary legislation were passed in an attempt to prevent overt demonstration of wealth. This new atmosphere was particularly conducive to an emerging group of gourmets: men competed over the size and variety of fish they could breed, over knowledge of the food itself and how it was prepared, and over the size and number of dining-rooms they could use. Some who were unable to control their appetites were publicly condemned and often satirized. We learn that even Cicero developed a smattering of knowledge about the sauces that were being served. In 46 BC, he took up this "art" of dining, and wrote a number of letter to friends and describes how he was able to have a sauce he had eaten duplicated in his own kitchen: "I even gave a dinner to Hirtius, but without a peacock; in that dinner my cook was able to copy everything except his hot sauce."(Cicero *Ad fam 9.20.2*). It is not clear what this sauce was like but I would hazard a guess that it contained *garos*.

The first indication of any kind of Roman elite interest in fish sauce comes in 35 to 33 BC, and the first recorded use of *garos* is rendered as *garum* in a Horace satire. The passage ridicules the idea of an elite gourmand who has become a bore about the food he serves to his guests. The host, Nasidienus, describes at great length the dish of eel and the sauce served, which is an *oenogarum*, a blended sauce made with oil, wine, vinegar, spices and *garo de sucis piscis Hiberii*, "*garum* from the juices of a Spanish fish" (Horace, *Sat. II 8.42*). This dressing of fish sauce with other liquids, which is similar to a vinaigrette in many respects, will appear ubiquitous in later sources, a standard accompaniment to all manner of vegetables and salads and across the entire social classes in Rome and the wider empire (see page 102). At the time it was clearly relatively new and fashionable, but it rapidly became commonplace.[11] The Spanish fish is almost certainly a mackerel at this time, but what kind of sauce this was is not easy to see. Nasidianus also serves *allec* blended with the lees from Coan wine as an appetizer alongside other relishes that are designed to stimulate the palate (*Sat*.8,2,9). I had long assumed that these Spanish mackerel "juices" were blood and viscera and that this was therefore a very early reference to a blood viscera *garum*, but now I think that this is in error. Leon (2001:175) has suggested that we should expect *sociorum* indicative of elite *garum* here, but this product had not yet become fashionable and as we have noted (page 68) may not have been even invented, as Apicius invented the very idea of *sociorum*. There are certainly no firmly datable *tituli picti* for *garum* this early: they are all from the Imperial period. Elsewhere in Horace's *Satires* a gourmet-philosopher recounts the precepts of fine living to a passer-by as a form of philosophy. The details are trivial nonsense about what to eat, including another indication that wine lees and *allec* were blended and served at Roman dinners. He also tells us more about these vinaigrette-like sauces:

> "It is worth the effort to get to know thoroughly the nature of the double sauce. Simple sauce is made from sweet olive oil, which is worthy of being blended with fragrant pure wine and muria provided that it comes with a powerful whiff from a Byzantine jar."
>
> (*Satires* ii.4.63–71)

The sauce is boiled with herbs and saffron and more oil is added. The identity of this *muria* is greatly disputed for numerous, complicated reasons. These issues are discussed at various stages throughout the book (pages 31;46;108 and specifically when Ausonius' letter 21 is discussed at page 104). For the present, it is sufficient to note that at this early stage, this product can only be a Greek *halmē* fish brine, as the vague and disputed association of *muria* with a bloody *garum* or *liquamen* does not manifest itself until Martial refers to it in c. 90 AD. There are other views (Studer 1994; Curtis 1991:7–8; Corcoran 1962:205; Rodríguez-Alcántara *et al.* 2018:150), but it is easiest and simplest to regard *muria* as being predominantly a brine in the first instance and a fish brine particularly derived from a cleaned salted fish. Its presence here indicates that fish brine was a desirable sauce to blend with oil and wine. It does not seem likely, given the earlier evidence for the use and value of fish brine in Greek sauces, that Nasadienus is using *muria* out of ignorance of its apparent low status. One may consider that, as *muria* seems to develops a lower ranking in later sources, it always had one in Rome, but this would be unfounded (see page 46). Using *muria* recalls earlier Hellenistic cuisine, and it was the Byzantine, Spanish and Sicilian salted *Scombridae* trade in the 3rd to 1st centuries BC that provided this sauce for the Greek markets. *Muria* was also associated with elite 4th century BC gastronomic tastes of Archestratus on Sicily (Wilkins and Hill 2011). This suggests that the culinary culture of Horace's poetry is still bounded in the culinary practices of the Hellenistic cuisine that arrived in Rome. The Hellenistic flavor to Horace's cuisine at this time reinforces the idea that the "*garum* from the juices of a Spanish fish" was actually still an early Greek *garos*, later renamed *liquamen*, and not the elite expensive blood viscera sauce which, as we have noted, (page 27) was probably not invented until the end of the century under Apician influence. Horace has painted quite a comprehensive picture of fish sauce in Rome in the 30s BC. *Garos*, now *garum*, was a familiar product, and it must have come in at least two qualities, the standard small-fish form and the one made for the elite market from larger fish, and the residues of these sauces were also consumed. The small-whole-fish sauce that arrived from Greece has become fashionable and desirable and as a result a gourmet product made with a much larger fish, that is, mackerel, has become the new taste at Roman tables, and it is possible that the *allec* from mackerel was the product that was more desirable and served as an appetizer. Experiments in making mackerel *allec* have demonstrated that it is exceptional in taste and texture. These references would seem to demonstrate that initially the transition from *garos* to *garum* was easy and simple as the term was logically and simply transliterated. Fish sauce at this time was essentially a single substance: dissolved fermented fish of varying qualities depending on species and size.

We are able to understand the uses of fish sauce at this time through an understanding of the concept of the *oenogarum*, a blended dressing of oil, wine, and or vinegar with fish sauce. The majority of our evidence for these sauces, bar what we have just seen in Horace, comes from Late Latin sources and it is rightly difficult to extrapolate back to the late republic/ early empire to interpret fish sauce use then. Recipes for *oenogarum* sauces are found throughout the *Apicius* collection as well as many other ancient sources such as Columella, Celsus, and Galen; these will be dealt with below. In each occurrence of *oenogarum* in *Apicius,* the recipe goes on to add *liquamen*. This is understandable only when you realize that the term is a transliterated Greek word and not a translation. *Garum* alone hardly appears at all in *Apicius* (there are just two uses of the term in it: see page 84; for the dating of *Apicius* see below). It therefore follows that every occurrence of the word *garum* within the term *oenogarum* was not an indication that *garum* and *liquamen* were the same, but rather the common use of a Greek term for these compound sauces, which it was not necessary to translate. This is an example of the "hybridization of the two languages" (Dalby 1996:180). Ultimately the term *oenogarum* is rendered into Latin in the late empire in the Gargilius Martialis recipe for fish sauce (see page 22), where it becomes a *confectio liquaminis*. The fact that the word is finally rendered in Latin is a part of a popular rejection of Greek terms for habitual things within the Latin language in the late empire, discussed on page 106. We have seen that these dressings were already part of the Greek cooking style. They were predominantly used as a dressing for vegetables but also for meat and fish. Galen, writing in the 2nd century AD, consistently recommends light, simple dressings of oil and fish sauce for vegetables in his dietary advice, each time using *garos* to mean the whole-small-fish sauce. In discussing the enforced vegetarian diet of a very poor student, Galen says "He ate (the vegetables) with *garos* of course, sometimes just adding oil to the *garos*, sometimes wine or again vinegar" (Galen, *On the Properties of Foods* 1.25.2). Contemporary with Galen, Martial describes a simple dinner in which the *gustum* consisted of lettuce with an *oxygarum*: oil and vinegar dressing, suggesting simple combinations like these were also consumed by elites but as commonplace fare (Martial *Epi* 3.50.4). When we see the people of Martials constructed world consuming *oxygarum*, we must not be fooled into thinking that *oxygarum* has been appropriated for elite consumption, or that its presence here in this place gives it some hidden meaning. This is equally true in reverse with references to the under-class consuming *allec*. These food items may have massages to tell us about social identity and class divisions, but they are first and foremost real food and food does not always communicate anything other than its prosaic existence. Galen's attitude to *garos* reflected in the casual "of course" demonstrates that fish sauce was ubiquitous in the diet of ancient Romans and Greeks. Here, as in so many other uses of *garos* in agricultural, veterinary and medicinal works, we are a long way from the elite, so-called "*nobile*" and arcane *garum* (Martial, *Epi.* xiii.82.2). Galen recommends *garos* without any adjectives of quality or variety on numerous occasions. The only indication of a different sauce, a *garon melan*, black sauce, comes in a remedy that cannot be attributed to Galen, and may be a later addition to the Galen corpus. This undated remedy has also been amended at some point in a medieval Latin translation incorporated into the Olms edition from 1823. The emendation includes the line *gari nigri quod Romani sociorum appellant*, "Black *garum* which the Romans call *sociorum*." Unattributable as this line is, it does nevertheless affirm that in the medieval period it was understood that *sociorum* referred to a type of sauce, not a geographical origin. The black sauce is used to make a medicinal remedy called an *oxyporium* (digestive), as we will see below. Despite the gourmet obsession with noble *garum*, reflected in the Martial epigram discussed below, it will be the medicinal uses of black *garum* that will dominate in the late empire.

Now that we can see what these simple dressings were like throughout the classical period, the single recipe that uses *garum* found in Celsus, who was writing during the early 1st century AD, can be seen as a Latinized form of *garos*, in a context suggesting the primary fish sauce (i.e., the later *liquamen*) used as a simple vegetable dressing entirely reminiscent of Galen: *holus ex oleo garove estur*, "vegetables eaten with oil and *garum*" (Celsus 2.25). Columella, who wrote on agriculture in the mid-1st century

AD century, uses *garum* in veterinary remedies, and in each case the product is used in substantial quantities for treating animals. It is poured into ears and down animals' throats, and there is little doubt that this is the simple small-whole-fish sauce that we have come to expect. The quantities stipulated alone prevent this being an elite or expensive sauce of any kind, including a blood viscera sauce. Columella's use of *garum* in food is a little harder to interpret. There is just one reference in food, in which an *oxyporium* (digestive) is prepared: pepper, parsley seed, silphium, cheese, and honey are pounded together and diluted with vinegar and *garum* (without adjective) when required (*De Re Rustica* xii.58.4.13). Given the way in which he uses *garum* elsewhere in the book it is certainly unlikely to have been a blood viscera sauce, though of course at the time of writing, c. 50 to 70 AD, the black and bloody variety had become the height of fashion. Given these considerations, many of the references to *garum* will have been simply transliterated from Greek sources by writers who, as we have seen with Pliny the Elder, did not necessarily comprehend that two sauces existed. We must therefore conclude that we cannot be sure which sauce is being referred to when we see *garum* with no other adjective in consumption texts, and this is probably true for all periods. The consequences of this are immense, as the context of the term would seem to be crucial in determining which sauce is meant. The remedies and recipes in Pliny are a case in point, as it is not always easy to see whether Pliny has copied a remedy from a Greek source that uses *garos* or a Latin one that uses a blood viscera *garum*. He includes *garum* in a dressing of oil and wine for parsnips (*HN*. 20.34.4); in a dressing of oil, wine, honey for *opsonia* (meat pieces, *HN*. 27.136.6); and for snails served with wine (*HN*. 30.44.4). This is reminiscent of the noble *garum* served on raw oysters that appears in a Martial epigram, discussed below. In a remedy for ears, Pliny lists *gari excellentis*, which was boiled down with honey and vinegar (*HN*. 32.78.1). The added adjective typical of amphora *tituli picti* is certainly reminiscent of the later use of *garum*, when it is traded alongside *liquamen*. We lack a point of departure for the introduction of the new term *liquamen*. This doubt as to which sauce is being referred to can be seen in all the subsequent evidence for fish sauce use, in texts and on amphora labels. This ambiguity compounds the essential debate about the nature of fish sauce and makes the process of untangling the evidence and providing clarity for archaeologist so difficult. In many early instances when the term *garum* is used, we must assume that the essential substance, the small-whole-fish sauce, is intended, and it is only when an additional adjective is used such as *sociorum, melan, haimation*, or as we will see in Martial shortly such terms as *nobile, arcano*, does it correspond to a blood viscera sauce.

The very existence of another sauce called *liquamen* as a separate commodity is only apparent in the early empire from amphora *tituli picti*, which are discussed in Chapter 12. A rare bilingual *titulus pictus* for *garum* from an amphora from Masada and associated with King Herod the Great, whose life spans the late Republic may shed a little light on this issue, though there are many interpretive problems. The amphora form is unclear and so therefore is the date of the deposit, and this influences how the *titulus* is interpreted. The vessel in question is fragmentary and identified from petrological studies as a Dressel 12/13, a type which is typically associated with elite fish sauces, particularly *garum*, and its identification as this form would strongly confirm an association with Herod the Great, as the form emerges in the mid-1st century BC. However, from a typological analysis, albeit fragmentary, the vessel has also been identified as a Dressel 38, also known as Beltran 2A/Pelichet 46 (Berdowski 2008:110). The earliest date of this form is greatly disputed.[12] There was a transitional period where the earlier Dressel 7/Pompeii VII type began to evolve through four stages (variants A-E) into the final formation of the Dressel 2A, and this early evolution is believed to have occurred either in the first few years of the 1st century AD or as late as the time of Tiberius, c. AD20. The form itself did not become settled until the mid-1st century. Even allowing for the potential early development of this form, it is very difficult, on typological or petrological grounds alone, to date the vessel to the time of Herod the Great.

However, the evidence provided by the *titulus*, while equally fragmentary, is more rewarding. In a single line, it reads *garum* in Latin letters, and then two unknown symbols, followed by Βασιλέω(ς), in Greek. The genitive ending means "*garum* of (or for) the king," but one has to ask, why it is bilingual, and when was it inscribed? It would seem very likely that the vessel (though its type is disputed) was certainly of Spanish manufacture, and the product it contained was also a Spanish one; it seems most likely, then that the inscription of *garum* was made in Spain. It is not however clear that the Greek was added at the same time, even though the editors had the impression (though that is all) that the inscription was made by the same hand (Berdowski 2008:112). Many wine vessels were also found at the site and in the same context, many with bilingual *tituli* stating that the wine was Italian, and in this case the inscription was very likely made at the point of origin. The link with Herod the Great is made much more likely as one of the *tituli* contains a consular date of 19 BC (Berdowski 2008:119). Herod the Great lived in Rome for some time during the early 40s BC and almost certainly acquired a love for Roman cuisine. The stores at Masada were notoriously luxurious, as noted by Josephus Flavius (*BJ* 8.4; Berdowski 2008:120), and a taste for Roman luxury would have followed Herod through his life. This inscription may represent the earliest securely dated *titulus* for any kind of fish sauce. I suspect that the label was bilingual because it specifically indicates that the sauce was the new blood/viscera sauce and not an ordinary *garos*, and the use of the Latin word signified this. As Βασιλέω(ς) is in Greek, we have to ask why they didn't use the term *garos* if the whole inscription were written in Palestine; it also seems more obvious to assume that the sealed container was labelled at its point of manufacture. The Greek term may have been added once the container had reached its port of destination. This is the most obvious explanation why this sauce was identified using the new fashionable Latin usage, that is, *garum* (Berdowski 2008:116). Herod was dead at some time between 1 and 4 BC, and while the link with him is not definite, this *titulus* may nevertheless be the earliest evidence for the "other" *garum* in the Late Republic.

Nevertheless, the date of the coining of the term *liquamen* is utterly obscure. It must, presumably, have coincided with the creation of the new bloody and or black sauce, otherwise there was no good reason to create a new nomenclature for such a simple product. We are in the dark of cause as to precisely what happened but we can imagine. In the period before the new sauce was developed *garos* was simply transliterated into *garum* and both terms represented the original small-fish sauce, which now began to be made with larger species such as the mackerel we see in Horace. The new blood viscera sauce emerges and ideally a new name would have been coined for the new sauce in Latin, but that is not what happens. Instead of a new name for the new sauce, the old name albeit Latinized was appropriated by the Roman gourmet community. Initially it was probably sufficient to use *sociorum* to separate the sauces but this was not sufficient as we can imagine a certain amount of confusion. For reasons that are not clear this appropriation of *garum* in Latin forced the manufacturers to coin another term in Latin to designate the original Greek sauce, *garos = liquamen*. This dilemma and it's solution must have been played out among the many parties in the salted fish trade, merchants, shippers, shopkeepers, cooks, and doctors. It was clearly not played out at the dining couches as Roman writers continued to use *garum* to designate the table sauces they saw before them, which ever type they were, without comprehending that another sauce (*liquamen*) was being used in the kitchen. It was essentially and still is a simple mistranslation issue but it was clearly not well understood outside of the industry hence the vast amount of confusion among ancient consumers and the ongoing confusion today. That this is what ultimately happened is without doubt as so many of the later sources keep trying to tell us. Diocletian price edict, Apicius' recipes, the Latin/Greek phrase books all state quite clearly that *garos* is *liquamen*. None of these Roman sources use the Latinized *garum* for good reason as they are not referring to the blood viscera sauce.

By the time Pliny the Elder was writing this new fish essence sauce had clearly evolved into something extraordinarily popular and by the time of the destruction of Pompeii the numbers of *tituli picti* that refer to it suggest a huge market for the bloody *garum* sauce. We have no textual sources for *liquamen* it this time, but in Pompeii it is found alongside *garum* as a separate and popular sauce. The fish sauce trader and potential manufacturer Umbricius Scaurus stakes a claim to make both *garum* and *liquamen* and the popularity of both as separate sauces rather than one being the weaker version of the other is obvious from the wording of the titulus. For the fish sauce trade in Pompeii, see pages 59; 195. Our first literary reference to *liquamen* can be found in recipes with an early provenance in the *Apicius* collection (see Chapter 3 note 4). It is otherwise remarkable that the next secure literary reference to a *liquamen* in Latin does not appear until the 4th century in the agricultural manual of Palladius, (*Palladius* Opus Agri 3.25.12) where, in referring to the idea of a fermented salty liquid derived from pears, he calls it a *liquamen* implying that the term had that generic function at this time. Other undated sources such as the *Colloquia of the Hermeneumata*: phrase books designed to teach Latin and Greek, which cannot be dated with more precision than 2nd to 4th centuries use *liquamen* to translate *garos*. These are discussed in detail on page 102 Dickey 2012. i. 119, *Colloquium Monacensia-Einsidlensia* 9d.

To sum up, our evidence places the transition from *garos/garum* to the use of *liquamen and garum* to sometime between the last years of the 1st century BC when *garum sociorum* was invented and the years preceding AD 69, which is not very precise or ideal. The length of time in which *garum* served to convey both sauces might have been only a few years of confusion or an extended period of chaos.

Fish sauce in the archaeology of the late Republic and early empire

The transformation in fish sauce manufacturing at Cadiz began during the last third of the first century BC, under either Roman control or influence. Typically, the number of processing tanks increases from two to around six to eight, in two lines parallel to each other, with a working area between them. The capacity of the tanks increases substantially and in the case of the Teatro Andalucia, 20 tanks were found with an estimated capacity of 10 m³ in total. An estimate of four to five times the productivity has been suggested (Bernal and Romero 2008:77). See Chapter 10 for more details. Previously the small Punic factories had been outside the urban area in the countryside close to the beach, but now they were constructed within the city of Cadiz itself, and the old factories were largely abandoned. We might link this to a switch from modest amounts of salted fish which is processed within a few days to huge amounts of fish sauce which took far longer and needed to be monitored on a regular basis. There appears to have been quite a dramatic change in terms of the products being made as well as the capacity output. At the height of the trade in fish sauces from the western Mediterranean, typical fish salting installations such as those at Troia I/II in Lusitania had a capacity of over 600 m³ (Trakadas 2005:61; Wilson 2007:4). The site in Gaul at Plomarc'h, next in size, had a capacity of c. 475 m³. These figures correspond to an un-imaginable quantity of fish and potential sauce as we are unable to visualize and equate the size and capacity of the salting vats to a product of a given consistency and appearance. At Cadiz the kiln sites, which were originally concentrated on the island of San Fernando, multiplied and were constructed throughout the countryside. They were making the traditional Dressel 7 to 11 amphora types closely associated with fish sauce (Bernal and Sáez Romero 2008:61). These changes are mirrored across the Spanish coastline in many sites including Malaga (Corralles *et al.* 2011:32). This new industrially made fish sauce appears in many new and complex amphora with segmented bodies from the mid-1st century BC, which we may conjecture corresponded to characteristics of the sauces and their consistency, discussed in Chapter 12. The reason for this apparently sudden demand for fish sauce has been attributed to military needs as the army expanded across Europe. Certainly, these new Spanish amphorae began to appear throughout Gaul and Germany and from the Claudian invasion into Britain in abundance (Martin Kilcher 2003). The cargo of many shipwrecks dating from the mid-1st century BC contain residues of both a salted fish and a *"garum/liquamen"* fish sauce made with mackerel, as illustrated by Horace, being traded across the Mediterranean into Roman markets. For residues and amphora Chapters 11 and 12.

The new and ever-increasing demand for fish sauce suggested by the archaeology is not mirrored in the literary sources, as these sources are largely subelite or elite in nature and rarely reveal information about the ordinary soldier, farmer, or tradesman, yet the increase in production suggests that all through the 1st century BC the use of fish sauce was commonplace and, in its cheaper forms, may have been ubiquitous in the Roman diet for some considerably time. The demand for it in the later period is only understood if the ordinary soldier had become so accustomed to consuming *liquamen* they could not contemplate a meal without it and military supply systems had to provide it either directly or ensure it was available (Davies 1971; Broekaert 2016:70; Ørsted 1998:25). The didactic sources discuss simple salad and vegetable dressings which, we learn from Galen, were eaten by everyone. A diet dominated by bread pulses and vegetables and small amounts of occasional meat is strongly associated with the majority of the population, and fish sauce must have been an integral part of that diet. We do not know when the change in tastes occurred, but It would seem not to have happened until a Greek culinary culture spread widely among the middle rank and lower classes within Italy at the turn of the 2nd century BC, a situation we can see reflected in the references to the use of fish sauce residues in Plautus. It is entirely plausible that the type of sauce that was used by most people at this time was a basic form of relatively cheap commonplace *garos* made with multiple species of small fish, always dominated by anchovy and sardine, but also made from any number of *sparids and clupeids* that were readily caught in estuaries and close to the shore. Martin-Kilcher notes that in early Augustan sites north of the Alps, Spanish fish sauce amphorae were often present, while oil was less common or even absent (Martin Kilcher 2003:79). We may reasonably conclude that the Spanish fish sauces traded at this time included a dedicated whole dissolved mackerel sauce, a pure mackerel blood viscera *garum* and a commonplace *liquamen* using mixed *sparids* and *clupeids*. Popular though fish sauce seems to be from amphora finds in the Northern empire, the discarded residue evidence, of its processing and harvesting, is largely an urban and semi-urban context and there is little to suggest that the rural poor engaged with fish sauce in the same way as the more Romanized urban masses (Clavel and Lepetz 2014; Locker 2007).

Nevertheless, the incidence of *garum* labels, compared to *liquamen* appears to be statistically significant on the northern military sites (Ehmig 2003) and in fact *tituli* for *garum* also dominate at the Pretorian camp and in Pompeii. I will argue in Chapter 12 that the statistical data derived from the presence and absence of particular types of sauce from *tituli picti* is not a good indicator of the actual amount of each sauce being traded and consumed. Most amphora do not retain a label and as the essential substance of trade from archaeological residues was a *liquamen,* we have to assume that most of the fish sauce amphorae without labels would nec-essarily be used to trade this rather than *garum*. There is in fact little to suggest that the special blood viscera *garum* was widely consumed in the military markets as it was largely a sauce consumed among commanders and elites in military settlements.

Fish sauce in the Late 1st century AD and beyond

We hear about the now well established *garum* from Seneca the Younger who in his *Epistulae Morales ad Lucilium* published in 64 AD wrote a stoic philosophical diatribe against the excessive indulgent and insatiable appetites of his contemporaries. Seneca demonstrates a desire to be a sat-irist and to "debunk the dining habits of the Neronian age" (Richardson-Hay 2009:74 quoting A.L. Motto). In Rome, Seneca claims, the elites' love of food and gourmandizing had led to such corruption that doctors were treating new illnesses brought about by gluttony. Seneca's images are graphic and sometimes gross in nature, but their intention is to shock and shame his readership. The *garum* that is discussed here represents the epit-ome of indulgence: "What? Do you not think that the so called *garum sociorum*, the costly extract of poisonous fish, burns up the stomach with its salted putrefaction?" (Seneca, *Epist. Mor.* 95.25.7). There is little doubt that Seneca is talking of something rather more exotic than the Greek small fish *garos* which had been part of an everyday Mediterranean culinary cul-ture for centuries. Seneca clearly disliked this *garum* as something inher-ently corrupt and corrupting; to him, it was more obviously the result of putrefaction, because it was made entirely with rotting/fermenting viscera and it was both, the epitome of and a kind of reversal of, the idea of indul-gent gastronomy as a kind of bodily poison (Richardson-Hay 2009:77). That blood was involved in its production is clear from the term *sociorum*, but also from the use of *sanies* (translated here as "extract") but which is technically the word for a bloody exudate from a wound. Pliny uses the same term in his description of *garum*, as does Manilius (see page 35). There is no commentary from gourmets on the common place small fish sauce, so one is limited in comparing the sauces, but it seems impossible to associate *sanies* with the ordinary small fish *garos* in the same way as can be done with *garum*.

Marcus Valerius Martialis, nowadays known more simply as Martial, was a Roman poet who hailed from Hispania is best known for his twelve books of Epigrams, published in Rome between AD 86 and 103, during the reigns of the emperors Domitian, Nerva and Trajan. In them, he satirizes the sophisticated life of comfortable and impoverished Roman elites alike, and uses the fashion for luxury food to good effect. He makes many references to *garum* which are mostly disparaging, but they also reflect that conflict in the Roman mind over *garum*: it is loathed and praised, feared and disparaged. The collection of little gift poems by Martial that act as a substitute for the gifts themselves are in books XIII, *Xenia*, and XIV, *Apophoreta*. In *Xenia* can be found a pair of poems, each of which replaces a gift of a fish sauce, one a *garum*, and the other a *muria*. Martial states in the introductory epigram of book XIV that he has presented gifts for rich and poor alternately, but the fish sauces are listed in book XIII, and it is not clear whether this ordering was true of this book too, and in any case this supposed order does not appear to have survived during the work's long textual transmission (Shackleton-Bailey 1993 i.1,2). In *Epi.* xiii.102 we are led to believe that *garum* is the valued sauce, as it is made "from the blood of a still breathing mackerel." In the following poem, *Epi.* xiii.103 the issues are rather more complex. Here, the poem replaces a gift of a jar of *muria* which is specifically described as having two qualities: it is a *muria* from an Antipolitan tuna (from Antibes in Gaul), but goes on to inform the recipient that "had I been of mackerel, I should not have been sent to you," the assumption being that this fish sauce made from mackerel would have been suitable for someone more important (Martial *Epi.* Xiii.103.2; Shackleton-Bailey 1993: iii.215). It is not impossible that the reverse is the case, and that the *muria* most valued was of tuna and the mackerel *muria* less valued, as the status of the first recipient is not clear. Many scholars believe that *muria* can have a meaning associated with the blood viscera sauce with reference to the use of tuna in the *Geoponica*, to a letter by Ausonius, and most of all because of the misreading of Isidore of Seville; the issues raised by these texts are discussed elsewhere: see pages 31; 46; 108. It is not necessary to see an association between the bloody sauce and *muria*. Were we to do so, it would mean that both gifts were an elite bloody sauce with *muria* reflecting the lower status one, for which there is no corroborating evidence. Other references to *garum* in Martial are at *Epi.* vi.93.6, where the body odor of an unfortunate character called Thais is likened to a number of offensively smelling objects, culminating in "an amphora polluted with putrid *garum*." Martial also uses *garum* as an example of a strong-smelling substance at

Epi. vii.94.2, It is true that *garum* does smell particularly strongly, whereas *liquamen*, while conventionally stinky, is by no means as powerful or potentially repugnant as a black *garum* (page 163). In describing a boar ready for the oven, Martial bemoans the fact that his cook "will consume a huge pile of pepper and add Falernian and *arcano garo*" (*Epi.* vii.22). The Shackleton-Bailey translation in the Loeb equates *arcano* with the "secret" or "private" *garum* of the cook, which is not a sound interpretation. There would be no reason for the cook to have a private stash of *garum*; *arcano* here is "arcane," mysterious and sacred *garum*. The mixture that is proposed here, wine, *garum* and pepper, is a *garum piperatum*, also found in in Petronius *Satyricon* 36.3.2, where statues of Marsyas stood with wineskins, from which a *garum piperatum* poured on to fish swimming about in a narrow channel. These fish were presumably still alive, and this is another indication that fish were permitted to live for a time in the dining room. Petronius is re-using the "season the fish while alive" trope attributed to Apicius. A *garum piperatum* and a *liquamen piperatum* are found in *Apicius* (see page 85).[13] These are our first hints as to what *garum* was used for. It appears that the concept of the *oenogarum* dressings, which were originally and continued to be made with *garos/liquamen*, began to be made with black *garum* too, but the use of it is more visible. The gift of oysters from the *Xenia*, xiii.82, includes a further reference: "*Ostrea*: a shellfish, I have arrived drunk from *Baian Lucrine*, now in my extravagance I thirst for *nobile garum*." The oysters are raw, and the *garum* is simply poured on. This is in actual fact quite a fabulous combination; were I not to have tried this myself, I might have retained my disquiet about this kind of *garum*. For those who already like oysters, it is a remarkable taste and is worth the experiment. In iii.77.5, disparaging the diet of a miser, Martial links onions and capers with putrid *allec* and dubious ham, a common trope in Plautus (Persa i.III.107); He also describes a concubine with an all-too-modest taste who "begs her lover for six *cyathi* (c. 250 ml) of *garum*" and has her slave walk behind her with a dish of *allec* so she can eat it with a spoon at as she goes (xi.27.2). It is not just that she eats it, it is that she eats it in excess that is condemned. *Allec* has clearly become something to looked down upon, in contrast to Horace's view in 33 BC, and it is now a common foodstuff that ordinary people consume and when social climbers continue to consume it they show their origins. Martial is no doubt reflecting a commonly held view about the oddity of fish sauce, a useful and in some form luxury product derived from an unappealing raw material and an equally unappealing production method. The very fact of *garum* and its bizarre production was an ideal topic for satire. It seemed to optimize the grossness and indecency of gluttony out of control and at its worst, yet it was also appealing, in a strange and incomprehensible way. It is noteworthy that few other writers comment so much about *garum* as Martial, and as we move into the 2nd and 3rd centuries it seems that *garum* when it is the "other" sauce rather ceases to be worthy of mention and seems to fade from the Roman consciousness. *Garum* no longer appears in literature, while the use of *garos/liquamen* was always so commonplace and ubiquitous that it hardly warranted mention at any time. In the late 1st century Dioscorides was writing from Anazarbus in Cilicia on medicines from plants and animals, and has a section on "Garron." The translation from Osbaldeston is "*Garum*, the liquid that comes out of salted flesh (of meat) or fish" (*De materia medica* 2.34; Wood and Osbaldeston 2000). There is little to dispute in the Greek. This translation and the Beck edition from 2005:404) simply translate *garon* into *garum* without question and cites Curtis for further reading. The interesting issue is that a *garon* can be defined as a sauce derived from animal flesh too. This is new, but essentially, we have the same substance that we have come to expect, fluid from liquefied fish or meat. The concept of a "black" or "bloody" sauce is absent and it is very possible that a *garum* was not relevant to the medicinal world in the 1st century. Greek medical and veterinary books do not register the presence of another and different sauce to *garos* before the 4th century, when we begin to hear about *garum* again from the veterinary texts. It is likely it took some time for the medicinal uses of this sauces to become sufficiently well-known to become included in remedy collections. As we have seen, the two documented instances of the use of a black and bloody sauce in pseudo-Dioscorides and pseudo-Galen are undated and obscure; nevertheless, the use of the black *garum* appears to be quite well documented from veterinary medicine, but only because this sauce appears side by side with *liquamen* in the same remedy collections.

There is one exceptional instance of interest in the other *garum*; this is the very perplexing letter by Ausonius which deals with the terminology for fish sauce in the late empire and is discussed page 104. Otherwise the remaining sources for *garum* and *liquamen* are largely from inscriptions, didactic recipe collections, and the practical sources, and as noted they are entirely dominated by *liquamen*. This phenomenon has led many to conclude that the term *garum* went out of fashion and that *liquamen* simply took over as the favored usage for the concept of fish sauce as a single entity. This theory is so well entrenched that when scholars cite recipes from the *Apicius* collection they regularly convert *liquamen* to *garum* in the belief that this is actually the correct term. As we have seen it is largely correct but technically not helpful due to the need to separate the two sauces. The *Apicius* collection of recipes is discussed page 82. A reassessment of the evidence can lead to a different scenario, in which the bloody *garum* simply became less popular, was made less often, and is therefore less visible in the inscriptional and didactic record as a result. In the inscriptional evidence, *garum* is found on the Zarai tax tariff (AD 202, *CIL* VIII 4508) from the Roman Berber town in Numidia. It is written in Latin on an imperial customs list, set up to record those items where tax was due in Africa, and it also records some items that were specifically exempt from tax (Asakura 2003:77). An amphora of wine and an amphora of *garum* are listed on a single line, with one *sesterce* being due on each. *Liquamen* is not mentioned, though it was definitely current usage, it is possible that it does not appear because it was such a mundane item that it did not incur tax and was not traded into Africa but from Africa. The sauce that was taxed was the luxury *garum* and that is why it is present on this list. It is true that there are numerous items listed including head of cattle and sheep, cloth, leather, nuts, dates, peas and fuel, but it is by no means a comprehensive list; there are many other everyday items that were surely traded that one would expect to see. A century later, a different picture is presented by Diocletian's price edict, inscribed on large stone tablets and set up all over the empire in an attempt to control inflation in AD 301. It is a far more comprehensive list of commodities than the Zarai tax tariff, and it includes the buying potential of ordinary people through the cost of labor provision. The edict was written in both Greek and Latin, and while the Latin versions seems to have rarely survived in the West, because they were pulled down as the edict utterly failed to control the inflation, examples of the Greek version survived well and can be read in full. The relevant part of the inscription in Greek reads γάρου γεύματος πρωτείου (*garos* food supplies of first quality) and δευτέρου γεύματος (*garos* second quality food supply) and in the Latin "first quality *liquamen*" and "second quality *liquamen*" (Diocletian price edict, Lauffer 1971:3.6–7). The inscription makes it quite clear that *liquamen* is seem as the equivalent to *garos*, and the absence of any other sauce or a *garum* is of interest because it may (and almost certainly does, in my opinion) imply that this kind of sauce was no longer sufficiently popular to warrant mention as so few people were buying the blood viscera *garum*.

The remedies of Serenus Sammonicus, physician and tutor to Geta and Caracalla (who died in AD 212) survive in a *Liber Medicinalis*. This records fish sauce use, but we do not know which sauce he is talking about. Many of his remedies are borrowed from Pliny the Elder and Dioscorides, and include various magic formulae (including the famous "abracadabra"). A remedy for a deceased liver involves grinding nine garlic cloves and nine pepper corns into a diluted *garum*, which you then eat and drink (18.331); snails are cooked with wine and *garum* (17.316); ashes are mixed with oil and *garum* and used to treat swine (40.762). *Garum* is simply used without characteristic or comment and there is no mention of *liquamen*. When we find *garum* in Pliny the Elder, Columalla and Celsus, we can legitimately consider that the *garum* is essentially a *liquamen*, because knowledge of the new sauce had not permeated into the texts, but here we cannot be sure as there is nothing to allow us to distinguish one from the other. The late Roman (but strictly undated) medical text *On Chronic Diseases* by Caelius Aurelianus is often quoted to assert that, "*garum* is commonly called *liquamen*." (2.3.70). This work is described by Caelius as a direct translation from the Greek of a work by Soranus from the 2nd century AD (Drabkin 1950). We have already seen how easily a mistranslation can occur. Caelius would simply see *garos* in the original Greek and transliterate to *garum*, not knowing that this would not necessarily be the accurate translation of the term. Some late Latin writers clearly knew that, technically and within the community of fish sauce users, *liquamen* was the correct translation of *garos,* and some clearly did not. It is a natural assumption to make and has been made ever since, for centuries and to this day. It seems perfectly possible that in the late 4th early 5th centuries AD, the bloody *garum* that was exclusively called *garum* in Pompeii and in satire had become so scarce that few writers and commentators would have been aware of it. Thus, the word *garum* had ceased to mean anything different from *liquamen.* We have in fact come full circle: *garos* was equivalent to *garum* when it was first used in Roman food, and for many it once again came to mean the same thing, that essential, dissolved, small whole fish sauce. This phenomenon has three stages: a convergence of *garos* with *garum* in the late Republic, then the divergence of *garum* from *liquamen* in the early empire, and a return to convergence in the late empire. This appears to be what the evidence is telling us, though given the twists and turns that the terminology seems to have undergone, it is not surprising that some prefer to take refuge in a more skeptical position. The blood black *garum* sauce characterizes the common image that so often persists of Roman fish sauce: all Roman fish sauce was rotten and putrid, because *garum* was; all Roman fish sauce was expensive, because *garum* was; all Roman fish sauce was a luxury, because *garum* was; yet this type of fish sauce almost certainly constituted a relatively minor role in the trade in fish sauce. It was *liquamen* that dominated the trade as we can see from the archaeological evidence, see Chapter 11 and 12. In the late empire, the blood viscera *garum* as a culinary commodity seems to fade from view, although it continued to be made in small quantities for medicinal uses; and we will see this particularly clearly from papyri from Egypt and from the veterinary and medicinal texts, discussed in Chapter 3. The Roman obsession with gastronomy and bizarre luxuries appears to fade from view as the empire proceeds. Macrobius looking back from the 5th century at what he considers the worst of the excesses comments "not that I am saying we should be thought superior to the ancients....people were keener on luxuries in those days then they are now" (Macrobius 3.13.12).

Fish sauce in the archaeology of the late empire

After the Augustan peace of 31 BC, ever-increasing numbers of kiln sites and salt pickling plants were built in the Bay of Cadiz and surrounding area. The standard fish vessels, Dressel 7 to 12 (Figure 12.2), were from the middle of the first century AD joined by new amphora, Beltran 2A and B forms, which have distinct segmented characteristics suggesting a direct evolution from the earlier vessels, in that they have wide straight long necks, ovoid bellies and long wide hollow spikes. This is a distinctive shape closely associated with fish products. They were clearly multifunctional vessels, able to accommodate both sauces in both forms whether as a liquid or as a fish mash, and for small solid salted product in a liquid brine (see page 245). Portugal entered the market, trading fish sauces using their abundant sardine from the Sardo basin at the vast site at Troia, which had 25 separate salting facilities at its height and represents the largest facility in the Roman world (Wilson 2007). The original site was founded in the Tiberian period, around the middle of the 1st century, and while many of the plants ceased to trade after the 3rd century crisis, (below) many continued to function salting fish until the 5th century. The facilities at Troia appear to have processed nothing but sardine, which has always been abundant in the Sardo, and while it was probable that salted whole sardines were processed as well, it is very likely that it was a sardine *liquamen* that dominated this trade particularly as with few exceptions the *tituli picti* for the local Dre. 14 form are for *liquamen* (Pinto *et al.* 2012).[14] The absence of Scombrid fishing, tuna or mackerel, at Sardo attest to the separation of *liquamen* from *salsamenta* in terms of production techniques and the need for large quantities of viscera. There is a strong indication that Spanish and African sites collaborated in fish sauce production. Products that were made in Africa were traded in pottery vessels of Spanish origin. These trading associations appear to have monopolized the industry. It has been determined that by the 2nd century AD, empty amphorae were being transported from Cadiz to processing sites in Tingitana and filled with African sauces or *salsamanta* (Bernal and Sáez Romero 2008:71). From the end of the 2nd century onwards we have what is termed "The third century crisis" in the salted fish industry in Spain and North Africa (Morales Muñiz and Roselló Izquierdo 2008:220). The salting plants closed in large numbers, as did the kiln sites, and a corresponding reduction in shipwreck evidence is apparent and amphora distribution in the north declines rapidly (Broekaert 2016; Wilson 2007). This crisis is understood to be the result of a combination of political upheaval (Bernal and Sáez Romero 2008:69) and a reduction in the stocks of mackerel resulting from overfishing to make *garum* and *liquamen* for the elite markets in Rome and beyond (Desse-Berset 1993). The crisis is most apparent in Spain where many sites were abandoned and did not reopen, while kiln sites were reduced by half during the 2nd century, and many did not survive into the third (Bernal and Sáez Romero 2008:69). In the mid-3rd century a new amphora type, the Almagro 50/51 A-C emerges in Roman deposits at Ostia. From wrecks such as Cala Reale at Asinara, we know that these brought a fish sauces and *salsamenta* from Portugal over the next century (Delussi and Wilkens 2000:53). Recent studies demonstrate that these forms were also made in Baetica and distinguishing fabric has proved difficult and as a result the origin of the product is less clear (Bernal 2016:300). In the 3rd century, new vessels known as Puerto Real 1 and 11 were designed for fish products from Cadiz. They were much smaller and were only made in two sites in the bay of Cadiz, demonstrating how much reduced the industry had become

(Bernal and Sáez Romero 2008:71). The bulk trade in fish sauce finally declines entirely through the 4th and 5th centuries in Cadiz (Bernal and Sáez Romero 2008:70). Other sites such as Rhode, near Barcelona in northern Spain, which operated until the 6th century, began to function only at this time and the precise nature of the decline in the bulk fish sauce industry in Spain is difficult to pinpoint. (Curtis 1991:58). Martin Kilcher suggests that the much lower distribution of fish sauce amphorae in the north from the 2nd century may be due to different refuse disposal (Martin-Kilcher 2003:81), however Broaekaert (2016:83) suggest that different production techniques such as barrels, and the use of locally produced fish sauces from Britany at Plomarche where the capacity of these facilities was immense, and considerably greater than the sites in Baetica (Wilson 2007:6), may have simply pushed the Spanish products out of the market. Evidence from many sites in northern Europe also attests to a locally made fish sauce product using typical North Sea fish such as herring and sprat and to local trading guilds identifying themselves as *Negotiatores allecarii* (see page 234; Curtis 1984; Van Neer *et al.* 2005; Broaekaert 2016:83).

In the late empire, it is apparent that when the fish sauce factories closed from the 3rd through to the 5th century and left their residues in place for archaeologists to find, they generally come from the typical *sparidae and clupiediae* of c. 5 to 20 cm, that is, a small whole fish sauce that we have come to expect. The few shipwrecks with amphorae and residues from this late period such as Randello are also dominated by small-fry sardine and anchovy (Table 11.3). The evidence for residues for the manufacture of fish sauce with a higher or total use of larger species such as mackerel is largely lacking, though from other evidence such as the Latium pots on the Rhône, shipwreck evidence such as Grado in the 2nd century AD, the *tituli picti* and the literary perspective, one can see that mackerel fish sauce residues reflecting a high-quality *liquamen* must have been deposited but have failed to survive. Morales Muñiz suggested that the switch from high-quality mackerel for sauces and *salsamanta* to the small fry represented not only a switch from good-quality, expensive sauces to cheap and abundant ones, but also a switch to low-quality sauces such as *allec* and *muria* (Morales Muñiz and Roselló Izquierdo 2008:220; Morales-Muñiz and Rosello-Izquierdo 2016:38)). There are many issues here that need unpicking. Morales Muniz is surely right in terms of size of species selection but not in terms of the low value of *allec* and *muria*. We have seen that *muria* could be valued when aged and also derived from tuna and mackerel (page 47). We will see that in the late empire, under specific Greek Orthodox dietary laws, *muria* becomes associated with the wider concept of a *liquamen* (see page 108). We will see that *allec* was not a separate, cheap low-value fish paste for the poor but a potential valued food recourse and a source of more *liquamen*. Unless *allec* was made with tiny fish or sea food it was always meant to generate a potentially good quality *liquamen* first, and repeatedly, before it was discarded.

There is evidence of the continuous production of fish sauces in sporadic archaeological sites across the Mediterranean. An example, the wreck Tantara F found near Caesarea in Israel which is dated c. 7th/8th centuries AD was found to contain eight amphora all containing very tiny (1–4 cm) entire fish bones of a local species *Cichlidae,* which have been identified as an *allec,* a fish sauce residue (Barkai *et al.* 2013:195). The report concludes that as there is only one species it must have been a valued and special fish sauce but as we have seen this was unlikely and in fact this would have made a standard/low grade *liquamen* of insignificant quality. Regardless, the sauce was clearly being made and traded with a potential market in ancient Palastine. Evidence such as this at the end of our survey of fish sauce throughout the classical period surely demonstrates that the small-fish sauce was the dominant sauce, the essential substance. It was traded from the Hellenistic period and throughout the classical period, and as we will see in Chapter 5, such production on a large and small scale will have been maintained, albeit largely archaeologically invisible into the Byzantine period all over the Mediterranean.

Notes

1 For a wider view point on social division and food consumption, especially fish Davidson, J.N., 1998. *Courtesans & fishcakes: the consuming passions of classical Athens.* Macmillan.
2 In modern techniques of preserving fish, fatty fish such as these do not dry well and can become rancid. This tendency ultimately led to the development of sauces rather than dried fish products in many areas of the world (Cuong Pham, of "Red Boat' fish sauce, pers. comm.)
3 Aristotle recounts that small fry were salted immediately after being caught so that they would hold until they get to the market. This is still what happens to whitebait on board modern trawlers today, apparently. Cf. Aristotle, *On Animals* 569b.30-570a.2.
4 http://oxrep.classics.ox.ac.uk/bibliographies/fish_industries_garum_production_bibliography/
5 Ellis defines fish sauce (*garum* and *liquamen*) as a by-product from the "mulching'? down of the "smaller parts'? of the fish while *"allec* and *muria* (both fish pastes?) were by-products of the fish sauce' (Ellis 2011:68). These definitions are somewhat unhelpful and contrary to the sources.
6 I believe the lining was due to be tested in 2020 but the results will not be published in time of going to press
7 Ellis cites the Byzantine edict from 6th century Palestine forbidding the manufacture of fish sauce within the urban area (*Hexabiblos* 2.4.22). Despite this injunction some fish salting factories were fully integrated into the urban environment, as at Baelo Claudia. I suspect that the manufacturer simply became accustomed to the strong smalls but the customer may not have!
8 Antonio M. Sáez Romero, Dario Bernal Casasola, Enrique García Vargas, José Juan Díaz Rodríguez, «*Ramon T-7433 (Baetica coast)*», *Amphorae ex Hispania. Landscapes of production and consumption* (http://amphorae.icac.cat/amphora/ramon-t-7433-baetica-coast), 10 July, 2016.
9 Bernal cites the 5th century site at Traducta, where hundreds of sheep, goat, cow and pig bones were found in and near *cetariae* (Bernal 2016:201).

10 The mullet, bream and wrasse were the kind of species that were used to make a basic fish sauce in later periods, though at Baelo Claudia the dominant species caught in the area in later periods to make sauces were in fact sardine and sprat (Bernal 2018:347).

11 Apart from Horace the only other indication of *garum* at this time come from Varro who is only concerned with the gramma of the plural form: *gara* (Varro *de lingua Latina* 9.40.66).

12 Enrique García Vargas, Daniel Martín-Arroyo, Lázaro Gabriel Lagóstena Barrios, «Beltrán IIA (Baetica coast)», Amphorae ex Hispania. Landscapes of production and consumption (http://amphorae.icac.cat/amphora/beltran-iia-baetica-coast), 08 July, 2016; Cotton *et al.* (1996:226).

13 *Apicius* 2.2.8; 7.9.3; 4.2.21; 8.17.14. Each time the sauce is made from a pepper mash, *liquamen* and a sweet reduced wine and they are often thickened.

14 University of Southampton (2014) *Roman Amphorae: a digital resource [data-set].* York: Archaeology Data Service [distributor] https://doi.org/10.5284/1028192.

3   Fish sauce in culinary, medical, and veterinary sources

In a prehistoric context the fermentation of meat and fish has been recognized as a fundamental part of the forager diet (Speth 2017:44). Fermentation acts to partially "pre-digest" protein and fats found in pre ceramic diets without the need for cooking. The lactic acid bacteria (LAB) present rapidly invade the decomposing meat and fish and effectively prevent pathogenic bacteria from forming, and the meat and fish so treated can be kept safe to consume for weeks if not months. These techniques do not involve large amounts of salt and would be dangerous if they did, as the LAB production would be undermined completely. The presence of aerobic bacteria in this kind of fermentation also reduces oxygen in the tissues, which results in the prevention of lipid autoxidation (rancidity). It is also apparent that vitamin C normally destroyed by cooking is preserved using fermentation. Psychological studies have also indicated that the disgust that is normally felt by modern western cultures to the idea of putrid or apparently rotten foods is not hard wired, but a culturally learned reaction. These strongly flavored foods were not "starvation foods" but were commonplace and both desirable, and almost certainly recognized as valuable nutritionally (Speth 2017:51). Over time these natural fermentation processors, resulting in controlled "rot" and a culturally acquired taste for strong flavors, began to be less desirable; meat was freshly cooked and the addition of spices was an efficient antibacterial agent, and so there was no need for fermentation of animal products. Nevertheless, fermented foods remained in the diet of many Northern and Southern European cultures, as there is great value to be had in the appreciation of the flavor enhancing qualities of umami, which is directly linked, despite our negative perception, to the very idea of deliciousness. A taste for fermented fish sauce is a little like a taste for stinky farmyard cheese, though it doesn't reach the levels of disgust engendered by *surströmming*, the Swedish low salt fermented whole herring, a new can of which, according to a Japanese study, is the most putrid food smell in the world.[1] *Garum* and *liquamen* could not have pleased everyone. They are today challenging to engage with and difficult to use. The standard *liquamen*-type sauce is far easier to use as it really does not smell that bad in comparison to many fermented foods and southeast Asian cuisine has spread so widely that it can be found in many households. Everyone who does own a bottle of Thai or Vietnamese fish sauce knows that it needs to be added to a recipe with some caution as two much can allow the fishiness to dominate rather than the umami.

Cooking with Fish sauce: Apicius

The cookery book known as *Apicius* is a unique collection of recipes that spans the entire Roman period. The recipes themselves reflect the long decades of Roman interest in good food from the 1st BC to the 5th century AD. The book survives in just two manuscripts from 9th-century France. Later copies of the manuscript have acquired the title *de re coquinaria* but crucially this is absent from the original manuscripts and is a medieval addition.[2] In our edition of *Apicius* we have concluded that the author and title of the book are combined in one word: *Apicius*. We concluded that the very concept of recipes and good food *per se* became associated with the name "*Apicius*" through the decades after the first *Apicius*. This *Apicius* is believed to be the famous and notorious Roman gourmet Marcus Gavius *Apicius*, though it is unlikely that the gourmet was responsible for the recipes (Grocock and Grainger 2020). *Apicius* and all gourmets would have relied on their slave cooks to prepare the food and it is the cooks themselves who recorded recipes and shared them amongst themselves, though some dispute their ability to do so on the grounds of literacy, as we shall see. The people who created recipes and first recorded them in collections were predominantly slaves, either of Greek or eastern origin or/ and trained by Greeks (Grocock and Grainger 2020). There is extensive evidence for literacy among the free Greek cooks, known as *magieroi*, who wrote and disseminated their recipes books (Dalby 1996:160). The extent of literacy among everyday Roman slaves is disputed, largely because the range of social functions, status and training of slaves in the ancient world tends to be underestimated, as there is inescapably a tendency to evaluate their activities and capabilities through the lens of more modern forms of slavery, which encompass a much narrower range of activities. There is an unfortunate assumption that slaves and ex-slaves lacked agency and motivation and even the intelligence to learn to read and write. Robin Nadeau has offered a critique of the assumption that they could do so, even at a very basic level of functional literacy, and claimed that cooks could not have written the recipes in *Apicius* as they were illiterate.[3] Nadeau claims that cooks relied on an oral tradition and simply lacked the motivation to read and write (Nadeau 2015:53). This is an *a priori* assumption which can readily be challenged, especially if we consider what the book was for, based on its contents, structure and style. This is not the place to discuss the arguments in full; suffice it to say that there does seem to be evidence for literacy in all areas of the empire among the lower orders, including cooks (Bagnall 2011; Woolf 2015). As long ago as 1914, C.G. Harcum recorded the funerary inscriptions from cooks in Rome which indicate imperial guilds of cooks, scribes of cooks and numerous detailed and technical funerary inscriptions, illustrating a proud and skillful profession which simply does not fit the picture of illiteracy, and the absence of both aspiration and agency among these skilled craftsmen. This issue is important in understanding fish sauce as we have to question why *garum* appears to be used hardly at all in the *Apicius* recipes, while *liquamen* dominates. The manuscripts of *Apicius* are Carolingian copies of a late imperial compilation, but the recipe collection itself has its origins in an earlier Greek version of the book. The chapter headings are all in Greek and the text is peppered with a bilingual Greek/Latin culinary language noted by Dalby (2013). Some of the recipes can be individually dated to the 1st century AD by reference to names of characters from that era, and by reference to food items that subsequently cease to be available in later periods (Grocock and Grainger 2006:361).[4] The collection was complete and effectively "published" by the late 3rd/4th century, but individual recipes were collected and re-collected over centuries. Early scholars on *Apicius* imagined that the book was rewritten for the Late Roman market from an earlier collection, at a time when *garum* was replaced by *liquamen,* but actually there is nothing within the text to suggest that happened, and it seems very unlikely that fish sauce terminology was changed while the other Greek terms were not (Grocock and Grainger 2020).

Garum and liquamen in use

It is very clear from a review of the didactic recipes that the two fish sauces were, in practice, used in very distinct ways.[5] What is most apparent is that blood/viscera *garum* had no role in the cooking process itself, while *liquamen* was predominantly used during the cooking. Without exception in *Apicius, liquamen* is added during the cooking process in order to blend with the other ingredients and allow the development of complex flavors. *Garum* appears as part of the names of finishing sauces or dressings, that is, *oenogarum,* but the fish sauce ingredient which is used in making them is *liquamen.* The older Greek terminology has been preserved in the "culinary vocabulary" but the Latin is specific about what ingredient/product is being used in it. When *liquamen* was blended, mixed with other ingredients, it had the same role in consumption practices as the complex mix of salt, pepper, vinegar, brown and red sauce still does in cafés and bars today, it enhanced and improved the experience of eating. The fact that *liquamen* and *garum* had different roles within Roman cuisine is also indicated by the distinct wording of the Manilius description of fish sauce (page 37). In this text, the blood/viscera sauce made with the juices removed are described in the words *mixto gustum sale temperat oris,* "mixed with salt balances taste in the mouth," which suggests a taste-experience that happened at the point of eating, while the product that is made with the juices retained is described in the phrase *communemque cibis usum sucumque ministrat*, "and it serves up a common condiment (i.e., *allec*) and a useful juice *(liquamen) for* foodstuffs" (Manilius, *Astronomica* 5.675) which as we shall see was concerned with cooking generally.

There are two recipes in *Apicius* that directly mention *garum* as a single product or ingredient. It is present unambiguously in just one recipe:

> *Apicius 7.13.1 fungis farneis: elixir calidi exsiccati in garo, piper accipiuntur, ita ut piper cum liquamine teres*
> "Ash-tree fungi: boil and serve while hot and dry with *garum* and pepper, so long as you pound the pepper with *liquamen*."

The recipe is perplexing of course as we have mention of both a *garum* and a *liquamen*, but logically the pepper mash could be made with *liquamen*, and this is then served at table with *garum* on the side.[6] The idea that *garum* is visible in the dining room so that the diner can pour it on himself is indicated by a letter of Ausonius, where it is suggested that the sauce that is being discussed could be poured onto a *patina* (frittata), and it is also apparently poured into spoons by the diner: "But I, by whatever name that liquor 'of our allies' is called," and at this point he begins to quote from an unknown poem "will now fill my *patinas* so that that juice, more sparingly used in our ancestors" tables, will flood the spoons" (Ausonius, *Ep.* 21; see page 104). The *patina* is a cooking vessel and also the name of the food, a kind of frittata, which was made inside these vessels. In *Apicius* these frittatas are made with *liquamen* as a matter of course. The implication may be that the diner adds *garum* via a spoon onto the cooked egg dish at table before he eats it. There is no absolute clarity here, as a pretentious ambiguity seems to be what Ausonius is aiming at. It is also implied that the diner pours the *garum* onto his oysters himself in Martial (*Ep.* 13.82). The idea of pouring the glossy black garum onto food at/or during service is not without culinary logic. The use of a table soy sauce which the diner handles is a familiar culinary practice in modern cuisine from South East Asia. The dark ishiri squid viscera sauce unique to Japanese cooking is also used in a similar way as a seasoning to finish a dish and it is never added during the cooking process. *Garum* might possibly be pre-mixed with *caroenum* (reduced grape juice) in the term *garoeni* at *Apicius* 4.2.29. This is a compound sauce that otherwise does not contain fish sauce and the absence is very unusual. There is also a possibility of *garum* being used in one recipe in the other *Apicius*, the *Vinidarius* collection.[7] Otherwise, as we have already noted, all mentions of *oenogarum* were in fact made with *liquamen*. The use of *liquamen* in cooking is understandable. It was relatively mild in comparison to *garum,* and, if the salt was used at the lower level found in the Geoponica (15%), it would not be excessively salty and could be used as a liquor, contributing to the volume of fluid required in a sauce rather than as a tiny percentage of it. Both sauces were used to make the compound sauces called *oenogarum*, a blended dressing of oil, wine or vinegar and fish sauce, which resembled a vinaigrette. These sauces are best understood in relation to *liquamen*, as is explained below. In Petronius' *Satyricon*, statues of Marsyas stand with wineskins from which a *garum piperatum* is poured on to fish swimming about a narrow channel (Petronius, *Satyricon* 36.3.2.) Given the Satyricon focus on satire of "elite" and "nourveau-riche" pretentiousness at this point, we might expect this to be an expensive *garum* being used to flavor an *oenogarum* which is then in theory used to season a *hydrogarum*, that is, a cooking liquor. There is ambiguity here, as the *garum piperatum* could equally be made with *liquamen*. The *piperatum* might not be made simply of fish sauce and pepper, though it could be, as at *Apicius* 8.7.14. At *Apicius* 4.2.21 we find the instruction to "pound pepper, lovage, pour on {*liquamen* and *passum*} so that it is sweet. Pour this *piperatum*..." This recipe demonstrates that these were more complex sauces, and at 9.1.2 a *piperatum* has coriander added too. Martial's complaint that his cook will bankrupt him mixing a "huge pile of pepper with *Falernian* and *arcano garo*" (Martial, *Sat.* 7.22) would seem to be a strong indication that that a blended sauce collectively known as an *oenogarum* could be made with the black and bloody *garum*. We have seen a form of *garum piperatum* in consumption texts and a *liquamen piperatum* in *Apicius*, and both represent sauces that were prepared at the last minute by the cook to serve as either dipping sauces or alternatively to pour over a dish before service.[8] What is clear to me is that these sauces would have tasted different depending on which fish sauce was used.

This is all the direct evidence we have for black garum in culinary use in the 1st century AD. Given how popular it seems to have been in Pompeii, on the basis of archaeological finds (mostly *tituli picti*), it is certainly strange that we are so much in the dark over its uses. We may conclude that if the diner chose to pour it on himself there is a very good reason for this sauce to become a topic of discussion and gastronomic competition. *Garum* was a precious highly valued fluid, more expensive than unguents according to Pliny (*HN.* 31.94), and this only makes sense if this sauce and its taste were unique and also too rich, too dark, too intense, and too expensive for general cooking. *Garos/liquamen* on the other hand appears to have been used predominantly in the kitchen and it is therefore quite understandable that the elite failed to recognize its existence as they never saw it in use separately. When it appeared at table it was either in the cooked food or blended into *oenogarum* already and indistinguishable from *garum* proper. We have learnt that when present in a premixed *oenogarum*, which sauce was being used was unclear. The nature of these dipping process is also illuminated in Chapter 5 when we discuss the phenomenon of the *Colloquia of the Hermeneumata*, Latin and Greek phrasebooks which describe everyday behavior in the dining room. Diners are encouraged to "dip it in," the "it" being bread, vegetables and meat depending on the context. The cup that is regularly used in *Apicius* to blend and serve these sauces is called an *acetabulum* little vinegar cup and these are regularly used to blend sauces in the phrase books too (Grainger and Buddulph in press). Later Byzantine evidence suggests that vegetables were very common and particularly the radish, often illustrated on banquet scenes, which was dipped into hot *oenogarum* as an appetite stimulant at the beginning of the meal (Anagnostakisis 2013:156).[9] In a 3rd century Roman wall painting from the Insula of the *Thermopolium* in Ostia the phenomenon of dipping is also made clear. The image is of a plate of food beside a glass and a form of maracas indicating that wine and entertainment is available. The plate of food includes a small cup imbedded in an unidentifiable food item that may be pulse or olives. There is a skewer with chunks of meat and next to this is a root vegetable which is almost certainly a radish given the later strong association with this dipping this vegetable. It is also almost certain that the small cup contained a sauce and one made with *liquamen*.

A few simple sauce recipes from the *Apicius* collection will suffice to demonstrate its utility:

> 5.4.1. With beans: cook them. Pound pepper, lovage, cumin, green coriander, pour on *liquamen*, balance their flavor with wine and *liquamen*. Pour into the pan, add oil. Bring to heat on a gentle fire and serve.
>
> 2.1.5. Another forcemeat recipe: put in a mortar pepper, lovage, oregano; grind it. Pour on *liquamen*, add cooked brains, pound carefully so it has no lumps, add 5 eggs, mix in carefully to make a smooth emulsion. Flavor with *liquamen* (*liquamine temperas*) and pour into a bronze pan. Cook it. When it is cooked, turn it out on to a clean board. Cut into squares. Put in a mortar pepper, lovage, oregano; grind them together. Add *liquamen* and wine. Put all together in a pan. Bring to the boil. When it has boiled, crumble *tracta* (dry pastry), stir it so that it thickens, pour out on to the serving dish. Sprinkle with pepper and set it forth.

This demonstrates the use of *liquamen* at three separate stages of the recipe: it is added during the pounding of the brains, then again to balance and correct the flavor of the forcemeat, then finally in the *oenogarum* that is served with the forcemeat. Of importance here is the knowledge that had fish sauces been made with the levels of salt indicated in the Gargilius recipe (page 22) then this would have been quite intolerably salty and uneatable. The concept of a vegetarian *liquamen* is found in the discussion on fish sauce in Paladius where a sauce made with pears is deemed a *liquamen* and we may conclude that this was used in basically the same way as a general seasoning to enhance the flavor of other foods.[10] A number of recipes use salt instead of *liquamen*, and in the case of the salt fish *patinae* naturally no *liquamen* is used (4.2.17 *patellam thirotarricam ex quocumque salso uolueris*: Salt fish and cheese patella using any salt fish you like), demonstrating a dislike for over salty dishes. Thickening is common with *oenogarum*, and we can see this concept most clearly in the following two recipes in the book on fish cooking in *Apicius*:

> 10.2.17 *in pisce oenogarum: teres piper rutam, mel commisces, passum liquamen caroenum et sic igni mollissimo calefacies.*
>
> 10.2.17. Oenogarum for fish: pound pepper, rue; stir in some honey, *passum* (dessert wine),
>
> *liquamen, caroenum* (reduced grape juice) and warm it through over the gentlest of fires.
>
> 10.2.18 *in pisce oenogarum: ut supra facies. cum bullierit amulo obligabis*
>
> 10.2.18. Oenogarum for fish: make as above. When it has come to the boil, thicken it with starch.

There is a description of "simple sauce and an allusion to a "double sauce" in Horace which has a bearing on the idea of thickening a sauce:

> "It is worth the effort to get to know thoroughly the nature of the double sauce. Simple sauce is made from sweet olive oil, which is worthy of being blended with fragrant pure wine and *muria* provided that it comes with a powerful whiff from a Byzantine jar"
>
> (Horace, *Sat.* 2.4.63–71).

There are different ways of interpreting this passage. Fairclough's Loeb translation, revised in 1929, considers the ingredients listed as potentially illustrating the *simplex* and the *duplex* sauces. Fairclough (1929:190) reports that "others" contend that the herbs and spices listed correspond only to a *duplex* sauce. This seems improbable, as all these sauces contained some extra seasonings. Though a fish *muria* is the sauce utilized I think we are still basically talking about an *oenogarum* in the sense that a form of fish sauce is blended with wine and oil. It may be that "simple" corresponds to *muria* while "duplex" corresponds to *garum*, though it is not sufficiently clear. The term *oenogarum* becomes synonymous with all compound sauces in *Apicius* even when made from vinegar, oil, wine, sweet wine and honey; as long as fish sauce is used and blended in some way they were classed as an *oenogarum*. *Duplex* means "double" or "twofold", but it also has a meaning in Lewis and Short of thick, strong or stout and I take that to mean that a *duplex* sauce would refer to one that is thickened, as in the *Apicius* sauces above (*duplex:* 3 Lewis and Short: 619). The *simplex* sauce is also found in *Apicius*, at 4.2.21 and 26, where an undefined *"oenogarum simplex"* is poured onto a *patina*, and at 8.1.5, where a mixture of wine, vinegar, *liquamen* and oil is classed as a *ius simplex* which is then seasoned with a pounded spice ball made up of pepper, lovage, celery seed, mint, thyme, roasted pine nuts which may correspond to this *simplex* becoming a *duplex*. Many *oenogara* such as 4.5.1 and 4.5.3 are also thickened and poured over various kinds of cooked food but they are not termed *duplex*. There are few indications of the uses of *allec*, apart from those cited from Plautus and Horace in Chapter 2. These imply that it was an appetite stimulant and was used as a dip with ham and vegetables. A recipe in *Apicius* for a sauce for boiled meat made with *allec* instructs the cook to grind pepper, lovage, carraway, celery seed, thyme, onion, date, strained *allec*. Honey and wine are added and oil poured on the top. (*Apicius* 7.6.14). The process of straining the sauce would appear to render it into a thick paste with the residual liquid removed. This technique is illustrated in Chapter 11 when we discuss *allec* more fully. Clearly *allec* had a role to play throughout modest and elite households, and was a product potentially utilized in the kitchens widely and this is reflected in the way Pliny describes *allec*. The sauce in *Apicius* resembles in many ways the modern spiced fish pastes, *pater piperium* and also the seasoned Pissalat of southern France (see page 230). There are few truly culinary recipes outside of *Apicius*, and this is largely due to the difficulty in distinguishing a culinary recipe from a dietetic one primarily designed to cure or improve health rather than please. Many dietetic recipes can be found in Pliny the Elder and Columella. Pliny suggests that wild parsnip can be taken with vinegar and *silphium,* or with pepper and *mulsum* (honeyed wine), or if you like with *garum* (*HN*. 20.34.3); snails can be boiled or grilled and taken with wine and *garum* (*HN*. 30.44.4); an *opsonia* (morsel?) is taken in a dressing of *mulsum*, oil and *garum* (*HN*. 27.136.5). We might legitimately conclude that these recipes were taken from sources of Greek origin, and that what Pliny has copied were recipes using *garos*, that is, *liquamen*, though there is no way to be certain. Columella's use of *garum*, as we have noted in Chapter 3, is also hard to interpret: the remedy *oxyporium* (digestive) which uses *garum* (without an adjective: *DRR* 12.58.4.13) is a complex blend of spices and liquids. A recipe in *Apicius* suggest that fish sauce could be deemed as to have "gone off" though we may conclude that this would mean becoming oxidized through exposure to the air rather than actually becoming rancid or decayed. The smell of an oxidized modern fish sauce is rather off putting but not unfit for use.

Apicius 1.7. On restoring *liquamen*: if the *liquamen* has produced a bad smell, fumigate an empty upturned pot with bay and cypress, and pour the *liquamen,* having whisked it, into the pot. If it is too salty you put in a pint of honey and stir it. You seal it with pitch and you have restored it. Fresh must also has the same effect.

This precise combination of burning aromatic herbs and adding honey has been found in residue by evidence from a *doliola* containing similar compounds at Boca do Rio in Lagos though the latter is 4th/5th century AD (Morais *et al.* 2016:107; page 126).

Cooking a quick fish sauce: Gargilius' *confectio liquaminis*

The idea that a sauce could be made quickly and simply by boiling the fish in brine survives in the *Geoponica* and is also found in a number of late and medieval manuscripts, as is discussed in Chapter 1. Small fish, salt to make a brine and wine are boiled together and reduced and then filtered and bottled (Curtis 1991:192). These "*confectio* of *garum* are also highly complex with numerous (25+) spices and herbs included in the recipes which are reminiscent of the original Gargilius version. The recipes from *Apicius* and other sources suggest that the majority of herbs and spices tended to be added to the fish sauce along with the oil, wine and vinegar when these blended composite sauces were made by the cook just before the point of service. Clearly, if one can boil up fresh fish and make an instant fish sauce, then one can also incorporate spices and herbs at this first and only stage of the process. The Gargilius recipe (page 22) presents another scenario where a fermented sauce is heavily flavored with herbs during the fermentation stage and then further spiced when it is diluted with wine and reduced to concentrate the flavor. There is certainly nothing in *Apicius* or recipes and remedies of a similar age that resembles the complex formulation of this recipe in Gargilius. We have already noted the medieval spices and north European species of fish utilized, which certainly places this recipe outside the Roman world. The reduction in the volume of liquid from c. 3 pints down to 1 pint is unusual and not found in any of the *oenogarum* recipes from *Apicius*, where sauces are brought to the boil but not for an extended time in order to reduce the volume. The Gargilius recipe was, I would argue, formulated at a time when the bulk "essential" fish sauce, that we see traded so widely in the 1st century AD, was no longer made in the same quantity, and cooks were therefore often forced to make their own, either by fermenting if they had the time or cooking if they needed it quickly. The widely traded essential substance of the 1st century AD was, I will argue, a largely unadulterated product comprising fish, salt, possibly wine or brine, and very occasionally a few herbs (oregano was used in the Geoponica), but certainly not the huge quantities of herbs in this recipe. I would also argue that the uniqueness and quality of any subsequent *confectio/oenogarum* recipe was the responsibility of the cook, rather than the fish sauce manufacturer. The cook or *opsonator* (steward) wanted to be able to know he could buy a consistent product of high quality, so that the blended sauces made in house were the best they could be.

There has however been some confusion concerning whether fish sauces were ever fermented and cooked at the same time in order to generate or accelerate a finished sauce. Researches understand from the Gargilius recipe that the two separate processes fermenting and cooking were often combined in the one process and at the same time and they consider that this was potentially a common practice in the early Roman period. Vargas (*et al.* 2014) understand the date of Gargilius recipe to be that of the original text, that is, the 3rd century AD, but we have noted this fish sauce recipe is almost certainly a later addition to the Gargilius text and cannot be used to illustrate a 1st century AD fish sauce (Vargas *et al.* 2014:70). Vargas *et al.* suggest that the *confectio liquaminis*, that the text calls an *oenogarum*, is actually incorrect and that the word should be *omogarum* meaning a "raw" and in fact also "simple", with reference to *simplex*. I do think that the initial fish sauce is technically "raw" as opposed to a cooked from scratch but I am less sure of this justification as it stems from a belief that the sauce created by the first stage of the Gargilius recipe was in some way incomplete as it took such a short time compared to the *Geoponica* (cf. *Geoponica* 20.46.4). The second stage, the cooking, was in effect an acceleration of the process of making the basic fish sauce and that this acceleration was a regular process in fish sauce production (Vargas *et al.* 2014:70). This would seem to be unnecessary in my view as it does not appear that there would be any benefit to accelerating the fish sauce process with cooking as the *allec* which contained the remaining potential nutrition had already been filtered out. It can be argued, after observing the sauces in production (discussed in Chapter 8), that one month of fermentation was sufficient to generate sufficient sauce of reasonable quality. There would not be any need to accelerate the process of making the fish sauce by cooking as it would have already been made effectively by fermentation and the long boiling was a process concerned with making *oenogarum* not the fish sauce. It could equally be inferred that the cooking and fermenting are separate processors which only occur here in the one recipe because, as it is a late addition to the text, it corresponds to a time when fish sauces were not traded widely and cooks had to make their own.[11] There is also every expectation of a time laps, not reflected in the recipe, between making the fish sauce and turning it into a *oenogarum*. It is often assumed that the basic bulk fish sauce of the 1st century AD was potentially cooked/heated/ boiled? after it left the *cetariae*. Not only is it very difficult to see how these fermented fish sauce could be further cooked at the production sites as there are no structures or vessels that could be identified as able to allow heat to be applied to the vast quantities of sauce that is envisaged (Chapter 10), but we must also acknowledge that cooking would have destroyed the enzymes which allow the protein to be liquefied. In our hunt for the essential substance in the early empire, it might be better to separate the quick cooked recipe from the fermented recipe and from the subsequent preparation and cooking of the *oenogarum/confectio liquaminis*. It is however, of note that the use of hot brine at the end stage when residues are almost exhausted is very probable as a means of extracting the last dregs of fish sauce from the residue, discussed in Chapter 8.

Medicinal uses of fish sauce

Robert Curtis' thesis was that the medicinal qualities of salted fish and fish sauces were fundamental to the Roman way of life (Curtis 1991:27–37). While this is true in principle, when one looks at precisely what fish sauces were meant to treat, the list of bodily ailments is pretty small. These salty liquids would naturally have antibacterial qualities in the same way that brine today (saline solution) is used to treat small infections in the eyes and ears, and this is how it was used in application in Roman times. In Pliny the Elder we find leprous sores treated with ash mixed with *garum* and oil (*HN*. 20.55.4) and there is the somewhat humorous idea that *garum* could treat fresh burns but only if the patient was not told (*HN*. 31.97.1). Curtis has questioned why so few remedies survive. It may be that fish sauces were more esteemed for the incidental therapeutic qualities provided by using them to stimulate the appetite, to accompany foods that were valuable and to aid digestion and promote the natural digestive transit processes. Fish sauces were classed as bad juices by Celsus (*med. 2.21*), as alien to the stomach (2.25), and liable to move the bowels (2.29). Galen describes *garos* as "heating" and "drying" and says that it should be used to treat putrefying ulcers and for dysentery (Galen *de simp. medic. temp. ac facul.* 12. 377k) and also notes that *halmer* (*muria*) had the same properties. The mixture known as a *liquamen catharticum* is found in two medical treaties (*Marcellus Empiricus De med* 30.52; Aetius *lib Medic* 3.83–84; 86–88). This is a medicinal concoction made of numerous spices and herbs bound with *liquamen* and sometimes a sweetener like honey. They have purifying qualities, and are hence a cathartic or laxative. A similar preparation is called an *oxyporium* in *Apicius* and Columella (12.59.4 cf. page 66) and is derived from *oxyporus* also from the Greek, for that which passes quickly! This is *Apicius*" recipe for *liquamen catharticum*:

> "1.32. *Oxyporium:* 2 oz. cumin, 1 oz. ginger, 1 oz. fresh rue, 6 scruples soda, 12 scruples date flesh, 1 oz. pepper, 9 oz. honey. Soak Ethiopian or Syrian or Libyan cumin in vinegar, dry it and grind it. Afterwards you mix everything with the honey. When necessary you use it with *oxygarum*."[12]

The occurrence of *garum* and *liquamen* together in a remedy collection is rare. We do find such evidence both in *Apicius*, albeit rarely (*Apicius* 7.13.1 discussed above) and it is common in the veterinary remedies sited in notes 12 and 13 below. The work of Marcellus Empiricus, a contemporary of the father of Ausonius in late 4th/early 5th-century Gaul includes, in his *De medicamentis*, the use of both terms, though never in the same recipe, and the same is true of a collection of medieval remedies which was derived largely from Pliny the Elder and other, unknown, sources.[13] It seems unnecessary at this point to consider that these references were anything but an indication that the bloody sauce and the essential whole-fish sauce were utilized independently of each other in the different remedies. The alternative is that compilers of these remedies were indifferent and copied recipes from different eras, in which early versions used *garum* and later ones *liquamen* without noticing the anomaly of the two terms and either took it for granted that they were the same or did not care. However, given the clear and precise separation of sauces in earlier text this would seem to deny too much of the evidence for the dual nature of ancient fish sauce. It is also quite possible that compilers lacked an interest in the minutia of the ingredients as they valued the recording of knowledge but not the actual empirical knowledge of the preparation process which was necessarily a slave occupation,

Veterinary texts and fish sauce

Columella uses *garum* copiously for veterinary purposes and it is fairly certain that it was unlikely to have been a blood viscera sauce, though of course at the time of his writing, c. AD 50 to 70, the black and bloody variety had become the height of fashion. There is no acknowledgment of any difference in the sauce types. I suspect that the cause is simply that compilers like Columella and Pliny were collecting remedies, but had no interest or comprehension of the empirical and practical aspects of the remedies themselves. That was the role of those who concocted them in the first instance: Greek practitioners and those who made them up: slaves and freed men. The later veterinary collections can tell us very little about the sauces that were used apart from the fact that now both terms were used. There is no direct mention of a black or bloody sauce, just the regular use of both *garum* and *liquamen*. The veterinary writers whose works survive largely date to the late 4th century AD. There are three key Latin writers: Pelagonius" *De veterinaria medicina*, Vegetius *Mulomedicina*, and the anonymous author of the *Mulomedicina Chironis*. Vegetius' text is largely made up of material taken from the other two. Pelagonius set himself up as a kind of consultant in imitation of his source, the Greek veterinary writer Apsyrtus, who was a horse-doctor of some repute. Apsyrtus was consulted by letter and the replies, imitated by Pelagonius, constitute the format of his work: a letter, often addressed to a named person, offering advice and a remedy. Adams considers that Pelagonius' skill is limited: "He is the sort of literary layman who might have found remedies in earlier handbooks while not necessarily performing treatments himself" (Adams 1995:663). The remedies themselves utilize *garum* and *liquamen* with equal regularity. Curtis had noted, what must have been an unusual occurrence to him, of both *garum* and *liquamen* within the same text, but makes little of this anomaly (Curtis 1991:37, note 37). He does point out that the practice among veterinarians is not just to use *garum* but to suggest *garum primum* and the *liquamen* is often classed as *optimum*, phrases associated with amphora *tituli picti* (cf Chapter 12).[14]

Notes

1   Excerpt from Hakkou ha chikara Nari, Takeo Koizumi, NHK Ningen Kouza 2002 ISBN 4-14-084183-4
2   I along with my husband Dr Christopher Grocock have been responsible for the most recent edition of *Apicius* (Grocock and Granger 2006; revised ed. 2020).

3 Wilkins, J. and Nadeau, R. eds., 2015. *A Companion to Food in the Ancient World*, John Wiley & Sons.

4 At Ap 1.13 there is a recipe for how to make silphium last for ever. Cyrenaican silphium had ceased to be traded by 50 AD according to Pliny (HN. 19.35–38, 22.100-6) and was replaced with the resin asafoetida, Syrian or Parthian Silphium. Recipes suggest that both were available at the time they were composed while in Columella (12.59.4) a similar availability of both Syrian and Cyrenaican silphium is found in remedies. The fact that silphium was potentially available when the Apician recipe was created suggest and early date. see also Theophrastus, History of Plants, 6.3, 9.1.7; see also Dalby, Food in the Ancient World, pp. 303–304.

5 For a general review of Roman food tastes see Banducci 2018: 138.

6 The concept of a *liquamen* pepper mash occurs elsewhere (2.2.8) where the mash is then sweetened with defrutum and thickened. The majority of Roman sauces begin with pepper in varying quantities.

7 Vinidarius 29 where a *cum inogaru* is variously interpreted as 'with cumin and garum' (Milham) or as we have rendered it '*cum inogaro* with *oenogarum* i.e. a *liquamen*, as is usual (Grocock and Grainger 2006: 325)

8 The term *oenogarum* and the other associated terms *oxygarum, hydrogarum, oleogarum* have never appeared on amphora labels. However, in Egyptian papyri there have been occasional references to these compound sauces in lists and purchase documents which suggest that they could be purchased pre-blended in smaller containers. See Chapter 4.

9 See, for example, the 'Heavenly ladder' detail of the wall-painting in the *exonarthex* of the *katholikon,* Votopedi monastery, Mount Athos (Anaknostakis 2013:197).

10 Liquamen ex piris, Paladius, Opus Agricultura 3.25.12.

11 There are a number of incidences of apparently 'raw' fish sauces from Egyptian sources using the term ὠμογάρον, which term, we might conjecture was used to designate a fermented sauce from a quick cooked one, in the late empire when the former was less readily available, though, as always, it is not entirely clear. (Cf chapter 5). For a different view: Varges et al 2014:70).

12 A very similar recipe for an *oxyporium* attributed to Galen, is sited in chapter 2.

13 Marcellus Empiricus *De medicamentis: liquamen,* 20.119; 23.51; 26.93; 30.52; *garum:* 9.14; 9.36; 9.112; 16.90; 30.1. The medieval medicinal text attributed ultimately to Pliny (Winkler 1984) is called *Physicae quae fertur Plinii Florentino-Pragensis liber primus: liquamen:* 1.5.17; 1.10.19;1.50.19; 2.9.1; 3.14.45; 3.15.17; 3.12.14; 3.43.6; 3.43.5; 3.43.9; *garum:* 1.9.1; 1.43.6; 2.18.18; 3.34.13.

14 Pelagonius, *De veterinaria medicina, liquamen:* 9; 11.2; 13; 98; 455; 457; *garum:* 428;13. Vegetius *Mulomedicina: liquamen:* 1.10.1; 1.17.10, 16; 2.91.2; 2.108.2; 2.132.4; 4.6.1; *garum:* 2.28.8; 3.28.10.

4   Fish sauce from Papyri in Greek speaking Egypt

Their value as evidence

There is a vital body of evidence about ancient fish sauce from papyri pre-served in the exceptionally dry conditions in Egypt. This material is unique, as the information found within these texts is rarely found in ancient literature from the manuscript tradition. On papyri we find the kind of everyday prosaic written material that can aid our understanding of how Greek speaking people engaged with fish sauce on a daily basis. Literacy appears quite widespread in Egypt despite beliefs to the contrary (Bagnall and Cribiore 2006; Bagnall 2011). The Egyptian documents include a legal archive of trading activities, letters concerning private deals, and simple records of foods purchased and their cost on a daily basis which include numerous different kinds of fish sauce.[1] It is important to stress that the ter-minology used in papyri is exclusively Greek. It might seem unnecessary to point this out but as the crux of the fish sauce nomenclature issue is derived from a translation/transliteration conflict, what terms the ancient Egyptians used for fish sauce would seem to be crucial. We also have a unique opportunity to understand fish sauce from a Greek perspective, albeit through the spectrum of many centuries of Roman influence over fish sauce technology, trade, taste, and consumption rituals. The modern sources for this chapter (Curtis 1991; Drexhage 1993) render the Greek *garos* into "garum" when they discuss fish sauce and this is normal and understandable as so many ancient authors did to but it is strictly speaking, not accurate as a precise translation of *garos* is clearly *liquamen*. To prevent confusion and maintain the system devised in the introduction, I retain the Greek form and use *garum* only when it seems that the sources refer to the high-status blood/viscera sauce.

The evidence from the 1st and 2nd centuries AD is scare in Egypt and would rather suggest that it took a while for an interest in "Roman" style fish sauce to percolate throughout the community. That is not to say that a local variety of some kind of fermented fish product was not made and locally used. Drexhage has noted that there was an abundance of fresh and salted fish from the Nile and it would seem unlikely that this would not be the case, though we have no knowledge of it.[2] When this Roman-style fish sauce first arrived in Egypt it was undoubtedly through elite consumers, and it was not until the 3rd and 4th centuries that *garos* had become sufficiently commonplace to be mentioned regularly.[3] There are many restrictions to our understanding of fish sauce in Egypt as we often cannot determine whether, when *garos* or the term *gararia* is used, it refers to the sauce or the vessel it was served in. We cannot determine the volume of the vessels referred to and, when the price is included, we are unable to determine its value and quality compared to other evidence. This allowed, the material is rich and informative. The earliest indication of *garos* comes from an Arsinoites source from around 180 BC (BL 3/89). In col. 3 line. 3 the entry reads: ὄξους γάρους, "oxous garous": probably a mixture of vinegar and *garos*, which is later collectively known as an *oxygaron* This may also refer to a *garos* made and initially diluted with vinegar rather than wine (see page 272, note 2). Athenaeus refers to ὀξυγάρου being served in a special vessel called an ὀξυβάφον (Ath. 3.67.e). In Byzantine sources the γαράρια was a vessel for heated fish sauces (see below). As with *oenogarum*, these mixtures were not necessarily just vinegar and *garos*, but seasoning and other liquids such as wine, sweet wines, and honey to balance the sourness and possibly thickened too (see page 87). There are a number of indications of ἐλεογάρου, "eleogaron," oil and *garos* with various spellings (ἐλαιογάρου: P. Wash. 1/59, 5th C.; ἐλεοκάρου: P. Hasitzka 25b, 7th C. and the more common οἰνόγράου (SB 16/12246, 4th century; P. Hasitzka 25b, 7th C.). From the *Apicius* we see that these blended sauces were easily made domestically, and it is often not clear from the context whether these references were to domestic production or purchased versions of these sauces. There is a diminutive form γαρίδία (P. Oxy. 8/1158, 3rd C.) and this has been taken to mean that the sauce is made with extra small fish, but we can probably reject this as most fish sauce in early Greece and later was made thus when of a commonplace nature (Drexhage 1993:34). I suspect this is probably a vessel for *garos*. These vessels are well known from Byzantine sources as special integrated braziers, also known as a chaffing dish, so that the *oenogarum* can be served and kept hot (see below; Vassiliou 2016).

The most valuable, for our purposes, is the record of purchase of five bottles (λάγυνος = glass) of "light" (λευκοῦ = white/light) *garos* and two bottles of "black" (μέλανος γάρους) which I will term *garum*, (P. Herm. Rees 23.6, 4th C.). This is probably the same product in the λαγύναιον (vial/small bottle) of bloody (αἱματίτου) *garum* (SB 14/11340, 6th C.),[4] both of which we can associate with the αἱματίου from the *Geoponica*. This juxtaposition of a light bright sauce compared to a black one reinforces and confirms my original premise that there were two distinct sauces. The "light bright" sauce is here contrasted with the black, and one must see this distinction as pertinent only in this context, that is, that the trader is demonstrating that he has both varieties. We do not hear of the term λευκός to designate fish sauces elsewhere, and "black" is also uncommon and only found to my knowledge in Galen (see below, and note 6). Color is crucial to a quality sauce but the darker is not, as one might expect, the better. Color is initially determined by protein content, but this is altered by the storage and production method. A sauce of high nutrition may be dark amber in color and only darken to brown with age. A fine aged mackerel sauce made by a colleague in the United States was golden amber and very fine in quality after 6 months, but after a year of storage in the presence of some light and in a glass vessel, it had darkened to a deep brown and the taste had altered to a more pungent flavor. This is the natural oxidation process at work. Cuong Pham from Red Boat has clarified that 'higher protein sauces tends to get oxidized faster therefor it gets darker faster' (pers. comm.) One might define the light bright sauce as a fresh one while the black one could represent an aged one. I am not convinced and find it hard to understand what might be sufficiently distinctive between a bloody sauce and a black one to warrant another sauce with another name. In pseudo-Galen, we do see a direct and clear association between the *garos melan*, "black *garos*" and the bloody sauce, though the dating is difficult.[5] The bloody sauce that I made according to the recipe in the *Geoponica* was a deep dark black/brown, even with minimal exposure to light. A freshly made *liquamen* of good quality by protein content was a dark amber and the light was able to penetrate fully and when held up to the window so that when in a glass bottle it looked amber. To my mind it is hardly likely that a distinction between dark brown and black would be made, but we may have to concede the black *garum* in this incidence may have corresponded to a dark, aged *liquamen*. In defense of my conclusion that the black and bloody are the same I would point out that even when a *liquamen* was darkened due to oxidation, light could penetrate the sauce, while the blood *garum* did not allow light to penetrate at all (page 165).

The diminutive for bottle rendered as "vial" by Drexhage is very interesting as it implies that these expensive blood/viscera sauces were bought and used in relatively small amounts. Most fish sauce, however, is discussed in relation to relatively small vessels such as jars, flagons, bottles of glass or ceramic. We find the earliest reference to a fish sauce is to a βῖκος γαριτικός, "a little jar of *garos*," from the Ptolemaic period (from 305 to 30 BC), which we may associate with the earliest form of small-fish *garos* (PSI 5/535. Col 2.l. 36). Drexhage notes however that the volume of fish sauce was always relatively small, regardless of the terminology of the vessel, and this was partly to do with the slow acceptance of the product as well as the fact that it was never really used in great quantities, despite the seeming abundance of the product in the Western empire during the imperial period (Drexhage 1993:38). It is certainly true that fish sauce of either kind was used relatively sparingly in terms of each individual meal. There is a reference to a woman asking for a a γάρος μικκόν (P. Giss. Univ. 25, 4th C.) for her chicken in an otherwise unclear passage and I would hazard a guess this was a conversation made either at table or at a *caupona* or bar. Here I think that both fish sauces were available to pour on to your food once it had been served. It is still noteworthy that the number of amphora referred to is quite small: the first reference to an amphorae for *garos* is from the 6th century (P. Mich. 15/740; Drexhage 1993:35).[6] We might consider the amount of foreign fish sauce was also scarce in these circumstances and this would seem to be confirmed by the absence of references to Spanish product, an issue discussed below. The assumption is that local and relatively small-scale production of fish sauce dominated the trade. A recent discovery of a 5th or 6th-century Byzantine papyrus from the Ashmolean records a shopping list and this includes γάρου σπανόν, "Spanish *garos*," and γάρου χρηστοῦ, that is, good quality (Maravela-Solbakk 2009:134)[7] which demonstrates that Spanish products did travel east at this time. In two Oxyrhynchus papyri, *garos* is also given the adjective χρηστός, that is, the best (P. Oxy. 14/1759; 41/2983). A ship manifest from the late 3rd century includes a γάρου ξενικόν, "foreign" (P. Fay. 104 verso l.28). The ship had sailed up the Nile from Nicopolis; it is not clear where this fish sauce was made (1993:48).

Garos is described as αὐστησιων, "sharp," "bitter" or "particularly intense" (P. Oxy. 14/1656). This may have a parallel with the *titulus argutus* which is generally interpreted as sharp, strong and often associated with aged salted products such as *cordyla* rather than sauce (Ehmig 2003:64). An account of food purchased from the 5th or 6th centuries (P. Erl. 111) includes among the food items two distinct qualities of sauce. At col. 2 line 14, the simple term γάρου is used, while at line 15 we have διχῶρου γάρου χηδεου from χυδαῖος meaning "ordinary" or "standard." Drexhage (1993:40) suggests that the first is an ordinary sauce while the second is of even lesser quality. We have at least three basic qualities to a *garos*, a dissolved small fish sauce, here: a low quality one, a standard variety and a high quality one, and we must add the μέλανος and/or αἱματίτου garum above as the top quality. A 7th century food list includes a γάρου πρωτιου and γάρου δευτεριου (ll. 4 and 5 in P. Ness. 3/87), that is "first and second quality *garos*," and this formula is mirrored in Diocletian"s Price Edict.[8] What we do not find in Diocletian is any indication that a blood/viscera sauce was sufficiently popular to warrant being listed and the relative rarity of this kind of sauce in papyri accords with this. The term ὠμογάρου in a 6th-century source is obscure (P. Vindob. Worp 11. 7); "fresh" or "raw" or "uncooked" would be the best guess for its meaning. This kind of sauce is also recorded in the Byzantine purchase list above, which the editor considers a physician or apothecary's list (Maravela-Solbakk 2009:127). Here we also find an ὠμογάρικιου, some kind of vessel for the serving or storing of this kind of sauce (Maravela-Solbakk 2009:136). It is clear that from later sources the idea of cooking a *garos* becomes quite common and we can conjecture that access to a raw fermented good-quality traded *garos* was limited and this was why cooking became more common, but this theory does not fit the evidence well, as we will see below that it was possible to make a small-scale local domestic product in Egypt. We may imagine that access to this inland was more limited and, as the product had become more popular, the need for a quick-cooked method became a regular way to make it. It is possible that the cooked version was more readily available, purchasers needed to be able to distinguish a "raw" fermented fish sauce from a quick-cooked variety and the color of the two products would be very similar. The alternative idea of freshness may correspond to a non-aged and therefore less powerfully pungent sauce. We have noted that others have a different understanding of the meaning of this raw *garos*, discussed in chapter 3 (Varges *et al.* 2014:70).

An official, Theophanes, made a journey with his staff from Egypt to Antioch and back in the early 4th century and in the process recorded his daily purchases of food, including considerable fish sauce and other supplies. In the majority of references *garos* is bought in various quantities and qualities, but on one occasion it is stated that he buys mullet in order to make *garos* clearly while on the journey (Pauni 6 lines 84–102; Mathews 2006:110)[9] He spent 700 drachmas on the fish, but he also later bought about half a litre of already-made *garos* for 200 drachmas. These were small mullet (*Mugilidae*) and were a regular fish-sauce fish. That *garos* was home-made is clear from the letter in which a mother reassures her son that the home brewed *garos* will be ready when he returns (P. Oxy. X.1299, 4th C.). This is reminiscent of the South East Asian domestic method in which small scale products were made in jars and bottles at home, and as they were rebrined they served to provide fish sauce over quite a long time (see page 132). The professional manufacture of fish sauce in Egypt was slow to develop, but eventually resulted in small numbers of traders' and fishermen uniting in guilds to make local products, though they were rare. In Oxyrhynchos in the 4th century there were so many γαροπῶλαι, "*garos* makers," that they formed a guild.[10] Curtis (1991:144) recounts the document from Oxyrhynchus that records the leasing of fishing rights at a private water course: the signee undertakes to pay a sum of money plus the provision of thirty *ceramia* jars of *garos* as well as various salted fish in jars (P. Lond. inv. 2143), while a fisherman from Panopolis is known to have undertaken to provide fish sauce for the visit of the Emperor in 215/16 AD (Curtis 1991:140; P. Got. 3).[11]

Calculating the prices of these products is problematic, not least because all prices fluctuated considerably in the late empire. Price determination is also difficult because it is not possible to be precise about volume, capacity and the value of currency in a given context. Drexhage determines that the maximum price was generally below that for good quality wine, oil and honey, and this is mirrored in Diocletian's Price Edict. It is of interest that the price ratio of *garos* to salt was about 6:1 (Drexhage 1993:52) making the purchase of salt relatively easy on the coast. It was also unfortunate that the vial of bloody *garum* was not priced.

Fish sauces were blended into various kinds of *oenogarum* in later consumption and didactic texts, and the vessels designed to serve these sauces changed over time. In Italy and the Western Empire, little Samian ware dishes called *acetabulum* were used to blend and serve dipping sauces based on fish sauce. We see these in use in the Latin and Greek phrase books, discussed on page 102. In Byzantium the vessel of choice was a γαράρια: known as a "chafing dish," these were a kind of integrated brazier and bowl, so that a small amount of charcoal could keep the sauce hot at table (Vassiliou 2016) (Fig. 4.1). A silver γαράρια is also referred to in papyri from the 6th century (SPP 20/151).

In sum, we seem to have a great deal of evidence for the purchase and consumption of fish sauce in Egypt, largely from the late 3rd and 4th centuries. There were at least three grades of *garos*, a low grade, a middle grade and a higher grade, plus the black and/or bloody *garum* variety. There were color differences, as *garos* could be light or bright, and from experimental approaches it has been observed that that fish sauce can be many shades of yellow/amber through to brown, depending on age and recipe. We have evidence of the black and or bloody sauce and we can say that it was quite a rare sauce compared with ordinary *garos* and that it was sold in very small amounts. The sauces were made by local fishermen in relatively small quantities and sold in small jars and bottles, while the black and bloody sauce was sold in particularly small vials.

Figure 4.1 Replica chaffing dish (*gararia*) that was used in Byzantine cuisine to serve a hot *oenogarum*. Copyright Yona Waksman.

Notes

1 This chapter is largely dependent on the work of Drexhage, H.-J., 1993. Garum und Garumhandel im römischen und spätantiken Ägypten. Münstersche Beiträge zur antieken Handelsgeschichte 12, 27–55. And Curtis 1991:131–141. See also Bagnall and Cribriore, *Women's Letters from Ancient Egypt, 300 BC to AD 800* (University of Michigan, Ann Arbor, 2006).

2 The Egyptian fermented fish known as *Faseekh*, which involved lightly salting whole anchovy, sardine and mullet to preserve while retaining the integrity of the fish has a long tradition and while it is clearly not a sauce as such the taste is similar (https://www.economist.com/middle-east-and-africa/2019/04/27/feseekh-an-egyptian-delicacy-that-is-sometimes-deadly. 05/10/2020

3 These are in chronological order: SB 16/12764, 1st /2nd C.; O. Elkab. 3/196, 1st /2nd C.; P. Alex. 21, 2nd 3rd C; P. Oxy. 14/1727, spät. 2./früh. 3rd C.; BGU 1/14, 255 n. SPP 22/75, 3rd C.; P. Corn. 35 recto, 3rd C; SB 6/9095,3rd C.; P.herm. rees 23, 4th C.; SB 16/12246, 4th C.; CPR 10/52, 4th /5th C.; SB 12/11077 verso, 4th /5th C.; P. Oxy. 14/1656, spät. 4th /5th C.; P. Wash. 1/59, 5th C.; P. Erl. 111, 5th /6th C.; P. Oslo 3/147, 5th/6th C.; P. Vindob. Worp 11, 6th C.; P. Prag. 1/90, 6th /7th C.; P. Lond. inv. 2840 = ZPE 78 (1989), 112 f., Anf. 7th C.; P. Hasitzka 25 b, 7th C.; P. Ness. 3/87, 7th C., BGU 2/377, 7th/8th C.; CPR 8/85, 7th /8th C.: P. Apoll. 93 A, 703/15 n.

4 Whether we can say for sure that the black and bloody *garos* are the same product is unclear. I consider they were, but see J. Sheldon *Garum* rouge sang' Zeitschrift für Papyrologie und Epigraphik 17 (1975), p. 156.

5 Galen: Kuhn 1833, Bk 12.637.

6 For the amphora in Latin papyri cf. P. Mich. 8/467, Comm. to l. 27.

7 The editor notes that the term σπανου may not necessarily refer to a genuine regional manufacture but more that it corresponds to a particular method of production.

8 Γαρου Γευματος πρωτιου, 'Garos food supplies of first quality; γαρου δευτερου Γευματος, "garos second quality food supply"; Diocletian's Price Edict (Lauffer 1971: 3.6–7).

9 Τιμης βοριδιων εις το γαρος (P. Ryl. 4/629. Col. 4, l. 88). The other references to *garos*, presumably purchased already made, are P. Ryl. 4/627 col. 4 l. 76; 629 col. 4 l. 88; 630-8 ll. 71, 81, 108, 122, 136, 154, 180, 230, 428, 495; 639 verso col. 2, l. 157.

10 P. Oxy.54/3749; Curtis 1991:141.

11 γαροπωλαι are also rare in the Greek inscriptions: there is only one, from Tyros (Drexhage 1993:47, n. 78).

5 Fish sauce in the late Roman, Byzantine and early medieval world

Fish sauce in the late empire

The impression one gets from text of *garum/liquamen* in this later period is that it was relatively insignificant and not worthy of much mention, and there also appears to be a reduction in interest, reflected in the rare derogatory mentions of the sauce and mirrored by the apparent reduction in production in the west. We see this reduced interest in *liquamen* in Anthimus' writing, between AD 511 and 534. Anthimus was a Byzantine physician working at the court of the Ostrogoth king Theodoric the Great and that of the Frankish king Theuderic I. His work, *De observatione ciborum* (*On the Observance of Foods*) is a unique account of a doctor advising his royal client what was best to eat to stay healthy. He proclaims that "we ban the use of *liquamen* from every culinary role" (Anthimus 9; Grant 1996:89;98). There does not seem to be much room for discussion here, as the use of fish sauce is rejected out of hand. Grant suggests that this was due to the pre-fermented nature of fish sauce interfering with the digestion (1996:86). However, Anthimus does relent, and uses it in a *hydrogarum,* which we have seen was a cooking liquor of fish sauce and water. This time, it was used to produce a flavored steam, and therefore a subtle flavor, to a meringue-like mixture made with pounded chicken meat. This apparent rejection of fish sauce in the western empire, while it maintained its predominance in the east, both in Egypt and the Byzantine empire, is intriguing. It is difficult to know how prevalent Anthimus' attitude was. There was an export ban on *liquamen* from the year 368 during the rule of Valens and Gratian (*Codex Iustiniana* 4.41.1: Curtis 1991:252), which is a rather obscure piece of legislation, but when one notes that barbarians outside the empire were forbidden to buy all three of the amphora-borne liquids (i.e., wine and oil too), it seems less important and it is likely that it had a political purpose related to maintaining control of the empire during invasions by the Goths. All three of these commodities were inherently valuable, desirable and also represented "the Roman way of life" which the Goths were desirous of obtaining.

Fish sauce from The Colloquia of the Hermeneumata Pseudodositheana: Latin and Greek Phrase books

The *Colloquia* were an "elementary Latin and Greek reader in the form of a phrase books" and were apparently a very popular means of learning Latin or Greek and many were compiled over much of the mid to late centuries of the empire and they exist in many formats and in many manuscripts though none of the manuscripts can be dated more precisely than to the 2nd to 6th centuries AD (Dickey 2012:3). They were clearly created in order to teach and inform those of Roman and Greek origin to converse and do business in each other's language. There is considerable repetition in the six known *Colloquia*, but in essence the kind of sentences that one finds are the same as those used today in any foreign language guide. There are numerous little dialogues on topics like how to buy food, borrow money, hold a dinner party, or have an argument' (Dickey 2012). There are conversations that one might have at a cafe and in shops, in school, in the market place, over a business deal, when purchasing clothes, tools, books; the list is endless. There are a number of conversations that occur in bars and private dining rooms that discuss fish sauce and how it was presented and how the guest engaged with it. The most important fact to be drawn from them, not I hope unexpected, that there is no mention of *garum*, with or without the qualifications *sociorum*, *haimation* or *melan*. All we find is *yápos/garos* in the Greek, rendered as *liquamen*, in Latin and the occasional indication of compound sauces such as *oxygarum* and ἐλαιόγαρον. In the context of daily eating and socializing, *garum* has no place.[1] The passages below (and their translations) are taken directly from Eleanor Dickey's edition of the *Hermeneumata*:

> εἰς τὸ ὀξόβαφον – *in acetabulo* – "in a vinegar cup" (ii. 105, *Colloquium Montepessulanum* 18d.6).

The *acetabulum* was a small vinegar cup which we learn from *Apicius* was used to measure, blend and serve *oenogarum* sauces at table (illustrated fig. 5.1, see page 86).

These cups provided personal portions of a sauce, into which the diner could dip small pieces of food such as bread, meat, or vegetables.

> Θές ὀξόγαρον – *pone oxygarum* – "put out fish sauce with vinegar" (ii. 104, *Colloquium Montepessulanum* 17d.3).
> ἄνοιξον ταμ<ε>εῖον, διένεγκε οἰνοφόρα, καὶ οἶνον, ἔλαιον καὶ γαρίδιον, σιτίνην – *profer vasa vinaria, et vinum, oleum et liquamen* (but γαρίδιον is a diminutive form not reflected in the Latin), *cervesiam* – "open the store room, bring out the wine jars, and wine, oil and fish sauce, beer (ii. 1800, *Colloquium Celtis* 49).

These two sentences suggest that vinegar or oil and fish sauce were available quite regularly and were in fact ubiquitous in a dining context. The fact

Figure 5.1 Various replica *acetabulum* vinegar cups: Drag. 27; Drag.24/25; Drag. 33. Copyright: Sally Grainger

that they are brought out separately along with wine also suggest that these items were part of the table accoutrements and as we shall see it is possible that guests could blend their own dressing and that these simple sauces were blended at the table at the last moment. They resemble a vinaigrette and as such need to be beaten to maintain a stable emulsion. Regardless as to whether the sauces were initially blended in the kitchen, they would also require rebeating regularly and it is conjectured that this was done by the guest.[2] In the next excerpt below, the ἐλαιόγαρον, rendered as an *impensam*, is one of these compound sauce, a joint oil-and-fish sauce mixture, and was potentially premixed and preseasoned:

βάλε ἐλαιόγαρον εἰς τὸ ὀξυβάφιον – *mitte impensam ad acetabulum* – "put some fish sauce+ oil sauce into the vinegar-cup" (i. 126, *Colloquium Monacensia-Einsidlensia* 11g).

ἐπίβαπτε – *intinge* – "dip it in" (i. 126, *Colloquium Monacensia-Einsidlensia* 11h).

Above, the imperative *intinge*, "dip it in," demonstrates just how these sauces were used. A little earlier in this text, in a list of "things which are necessary" which must be brought out, we find the terminology of graded sauces of 1st and 2nd quality familiar from Diocletian's price edict:

γάρον πρῶτον καὶ δευτέριον – *liquamen primum et secundum* – "first and second quality *liquamen*" (i. 119, *Colloquium Monacensia-Einsidlensia* 9d).

μετά γάρου καὶ ἔλαιον Σπανόν – *cum liquamen et oleum Spanum* – with fish sauce and Spanish oil (i. 127, *Colloquium Monacensia-Einsidlensia* 11K.5).

In the second excerpt above, the oil is definitely Spanish, but the fish sauce may not be, unless the adjectives Σπανόν/*Spanum* are to be taken with both nouns; this is a curiosity, but as we cannot date these sentences it tells us very little. Fish sauce appears fundamental to a meal in these instructions. In every scenario where food is served fish sauce appears in some form or other, and it is remarkably that this same simple oil/vinegar and fish sauce combination could be found in early Greek and Hellenistic cuisine and it could be found in the satires of Martial where with the first course of lettuce, *oxygarum* is served on the salad (Mart. *Epi.* iii.50.4). It is very apparent that very little has really changed over the centuries. The bloody black *garum* has no role here and it is therefore interesting to consider what it was used for at this time. Papyri from Egypt provide evidence of two reference to black and bloody *garum* in the 4th to 6th centuries AD discussed in Chapter 4. It is plausible that in the late empire this kind of *garum* was mainly used for medicinal purposes, though there are some exceptional indications that this *nobile garum* was still being desired and consumed with food in the challenging letter from Ausonius, which we will examine next.

Ausonius, fish sauce and the muria debate

There is a very complex letter by Ausonius (c. AD 310–393 or 394) written to his former pupil and close friend Paulinus of Nola in c. AD 390 which seems to be entirely concerned with the naming of fish sauce in Greek and Latin and the problems that arose from distinguishing them. It does seem to suggests that a true black *garum* was not that easy to obtain. The translation offered here is a new one, as the interpretation generally placed on the letter needs revising. Ausonius had begun his career as a *Grammaticus* in Bordeaux, and was later tutor to the emperor-to-be Gratian, before being given the rank of *comes* in AD 370. Ausonius begins by thanking Paulinus for a gift of oil and another gift of a fish sauce:

> *quanto me adfecit beneficio non delata equidem, sed suscepta mea querimonia, Pauline fili! veritus displicuisse oleum, quod miseras, munus iterasti, addito etiam Barcinonensis (muriae) condimento cumulatius praestitisti. scis autem me id nomen muriae, quod in usu vulgi est, nec solere nec posse dicere, cum scientissimi veterum et Graeca vocabula fastidientes Latinum in gari*

appellatione non habeant, sed ego, quocumque nomine liquor iste
sociorum vocatur,
    iam patinas implebo meas, ut parcior ille
    maiorum mensis applaria sucus inundet.

Paulinus my son, with what great kindness my complaint has been handled, even when it has not been delivered! Fearing that the oil you sent me had not pleased me, you repeated your gift and distinguished yourself more fully by adding a condiment (of *muria*) from Barcelona. But you know that I have neither the custom nor the ability to say the word *muria*, which is (or 'because it is") used by the common folk, although the most learned of our ancestors and those who shun Greek expressions do not have a Latin expression for the appellation *garum*. But I, by whatever name that liquor "of our allies" is called,

> will now fill my patinas so that that juice, too sparingly
> used in our ancestors' tables, will flood the spoons'.
> (Ausonius Ep. 21; Prete 1978:267).

The letter appears to corroborate strongly the commonly-held view that the term *muria* may have referred to an expensive sauce or one with associations with a bloody *garum sociorum* or *liquamen* sauce. Curtis says that the "Roman penchant for using the terms *liquamen* and *muria* imprecisely renders elusive a firm understanding of these products," and this is where that imprecision is most apparent (1991:7). We have seen that the original idea of *muria* as an elite *garum* was derived from a number of mis-readings that can now be effectively removed from the debate (see pages 31;46). As the discussion on the pages indicated has shown, it is far clearer not to regard *muria* in the first instance as anything but salted fish brine; and, more to the point, it accords with archaeological and experimental evidence if we regard them as different types of fish sauce. One is then left with the puzzling question as to why Ausonius called the sauce he has received a *muria*. He may have simply been given a *muria salsamentorum* which as we have seen was not as undervalued or underrated as he seems to think. Ausonius claims not to want to use the term, but has already done just a sentence before, which is certainly odd. It has been pointed out by Andrew Dalby (pers. comm.) that the first *muria* is actually superfluous to the grammar and description. By excising this word, the line then reads "a condiment from Barcelona," which would be enough to acknowledge the sauce as he is writing to the sender of the gift. We know from another letter, a tour-de-force in verse written to Paulinus under the title "he replied to everything but did not promise that he would come to see him," that Paulinus sent the sauce from his home "Saragossa, across the Alps and the rocky Pyrenees, Tarragona by the Etruscan Sea, and Barcelona, built above the oyster-bearing deeps" (Ausonius, *Ep.* 27. 87–89).[3]

It is possible to interpret this in a number of ways. Ausonius may be affecting (or demonstrating) an elite snobbery, admitting he cannot bring himself to use the term *muria*, which is in common use. It might equally show that he knows that the fish sauce he has received is not a *muria* at all, but is something quite different – though the only term he could use for this is of Greek derivation, for which "the most learned of our ancestors and those who shun Greek expressions do not have a Latin expression." The special gift and his delight may be indicative of a certain difficulty in finding this kind of sauce on the market in Bordeaux, Gaul at the end of the 4th century. The question we then ask is why Ausonius has not mentioned the existence of the term *liquamen* even if he did not want to use the term; it is a standard usage in Anthimus just a few decades later. Given how far we have come in this debate, it seems patently obvious that it is a *sociorum* (i.e., an elite bloody sauce) that he has received, given that *muria* is out of the picture, and he uses that term and cannot call it a *liquamen*. This reading of the letter enables us to lay to rest the debate about whether *garum* and *liquamen* were the same thing once and for all. This letter cannot be read to indicate that *garum* could be called a *muria,* it rather demonstrates that of the three sauces, *garum* and *muria* are suitable topics for discussion and debate among gourmets, because they were largely visible in the dining room in tables in blended sauces, while *liquamen* was not. He complains that he does not have an original Latin word he can use for the sauce he has received. This statement is on face value patently incorrect as we know full well that *liquamen* was an original Latin term and it was used to indicate the everyday fish sauce used by everyone else. The key of cause to this letter is that Ausonius is definitely not like everyone else and he does not want to be seen as common and use vulgar terminology. It seems that the use of the term *liquamen* was always far more vulgar than that of *muria* and this reflects an ongoing ambiguity about *muria*, as it always had a certain status from Hellenistic times, could be desirable when aged, and had specific purposes within the meal (see page 46). I have concluded that, in presenting himself as being oversensitive and fussy and acting the dilettante about his choice of language, Ausonius backs himself into a corner and has no other word in Latin to use but *muria*, and even that he distains to use. The Greek terminology was clearly at this time non-U and he knows perfectly well that what Paulinus has sent him isn't *muria* or *liquamen* but a *garum sociorum* but if you refuse to use words of Greek derivation, as he so patently says, what is there left? *Liquamen* is without doubt of Latin derivation while ironically *muria* is of Greek derivation, but this seems to have passed Ausonius by and may cloud the issue a little too much. The other dilemma created by this letter is the strange statement that the sauce in question was "too sparingly used in our ancestors' tables." We must remember this is written in c. AD 390, so which ancestors is he referring to? How far back should we look to find Romans who didn't use this particular sauce? (but see page 53 for a similar remark by Pliny) I would of course see this as indicating the relatively late introduction of the bloody *garum* in the late 1st BC/early 1st AD to a Roman cuisine that had probably been using some form of *garos* since the beginning of the 2nd century BC, though this is admittedly difficult to be sure. The letter purports to be all about the naming of fish sauce, but gives us very little real information and creates more problems than it solves.

There were various manifestations of *muria* at this time, and it is apparent that this word had not only become a generalized term for fish sauce but also taken on new meanings, particularly in the east. The first source of information is a Scholia to Horace written in the 5th century. Earlier consideration of this text would have lent support to Curtis' view that the meaning of the terminology is elusive, but when all the evidence is used to provide a context for the Scholia's contribution it may be susceptible to a clearer explanation. On Horace *Sat.* 2.4.65 (above page 63), when the *muria* is added to the dressing, the Scholia comments *muriam antique dicebant liquamen et Syrorum linqua sic dicitur,* "the ancients called *liquamen muria,* and this is what it is called in the Syrian language." We may conclude that the scholia had a limited knowledge of a true *garum* or the original meaning of *muria* as a saltfish brine. That *muria* had by this time actually become the word to designate a general fish sauce, that is, *liquamen,* is fairly well documented from Palestinian, that is, Syrian sources though, under a slightly different term, *muries* (Weingarten 2005:64). There is however much doubt about the identification of this *muries* with *garum/liquamen,* due to the ongoing assumption that there was only one general type of sauce. Weingarten says that it is not clear whether the term *muries* in the Talmudic literature was a name for a spectrum of fermented fish and grain products encompassing all the potential fermented sauces (Weingarten 2005:64; see page 229 for a wider discussion on the Talmudic sources). This implies that in Palestine *muries/muria* could have had a generic function in the late empire precisely when *garum* was largely out of the picture and a general fish sauce of the *liquamen* type was all that was available. This usage does not mean that this generic function applies in the late republic/early empire as well. The scholia understands these products as they appeared to him, rather than as they were. The Talmudic sources consider *muries* to be both a fish condiment and also a grain-based condiment made from fermented whey and barley, and it was that which became the more popular and commonly used sauce in later Babylonian Talmudic sources. It is of course possible that Ausonius, in Bordeaux, may have been aware of this developing usage for *muria* in Palestine as an all-purpose term, that is, a *liquamen,* but given his use of *sociorum* and its strong association with a bloody sauce, and the fact that he suggest that it could be poured onto cooked food, this is clearly not the way to understand the use of *muria* in this letter. For the information we may glean about the consumption of fish sauce from this letter, see Chapter 3.

On the fish sauce passage in Horace *Sat.* ii.8.46, where the *garo de sucis piscis'* is added to the *oenogarum,* the Scholia comments *hoc liquamen de sucis piscis fit*, "this *liquamen* is made from the juices of a fish" (Psuedo-Acronis, *Scholia in Horatium Vetustiora*). The Scholia takes it for granted that the *garum* in Horace was a *liquamen*, and there is no doubt in his mind or mine that it might have been a bloody sauce. The bloody sauce is so little used in the 5th century that is has no bearing on the scholia's reading of this passage. A factor that may have contributed to this confusing conflation of terms is the color of these sauces. In Chapter 8 on experimental archaeology I discussed a number of different kinds of *muria* that I had made. It is noteworthy that a six-month-aged *muria* from mackerel was the same color as a standard *liquamen* from sardine. It was practically impossible to tell the difference from the appearance. The taste was dramatically different as the *muria* was lacking in umami and anything but a mild salty fishiness. The pungency was also limited, and all in all it was pretty innocuous and ineffective (certainly in relation to *liquamen*), but this may have been the preferred taste. We can envisage a situation where it was no longer that important which sauce you used. Pliny says after all that salt was the taste that was most looked for in the consumption of what he calls *garum* (Pliny the Elder, *HN*. 31.93).

There may be an explanation for this switch in terminologies for fish sauce in the prohibition against consuming blood from any animal found in the Greek orthodox church. The original Christian church, following Jewish practices, initially prohibited the consumption of any animal blood, later becoming less strict. The Greek orthodox church maintained this prohibition and this clearly explains the apparent absence of a blood garum in Byzantine cuisine. This prohibition was particularly enforced in the Cypriote church which even banned the consumption of fresh fish as you could not guarantee to remove all the blood. A cleaned and salted fish was permitted, as it had been through a cleansing salting process to remove the blood before being submerged in brine and the subsequent fish brine, after a process of aging and desirable oxidation was quite dark amber and was identified as a *garos* in Cypriot texts (William Woys Weaver pers. comm.) This prohibition goes a long way to explaining why Syrian and Palestinian sources, as reflected in the Scholia to Horace and the rabbinic sources, seem to move away from the original idea of *garos* as a dissolved fish sauce towards the idea of it as fish brine, that is, the original *halmer/muria*. However, the debate about the consumption of *allec* in rabbinic sources in Palestine (see page 233) would seem to suggest that a dissolved fish sauce was utilized alongside these sanitized fish brines in Israel.

Morris writes that '(some?) Eastern Christians understood the injunction against blood in Acts 15 to forbid the consumption of blood, (any animal blood). Most early western Christians had understood the prohibition against blood to be a prohibition against murder, rather than against consumption (Morris 2018:159). Woys Weaver tells me that this canon law was established at the Council in Trullo (692 AD). Theodorus Balsamon's Commentaries explain it thus 'Canon 67: The divine scriptures commands us to abstain from blood, from things strangled, and from fornication. Those therefore who on account of a dainty stomach prepare by any art for food the blood of any animal, and so eat it, we punish suitably'. Woys Weaver (forthcoming) was not able to ascertain whether this extreme blood rule which forbids consumption of a dissolved whole-fish sauce was practiced more widely in the Greek world and particularly by the Syrian Christian Church. Clearly these sauces, which were once separate, have completely merged and given that visually it was often hard to distinguish one from the other this is not surprising. The evidence from the Byzantine sources gives no hint that the sauce named as a *garos* could potentially, in some circumstances, have been in reality a sanitised salt fish brine. However, what seems clear is that the merging of *liquamen* with *muria* can be attributed to orthodox dietary prescriptions from a late Roman context, dating from the 4th century onwards and specific to particular regions and it should not be attributed to an earlier period. The Cypriote evidence came to my attention at proof stage. I have not been able to pursue this more fully and clearly a great deal more work is needed on these issues.

An experiment conducted in 2019 to make a fresh tuna *muria* with 15% salt by weight was quite illuminating. The salt percentage was a good deal lower than the 50/50 w/w of salt envisaged by the salting method described in Columella and the Gargilius recipe. The result after three days was the production of a brine which had a delicate pink color with the taste that was essentially essence of tuna. Many of the people who subsequently tasted this liquor, blended as Horace suggests, thought it exceptionally delicious. We may also consider another quote from Horace concerning *muria,* which is rarely mentioned but which also places it on a higher culinary plane: "Curtillus would use (as a sauce) sea-urchins unwashed, insomuch as the yield of the sea shell itself is better than *muria*" (Horace, *Sat.* ii.8. 53). This delicate fresh taste of the sea which is wanted here, was clearly considered desirable in the ancient world and it cannot be compared to the umami pungency of *liquamen*.

Fish sauce in the Byzantine empire

The Byzantine empire based in Constantinople retained the ideas of "Roman" culture in the east and thrived on them while Rome itself and the western empire declined and faded from memory. Up until the 6th century, the cuisine found around the Roman empire was pretty much the same in terms of its flavorings and seasonings, but after this point a divergence happened. In the west, a far less "fermented" cuisine developed, which led to a certain conflict across the ambassadorial tables, as we shall now see. The cuisine of Byzantium was essentially Roman, with many Medieval and Arab influences (Dalby 2003; Regia 2018). A basic *garos* fish sauce was a commonplace ingredient, still consumed widely and in exactly the same way as we found it used in recipes in *Apicius*, as a dressing for vegetables and salads, for example, *garos* with wine, *oinogaros,* with oil, *elaiogaros* and *garelaion,* with water, *hydrogaros,* and with vinegar, *oxygaros.* These mixtures appear regularly in Byzantine literature and continued to be consumed right up to the 10th century and beyond among Greek-speaking peoples (Anagnostikas 2013:86). From the western perspective, the East was seen as different, strange and foreign. The combination of oil and fish sauce, which was the very first compounded sauce that we saw in 5th century Athens, was still widespread, yet in the west, oil had ceased to be available and had been replaced with animal fats. In the 11th century, the doctor and scholar Symeon Seth in his *On the Properties of Foods* recommends typical Galenic culinary uses of *garos,* even though he has a reputation for criticizing Galen and emphasizing Eastern medical traditions: a pottage made with pulses is flavored with oil, vinegar and *garos,* and just such a combination can be found in *Apicius* as well as in Galen (Anagnostakis 2013:99).

Liutprand of Cremona, the ambassador to the Byzantine court of the Emperor Nikephoros Phokas, wrote to his Emperor Otto I in AD 968 and complained bitterly about the food he received in his *relation de legatione,* an annual report. A meal was "foul with an exceedingly bad fish liquor" (Chapter 11), and later he received a gift of a kid "stuffed with leek, onion, and garlic, and steeped in *garos*' (ch. 20). This was sarcastically praised, in such terms as to make it clear that it would not have been to his emperor's liking (Koder 2013:105). Koder suggested that the frankly expressed dislike of this kind of typical traditional Byzantine food had political connotations, as Liutprand had visited in previous years and made no comment on the quality of the food (Koder 2013:107). We are informed that a liking for fish sauce was maintained in Constantinople into the 16th century, now under Turkish rule, in a most intriguing passage from a French traveler and naturalist, Pierre Belon. The passage is translated by Andrew Dalby (2003:68):

> "There was a liquor called *garum* which was once as widely used a Rome as vinegar is now. We found it as popular in Turkey as it ever was. There is not a fishmonger's shop in Constantinople that has not some for sale ... The *garum*-makers of Constantinople are mostly in Pera. They prepare fresh fish daily, sell it fried, and make use of the entrails and roe, steeping them in brine to turn them into *garum*"
>
> (Belon, *Observationes* 1.75).

It seems hardly credible that at this time a true blood/viscera *garum* could still be being made in Constantinople, yet the passage is quite clear that the fish were cooked and sold as fried fish, and it was only the roe and entrails that were steeped to make the *garum.* This is very perplexing. It is perfectly possible to make an adequate *garos/liquamen* from fish entrails alone, as this material is still high in protein and the resulting sauce has the same umami rich salty taste, and once fermented and filtered it would be difficult to tell apart from a type of sauce made with the fish flesh.[4] If *garum* was going to survive at all in such a place one would expect that it would be the original *garos/liquamen*, but this sauce seems to be an anomaly. One can only speculate that Belon, who probably witnessed the evisceration, had failed to notice that some of the fish were already included whole in the recipe. It is perhaps also noteworthy that in Istanbul the small fish species such as anchovy and known as *hamsi* are caught in abundance everywhere and are frequently cooked for sale to this day from many of the bridges of the Bosphorus and there is little potential for these to be bled to make a true blood *garum* (Demir 2007).

A gift of Amalfian garos on Mount Athos

There has been speculation that *garos* was known and liked in Byzantine Italy, as we hear that a gift of *garos* had come from Amalfi and was delivered to the Lavra monastery on Mount Athos in the late 10th century (Anagnostakis 2013:56). However, the description of this gift from *the Life of Athanasios* can be understood in another way as Ilias Anagnostakis has pointed out (pers. comm.). The passage reads that "old men from the Amalfians came to the abbot, bringing him a gift of *garos*" (Noret 1982:127–213). This is more likely to be the Amalfian monks from a neighbor monastery on Mount Athos. The gift was accepted and passed on to the steward, with the instructions to "to put some of this in when the occasion arose," as the old men (who brought it) had praised it highly. It is interesting that the instruction was to "put it in," meaning, I suspect, to add it to the pre-prepared sauces as usual. The steward, who resented this instruction, did not do as he was told and instead put in some of his own *garos*, which he had made himself. When the guests came to eat this *garos* they praised it as being the best. When the abbot said that it was the work of the old men of Amalfi, the steward resented this and spoke out, telling them that he had made it himself. The abbot then commanded that all the home-made *garos* of the steward should be thrown away, so as to correct the disobedience of the household disciple (Anagnostakis, pers. comm.; Noret 1982:127–213).[5] It is possible that the praised *garos* could have been a true fermented fish sauce and that the home-made one was a cooked variety, though the steward could easily have made his *garos* in the traditional way. As it turns out, the two sauces in this story were in fact similar and equally noteworthy (or of equally indifferent quality!) Perception is all, as many a wine taster has found to his discredit. The guests, who clearly were expecting a good sauce, praised it to be polite, while the steward, whether he had cooked his sauce or fermented it, naturally resented not receiving the praise which was lavished on the sauce.

Fish sauce in the Western Medieval sources

Fish sauce bows out gracefully in the west during the 5th and 6th centuries, and we have seen the beginning of this in the ban on using *liquamen* in cooking in Anthimus (Anthimus 9; Grant 1996:89;98). Curtis lists the remaining sources (1991:184–190), and I will briefly cite a few examples to illustrate this point, while adding a few culinary references to the later use of what was termed *garum*. Cassiodorus (AD 490–583) makes mention of a *garismatia* on the coast of Istria, which is understood to be a place of the manufacture of *garos,* as it is seemingly derived from a Greek word. Buonopane (2009) suggests that it may have a connection to what he calls the luxury sauce *haimation*, which is plausible, though little is known of its meaning. Gregory, Bishop of Tours in Merovingian France, (AD 538–94) records the theft of 70 *orca* of oil and *liquamen* from the port of Massilia. The continuing use of *liquamen* is interesting here as a Frankish monk from the mid-7th century records a royal purchase list for visiting dignitaries which included oil, *garum,* honey and vinegar along with quite few exotic spices. There is a question as to why the two terms are still being used separately. It's just possible that later transcribers, commentators and translators of this material used the more familiar *garum* when *liquamen* was used and vice versa, as occurs in relation to Greek and Latin sources elsewhere. We might conclude that the continued use of *liquamen* reflects the writers concern with the potential ambiguity associated with *garum* and prefer the precision that the use of *liquamen* affords. The use of Latin *garum* would then reflect a less precise usage and be typical of late Latin, where *garum* once again became the essential substance. In the 7th and 8th centuries the occasional mention of *garum* being imported from the coast continues: a cooked *garum* recipe from the Abbey of Echternach probably dates from this period, and is very similar to the Gargilius recipe in that it uses a huge quantity of spices and herbs and honey, but it is cooked from fresh fish, with wine and salt added from the beginning, rather than being fermented and then reduced. A correspondence in verse published in 1538 between François Rabelais (1483–1553) and his friend Etienne Dolet, suggests this. Dolet states in an epigram *de Garo Salsamento* that "salted *garum* has been restored from the ancient age by your genius" and that it has a "sweet smelling odor." It is difficult to know for certain whether this was a true *garum* or his imagination at work. Rabelais replies in turn, with a further epigram with the same title (*Oeuvres Diverses* xiv), and apparently sends Dolet some *garum,* which he suggests he blends with "sour wine and oil, however much you like." Dolet lived in Lyons; Rabelais was based in the south of France in the period 1537–1539, and occasionally lectured on medicine at Montpellier. Curtis concludes that the use and production of fish sauces may have been limited to a small area of the south of France, and we may be able to see the remnants of this production in the development of *pissalat* in Nice (Curtis 1991:186, notes 10,11,12).

Cooking in the Renaissance and beyond

From a culinary perspective, cookery books and books that discuss food generally record a renewed interest in fish sauce through the Renaissance in

France. Sarah Peterson has recounted this resurgence in her book *Acquired Tastes: the French Origins of Modern Cooking* (1994). The salt/acid flavor combination so typical of Roman and Greek sauces had ceased to be popular and had been replaced by a love of sweet and spicy Arab foods which used combinations of honey rose petals, almonds, far East spices such as cinnamon ginger, while salt and saltiness was not valued. As Peterson notes, "Roman salty fish sauces had been flushed from European cookery by the high Middle Ages" (1994:134). With the resurgence of interest in the classical world came a renewed interest in ancient food and that obscure and by this time little-understood sauce called *garum*. Rabelais' literary works contain many descriptions of food, and the rediscovery of salt and salted foods is reflected in the menus he creates. Gargantua's father Grandgousier has a penchant for salty foods to provoke his thirst (*Gargantua* iii) and Gargantua himself later prepares a dressing for his salad which is made from just salt, oil and vinegar, sadly lacking the umami of the ancients (though it accidentally included six pilgrims! Rabelais, *Gargantua* xxxvi). The newly discovered salted tuna roe known as botargo was popular, as was caviar, and both were associated with all three sauces of the ancients, *garum*, *liquamen* and *muria*, by Guido Panciroli, a 16th-century Italian antiquarian (Peterson 1994:138; Panciroli, *Rerum memorabilium* 25), who saw a similarity between the viscera used in *garum* and the fish eggs of botargo without fully comprehending their nature. In Italy at this time these liquids were hardly to be found beyond the embryonic survival of *colature di alici*, that is, anchovy *muria* on the Amalfi coast in the tiny town of Cetara, as well as other nearby towns. Today in modern cookbooks there is a natural assumption that *colatura di alici* and *garum* were the same thing (Spieler 2018:16). This association between *garum* and the Cetara *muria* can be traced back to Renaissance sources where the salty liquid from gutted and pressed salted anchovy (precisely what *colatura di alici* is still made from, Carannante *et al.* 2011) was described as a *garum*. These sauces were made from clean fish and the intense umami kick that would be present with the presence of fermenting viscera was absent. There is no question of not eviscerating, despite the labor involved in handling thousands of tiny anchovies. Bruyerin also describes a composite sauce of salted anchovies, vinegar, oil and parsley, which was then melted down over heat to produce what they thought was a superior sauce to the ancient *garum* (Peterson 1994:138; Bruyerin *de re cibaria* p 572–573). The sources at this time talk about the blood and viscera with some incomprehension while enthusing about their sanitized *garum*. Melting clean salted anchovy over heat became the way to add umami to food, a practice which can be found in Varenne's *Cuisinier* of 1651. In England the use of melted anchovy is mentioned in 17th- and 18th-century cookbooks as a *ketchup* but, as in the Bruyerin recipe above, they are not just salted anchovy but rather a complex concoction which has acquired this obscure name. One recipe includes "twelve to fourteen anchovies, ten to twelve shallots, white wine vinegar, two types of white wine, mace, ginger, cloves, whole peppers, a whole nutmeg, lemon peel, and horseradish." This is from the first known English ketchup recipe, published in E. Smith's *Compleat housewife* of 1727 (Smith 1996:12). Mushroom ketchups were made without anchovy, and the origin of Worcestershire sauce is surely to be found in these concoctions as it is principally a lightly fermented anchovy sauce, flavored with tamarind, vinegar, garlic and cloves, though the recipe originally came from India, not South East Asia (Smith 1996:177). A direct line can be surely drawn between these ketchup compound sauces and the reduced *confectio liquaminis* found in the Gargilius recipe and also those late Roman cooked fish sauces. These early modern ketchups were often reduced by boiling to make a concentrated sauce and, as is well known, eventually turns into tomato ketchup (Smith 1996:24).

Ketchup is believed to have been derived from various related words, *catsup* or *catchup*, and the origin is a matter of great confusion according to Smith (1996:4) as modern food historians compete over the numerous geographical areas from where this word may originate. The most plausible, in my view, is the one that gives it a Chinese origin, derived from *ké-tsiap*, an Amoy dialect word meaning "the brine of pickled fish." There is also speculation that the word came from Vietnam, or what was then called Tonkin (Smith 1996:4). These new sauces were, when they reached Europe, already compounded sauces made with many ingredients and therefore not directly related to the pure fermented whole fish sauce that we find today in Vietnam and Thailand. It is the *nuc nam* and *nam pla* from South East Asia which have that strongest links, in terms of their simplicity, with the *garos/liquamen* of the ancients.

Notes

1 It is rather strange therefore to note the anomalous late and/or medieval manuscripts that survive in Latin which continue to use *garum* to convey a basic quick cooked fish sauce: (Chapter 2). I am inclined to place these documents in the medieval period precisely because they use *garum* to convey a relatively everyday cooked fish sauce; see below.

2 The small Samian ware vessels known as *acetabulum* that are strongly associated with *oenogarum* sauces often have a ware pattern linked to stirring, grinding and beating and results of initial residue analysis also suggest that fish sauce was served in them (cf. Biddulph 2008; and Grainger and Biddulph in press.

3 It is of interest that at Rhode (Rosas), north of Barcelona fish processing facilities that we can specifically associate with a blood/viscera garum have been identified from a site dated to the mid. 3rd century AD. (page 202; Curtis (1991: 55) siting Nolla and Nieto (1980).

4 I am grateful to Ilias Anagnostakis for advice in re this passage and to Dr Christopher Grocock for the translation.

6 Fish sauce from an Archaeological perspective

The literature review: modern scholarship on garum and previous studies into fish sauce from an archaeological perspective

It is of note that, in the study of the classical consumption texts, (satire, letters), little attention has been given to the consumption of ancient fish sauces, which is not in itself unsurprising. There is general agreement among Classicists that the consumption of *garum* was a useful and convenient symbol of Rome's corruption, as it is perceived as "rotten" and bizarre and conveniently demonstrates the essentially alien nature of the Roman cuisine. As a result, the amount of modern classical literature dealing with the practical aspects of its consumption to review is virtually non-existent apart from Curtis (1991) and Corcoran (1962). The situation is very different in archaeology, as most debate is concerned with the interpretations of structures associated with the manufacture, storage, transport, trade and commercialization of the products as well as of the fish bone residues of fish sauce, and lastly the epigraphic study of labels associated with fish products. In order to condense the literature review and at the same time highlight the complexity of the fish sauce nomenclature debate, this section will record the views of the leading scholars in this particular area, which express distinct trends in research. In the introduction I introduced the concept of the "single sauce hypothesis," a term I developed to illustrate the idea, believed by many, that *garum* was a generic term signifying one basic commodity of various qualities in the early period and which was replaced with *liquamen* in the late empire. The review of the chronology of fish sauce in Chapter 3 has shed considerable doubt over this idea. It is clear that each of the leading researchers in this field have negotiated the anomalies and contradictions that led to this impasse in different ways in order to maintain the single sauce hypothesis.

Readers looking for publications with bibliographies covering the topic more generally should of cause begin with Curtis' 1991 work *Garum and Salsamanta: Production and Commerce in Materia Medica*, which is still the most comprehensive and truly interdisciplinary approach. The conference proceedings *CETARIAE* (Lagóstena 2007) also a good starting point, as well as the numerous works in the bibliography by Bernal-Casasola alone and with others. In recent years there have been many publications on the topic such as Bekker-NIelsen and Gerwagen 2016), *The Inland Sea: Towards an Eco History of the Mediterranean and the Black Sea*; *Fish and Ships: Production et commerce des salasamenta durant l'Antiquite* (ed. Botte and Leitch 2014); Botte (2017) *L'exploitation de la mer en Italie centrale tyrrhénienne (Étrurie et Latium): production et commerce durant l'Antiquité*. An excellent bibliography can also be found through the *Oxford Roman Economy Project* website.[1] In 2017 this author was involved with a conference entitled *The Bountiful Sea: Fish Processing and Consumption in Mediterranean Antiquity,* held at Oxford and published in 2018.[2] Many papers dealt with aspect of fish sauce and it currently represents the most comprehensive source of material on this topic.

The first and most important archaeological survey of the fish processing sites in Africa and Spain was produced through the University of Bordeaux by Ponsich and Tarradell in 1965. They set the criteria to identify the archaeological remains of fish sauce production that later studies rely on in interpreting the evidence. Ponsich and Tarradell also devised the first interpretation of the nature of these products, using Pliny the Elder and Isidore of Seville as their main source (see below and page 25 f). Thus, they concluded that there were many types of *garum*, and they do recognize the distinction between a pure blood/viscera sauce and a whole-fish sauce in a way that has since fallen away in recent interpretations (Ponsich and Tarradell 1965:98). Despite recognizing this distinction, they still conjectured that *liquamen* and *garum* may have been differentiated by the size of the mackerel utilized to make them (Ponsich and Tarradell 2002:50). Djaoui (2016:119) maintains the same idea in recent work (page 236). I am less convinced by this as discussed in Chapter 8. It seems to me that it is unlikely that either sauce, would be significantly different, depending on the size of fish utilized, certainly not significantly different to warrant a different name

Curtis's seminal work *Garum and Salsamenta in Materia Medica* of 1991 was ground-breaking, as it incorporated ancient literary and medicinal literature as well as the archaeology for the processing sites and incorporates ethnographic materials (information from south East Asia) to examine the similarities with modern fish sauces around the world and enhance our understanding of a product that is now extinct. Immensely valuable though that book is, it may have greatly exacerbated the confusion over terminology. Curtis suggested that initially, when *liquamen* was perceived to be different from *garum* with reference to *tituli picti*, it was still just a by-product of *garum* and then later became synonymous with it. He believed it was a liquid weaker, paler and of lesser quality, and potentially a second or subsequent washing of the residue from *garum*, but he did acknowledge that this was a guess, without any corroborating evidence, based on the anomaly that *garum* and *liquamen* appear to be synonymous in late Latin texts (Curtis 2009:7). This guess, as to the meaning of *liq-uamen,* has been taken as a legitimate definition and repeated and used to interpret new evidence, which can be problematic. In recent work on food ways in Roman Italy Banducci follows Curtis' definition of *liquamen* (Banducci 2018:125). In analyzing the Tanura F wreck Barkai and col-leagues (2013:190) conclude, following Curtis, that *liquamen* was simply a second rate *garum.* Curtis elaborated on his understanding of fish sauce in 2001 and 2007 and to some extent further exacerbated the way in which others interpret the evidence. Curtis imagines how fish were processed on the beach or workshop at the height of the industry in the 1st century, AD, which are documented by numerous excavations such as those at Neapolis (Slim *et al.* 2007) and Baelo Claudia (Bernal-Casasola and Sáez Romero 2008). He developed the idea that, as the manufacture of fish sauce was an integrated aspect of the production of *salsamenta,* the fishermen would have had access to the larger fish (tuna or large mackerel) for salting whole in order to generate the quantities of viscera that it was believed were added to the smaller shoaling species such as sardine or anchovy in order to make fish sauce (Curtis 2001:406). This view relies on the view that viscera was the primary and main component of fish sauce which can be disputed. This picture is mirrored in analysis from Edmondson (1987) in his *Two industries in Roman Lusitania,* where the concept of fish sauce being an integral part of the production of *salsamenta* is taken as a given. In Edmonson's view the large species were salted and the refuse and scraps of that process were added to otherwise worthless small fry to make the "sauce" which were essentially a poor food for the masses to flavor their pulse (Edmondson 1987:100). This interpretation is fundamentally at odds with the recipes but it also highlights a dichotomy. Fish viscera and blood was valued in expensive *garum* but was also essentially rubbish, and it is difficult for modern scholars to see how it could be both? One can see how the description in Pliny the Elder has informed Edmonson's interpretation, as he also stresses that fish rubbish is a major component. To hold that the very nature of fish sauce was simply a by-product of *salsamenta* is an error. We have seen that *garos* fish sauces were products derived from common-place small-scale fishing and later Greek culinary practices. Our review of the chronology of fish sauce (Chapter 2) has demonstrated that *salsamenta* and sauces were not necessarily manufactured at the same time from the same fish in the early fish salting industries in Punic Spain and they may never have been fully integrated in many facilities. There are other views, such as those of De lima (2018:121) who has suggested that the great fame of the early tuna industry from the Iberian Peninsula, which we see in the minting of coins with images of tuna on them, was in fact more to do with the production of *garum* from tuna rather than *salsamenta.* This simply cannot be understood from these images. There is very little in the archaeology to suggest that any kind of fish sauce *garos/garum* was actually manufactured in any kind of systematic way from tuna before the early

2nd century BC. These views of Curtis and Edmonson are taken up widely without questioning them sufficiently and they are then used by many as an interpretative framework when dealing with data from the processing sites as we shall see below. If one blends those ingredients that are perceived as elite and expensive such as mackerel or tuna and their viscera and blood with the commonplace small fish *garos*, one gets a universal "sauce." This generic notion of "sauce" is self-defeating as a means of understanding these products because it fails to expose their diversity.

The University of Cadiz publishes widely on fish sauce amphora and the processing sites under Professor Dario Bernal-Casasola (e.g., Bernal-Casasola, *et al.* 2007; Bernal-Casasola, and Sáez Romero, 2008; Bernal-Casasola, 2018).[3] In addition to the immense quantity and quality of the research in discussions of stratigraphy, architecture, pottery and small finds, they perform analysis of the fish bones residues found in the *cetariae* in Baetica (Bernal-Casasola *et al.* 2014; Bernal-Casasola *et al.* 2016) and also those found at the *Garum* Shop in Pompeii (Bernal-Casasola D., Cottica 2017).[4] They have also been experimenting with fish sauce, and perform chemical analysis of actual fish residues and pottery. These experiments, along with those of my own and others, will be discussed in Chapter 8 (Bernal, 2014; Palacios *et al.* 2016). In keeping with the conventional wisdom, the Cadiz team do not clearly distinguish between *garum* and *liquamen*, as they believe that they were effectively the same thing, and the evidence for clear differences between the two types is understated in their discussion. Palacios, in (Bernal-Casasola *et al.* 2016:92), following Ponsich and Tarradell 1958 and Curtis 2001, regards the distinction between the two terms as being concerned principally with the size of species rather than the presence or absence of fish meat, viscera and/or blood. They conclude that only when the larger scombrids such as tuna and mackerel were used, would the intestines **and** blood also have been used. It is concluded that when the flesh of the tuna or mackerel had liquefied with these extra bloody fluids and become a sauce it was only then called *garum haimation* for its blood red color. As we have seen the four recipes that survive do not put fish meat and blood together in sauce. The definition by Palacios takes no account of the clear and dominant evidence from the *Geoponica* that a true *garum haimation* was clearly made **without** fish meat while a *liquamen* was made **without** blood but clearly with a quantity of viscera whether it be added or naturally present.[5] The definition from Palacios (Palacios *et al.* 2016:92) takes no account of what distinguishes *garum* without a designation from one designated *haimation* or *sociorum* and from a *liquamen*. It also does not take into account the fact that a *liquamen* was clearly made with mackerel (as the evidence of the *tituli picti* from Pompeii shows), and this mackerel *liquamen* cannot be a *garum haimation* or a *sociorum* as well as a *liquamen*. They conclude that when small fish were used, then it is assumed that no extra viscera was necessary as autolysis was guaranteed (contrary to Curtis above) and in principle, they conclude that the Romans called the first product (i.e., large fish and the combined visceral juices), *garum* and the second, (little fish no visceral juices or blood) *liquamen* (Palacios 2016:92). Contrary to this, the recipes specifically add extra viscera to a mixed small fish *liquamen,* and as can be demonstrated, the mixture of small fishes utilized in sauces, from surviving residues, are invariably of multiple sizes, in the range 5 to 20 cm, and experiments have demonstrated, from surviving residues, that it is the upper size range of these that need the additional digestive enzymes, when they are too numerous to be individually cut open (see page 20).[6] In the most recent publications from Cadiz, in a joint project with the University of Seville (Rodríguez-Alcántara *et al.* 2018:150), *garum haimation* is defined principally as a product made with both small fish and the blood and viscera of tuna rather than mackerel. What is most at odds here, as elsewhere, is that this interpretation blends the characteristics of a commonplace small fish *liquamen* with what was essentially an elite and expensive blood viscera *garum*. It also makes the assumption that viscera and blood were added together, which is not necessarily the case. It is more than likely that the bleeding and eviscerating of the larger species occur at different times, as discussed in Chapter 9. If *garum* is made with small species (sardine, anchovy) and with the large fish (mackerel, tuna) with blood and viscera being added to both then you have a single sauce utilizing all the ingredients. As most commentators continue to want to think of *"garum"* as a single substance there is no problem with this, but it is fundamentally at odds both with the recipes and much of the archaeology. The evidence from Aila Aqaba in Jordan attests to the concept of blood and viscera from tuna being collected and fermented in a vessel (Van Neer and Parker 2008; discussed page 201) and while we cannot be certain that tuna meat was not added to this mixture, the evidence is not clear.[7] We have seen in Chapter 1 that a misreading of Isidore of Seville in relation to the definition of *muria* leads to a belief that this sauce was derived from *liquamen* (page 33). Consequently, the Cadiz team continue to consider *muria* to be the brine that rises to the surface when the fish used to make a sauce begins to dissolve, and this, they suggest, was taken off and utilized as a cheap sauce in its own right (Palacias et al 2016:92). This thinking inevitably leads to the idea that a *muria* could have be taken from a *garum*, which seems overly complex and effectively merging all the sauces again. The issue of the brine in *liquamen* production will be discussed in Chapter 8, suffice to say this kind of brine is the liquid that will eventually form the basis of the fish sauce itself, and to remove it would seem to reduce the volume of available sauce. *Allec* is defined by the Cadiz team as the by-product or residue derived from the manufacture of *garum*, representing the remaining, solid bony part of the *garum*. If one has determined that there is only one generic sauce, there is no problem with the idea that the residue or *allec* had to be derived from this generic sauce. If one accepts that there were two types of sauce then only one, made from whole fish or parts of fish containing bones and flesh can generate a fish paste residue, because blood and viscera alone would not generate a paste detectable in archaeology. In sum, Palacios (*et al* 2016:92) conclude that *garum, liquamen, allec* and *muria* could all potentially came from the one process using one batch of sauce. This idea of a fully integrated process was also advocated by Corcoran (1962:205).

As we have seen with the chronology in Chapter 2 there are other ways at looking at the evidence which points to *garum, liquamen,* and *muria* being made separately in time and space from different batches of fish using different techniques and recipes. This "fully integrated" approach ultimately implies that the terminology as reflected in *tituli picti* on amphorae and in texts were meaningless as the individual products were apparently so homogenous. While seeming to still hold to the idea that all four products could have come from a single batch of fish, the lead archaeologist at Cadiz, Dario Bernal-Casasola, has also moved further away from the idea of a unified essential substance in other ways. Bernal-Casasola believes that there were numerous different types of sauce depending on individual recipes, and that they may have had many other individual names which are unknown to us. He has highlighted the short acronym *tituli picti* associated with the pile of Dressel 21/22 amphora in the *Garum* Shop at Pompeii to illustrate this phenomenon. These labels have however been linked to the *salsamenta* which the amphora originally held before being reused for a standard small fish *liquamen* (see Chapter 12; Botte 2009; Bernal-Casasola *et al.* 2017:242). Bernal-Casasola also question what the "the alleged diversity of these products is due too when the fish composition of the residues (*garum/liquamen*) is so similar or the same" (Bernal *et al.* 2014:231). He is discussing the residue evidence in the form of *allec,* which displays considerable uniformity (see page 217). He has recently rejected what he calls the "*garum* effect" bias, which seeks to find scombrid meat and other fish sauce residues, and attributes them to the four known names *(garum, liquamen, muria, allec)* which he sees as missing out on the potential diversity (Bernal 2016:193). This is a very complex issue, but requires a close analysis of the existing information rather than speculation that these are only a few of what may have been many more types of sauce. The terms which survive deserve to be examined rather than being set aside because it is a challenge to see what each referred to.

One of the leading scholars of the fish sauce trade is Emmanuel Botte, a French archaeologist who has worked extensively in Sicily on the Hellenistic salted tuna trade and the amphora designed to transport these products (Botte 2009). Botte, contrary to our discussion on the description of fish sauce production in Manilius in Chapter 1, considers that a salted tuna product and the "single" fish sauce made with tuna meat and tuna viscera was being discussed in the poem. Botte then contrast this sauce with the one made from little fish in the next passage in Manilius at lines 676 to 680 (Botte (2009:19). Botte however does reject the Gargilius Martialis recipe as a medieval addition to the 2nd century text. The issue that he considers the most important to distinguish between the sauces is the one between cooking and fermenting (Botte 2009:20). Botte recognizes that *muria* is undoubtedly the brine associated with a salted fish product. His recent work on the fish processing evidence in Italy in the Republican period and early empire (Botte 2017) has been invaluable in composing the chronology of the fish salting trade in the Mediterranean (chap. 2).

Alfredo Carannante is the leading Italian archaeozoologist working in the ancient fish sauce, and in particular the fish sauce residues associated with the *Garum* Shop and those found at the villa of Marcus Fabius Rufus (Regio VII, Ins. 16, Ins. Occ. 22), in Pompeii (see page 230). In his most recent work (Carannante 2019) there is a recognition that we can no longer use *garum* as a generic term to designate all the various sauces and he also acknowledges my contention that the manufacturer and trader had a system for naming fish sauces that may well have been subtly different from the way in which the elite consumer named these products which is a valuable step forward. There lies our dilemma: do we, as archaeologists, ally ourselves with the consumer, or those who developed and managed the industry? Archaeology is if nothing else a discipline involved with non-elite and non-literary approaches.

This integrated generic idea of fish sauce is also to be found in the inter- pretation of the remarkable fish bone residues found in the little *Latium* pots from the Rhône at Arles, (discussed fully on page 191; Djaoui *et al.* 2014; Piqués *et al.* 2015). These residues are typical of fish sauce and a form of *allec* but the conclusion by Djaoui that it was both a cheap and low value *allec* food for a ship's crew as well as being made with mackerel blood cannot be sustained. A seminal French work of zooarchaeology by Desse- Berset and Desse (2000) has formulated the criteria for identifying the fish bone residues associated with either fish sauce or a salted fish. Fish sauce residues are identified as small or very small mixed species, dominated by the *Clupeidae* and *Sparidae*, such as sardines (*Sardina pilchardus,* ancho- vies (*Engraulis encrasiculos*), bream, *Pagellus erythrinus*, picerel (*Spicara maena*) all under 20 cm.; they are fragmented, disarticulated, chalky and deposited in large quantities (Desse-Berset and Desse 2000:91). Though these criteria are generally accurate, there are many occasions when they don't fit the evidence. The poor quality of the fish sauce bones has been attributed to the fermentation process or to the addition of cooking which together is assumed to soften, damage and otherwise erode bone. This is however extremely unlikely as it can be demonstrated that these actions do not damage bone and it can also be demonstrated that fermentation does not necessarily erode the collagen, which allows the identification of finds of articulated bones to be a fish sauce residue as noted with fish sauce residue from the vats at Nabeul (sardine anchovy and sardinella), tested by Garnier *et al.* (2018) which were still articulated (Sternberg 2000). This allows for the possibility that well-preserved and articulated bones identified as a solid salted fish product in amphorae from ships wrecks could also quite easily be a sauce residue, transported with its bone in a semi-fermented state (Grainger 2013). There are clearly considerable problems in distinguishing between what might be a purpose-made *allec* meant to be consumed as a fish paste, the residue of fish sauce at the production site and also discarded after it is used up, and the potential remains of the consumption of a salted fish products. These issues are discussed more fully in Chapter 11.

Desse-Berest and Desse (2000) identify the bones for a salted fish product as fish of larger size, that is, >10 cm. up to 40 cm., mainly mackerel, but also sardine and anchovy of the upper range; articulation is common and the bones are very well preserved, often with organic matter, particularly from shipwrecks where the amphorae are sealed. What is not made clear is that all the fish bones identified by Desse-Berset and Desse as a *salsamenta* are derived from shipwrecks, where (because of the unique conditions) preservation is often very good, while all the residues found on land are assumed to be a sauce. This is contradicted by Van Neer (2010, see below) who has a different view of the mackerel finds on Roman sites in the north. Vim Van Neer, undoubtedly the leading archaeo-botanist in Northern Europe, has made a comprehensive study of the fish sauce and *salsamenta* residues found outside the Mediterranean (Van Neer *et al.* 2010) and fish consumption and status of the consumer in Roman, Egypt and early medieval contexts, and has published widely in Roman zooarchaeology generally (Van Neer and Ervynck 2002; Van Neer *et al.* 2004; Van Neer 2008). Though Van Neer takes great pains to elucidate the fish sauce nomenclature issue, he falls back on the accepted "single sauce" hypothesis, a product which he interprets as being relatively cheap. He acknowledges that elite expensive versions probably did exist, but he has suggested that the expense is to some extent exaggerated by Pliny the Elder and others in order to criticize society's extravagances (Van Neer *et al.* 2002:208). However he does conclude that "although the precise nature of the products remains unclear, each name must correspond to a particular type of preparation and a specific set of ingredients; it is hardly believable that that these terms were used indifferently as is sometimes suggested" (Van Neer *et al.* 2010:163) Van Neer is one of the lead promoters of the idea that *allec* was a separate purpose-made sauce which naturally contained bone which the purchaser was expected to consume, and the presence of the bone is thus a demonstration of its low value. The residues of fish sauce production are found all over the empire at production sites, disposal sites and also in ship wrecks and there is considerable doubt as to how it was ever economically viable to transport and trade such an apparently low value product. Van Neer also agrees with Desse-Berset and Desse (2000) that the residues of fish sauce and *salsamenta* are to be distinguished largely by size of species, and also by state of preservation. In an extensive list of mackerel finds from Northern Europe (he provides the references to over 40 sites, many of which are military), he reveals that mackerel consumption appears to be quite common, particularly given that preservation of fish bones is highly problematic (Van Neer *et al.* 2002:207). It is perfectly feasible to imagine a situation where a mackerel or a mixed sauce with mackerel was discarded in a similar way to those residues that are normally recognized as the result of fish sauce production. There are, sadly, no reported sites of large piles of discarded mackerel. There are significant examples of mackerel bones in "sauce" amphora, cited by Ehmig (page 222; 2003:78) and in vessels other than amphorae in various sites in Europe and they strongly indicate the possibility of a fish sauce. It is not easy to understand how mackerel bones could end up in a vessel at the consumption site regularly if it was a salted fish product, as they would have been consumed individually and discarded in the normal way (Van Neer *et al.* 2010:169; Hüster Plogmann 2006; Lauwerier 1993: see Chapter 11). Semi-processed fish sauces made with all or some mackerel have undoubtedly appeared in shipwrecks such as Grado, Sud Perduto II, Cape Bear III, Port Vendres II, and St Gervais III (see page 218), and while most of the evidence is currently identified as a salted fish product by Desse-Berset and Desse (2000), there is every possibility that these products were sauce-related, especially when the shape of the amphora is taken into account. The characteristics of particular discard practices are used to distinguish fish products, but these behaviors can be complex, and a fuller understanding of the residues themselves and how they might be discarded will greatly improve our understanding.

The study of the *tituli picti* that designate the various forms of salted fish product is dominated by German scholars (Ehmig 2001; 2003; Berdowski, 2003; Martin-Kilcher 1994). Ulrika Ehmig classifies and quantifies the various terms and recognized the incongruity of *salsamenta* being traded in vessels clearly designed for a liquid with narrow necks and small openings in her seminal 2003 work on the amphora at Mainz. She had already posited the idea that mackerel shipped in narrow necked Dre. 9 couldn't possibly be a solid product and gave it the title *allec,* (Ehmig 2003:78), while also questioning the validity of transporting such a residue as a fine *garum scombri* was expected to be filtered (Ehmig 2003:79, her table 11). She cites the incongruous image of the Dressel 9 on display at the Le musée d'Histoire de Marseille in which the mackerel have been placed head first down the vessel in a way that is surely inaccurate, given the width of the neck. I was equally confused by the site of this amphora and include my own picture taken in 2010 (fig 6.1).

There are extensive online recourses for amphora epigraphy: *Amphorae ex Hispaniae* has detailed lists of the amphora typologies and associated *tituli picti* (http://amphorae.icac.cat/) and research group CEIPAC (Centre for the Study of Provincial Interdependence in Classical Antiquity) of the University of Barcelona coordinates a study of the amphora and their labels, http://ceipac.ub.edu/index_en.html. *Tituli picti* are dealt with in Chapters 12.

A recent work on marine resources and the economy in the Roman Mediterranean, *Harvesting the Sea* by Annalisa Marzano has greatly improved our comprehension of the wider economic situation. She uses the diverse epigraphic, ethnographic and iconographic evidence along with Egyptian papyri and the technical didactic literature, and particularly the judicial evidence for fishing disputes, to illuminate the wider fishing industry in the ancient world.[8] Marzano understands *allec* principally as a "purpose made" sauce from little fish which would naturally contain bone but, though she does not state it outright, infers that it would not necessarily generate a liquid sauce first (Marzano 2013; 2018:170). This is contrasted with the other kind of sauce which she calls *garum* and which she assumes was made both with butchered fish meat and blood/viscera and which therefore is not identifiable from bone residues. This analysis is fundamentally at odds with the evidence as it effectively relegates all the bone residue evidence identified as *allec* to a purpose-made paste without generating sauce and sees everything else as an elite sauce. The middle ground in this approach is lost, where good quality everyday sauces, that is, *liquamen* for all culinary uses should be found.

The new science of residue analysis

There are now increasing numbers of articles reporting on residue analysis to identify invisible biomolecular residue's inside the fabric of amphorae and in the walls of processing tanks. Attempts have also been made to identify the numerous complex amino acids and minerals in ancient remnants of fish sauce at various sites. These studies represent new and exciting ways to study this material, and there have been some hugely important and valuable results (see below). Much of what is done provides micro-scale detail about a product that we are still struggling to understood on a macro scale. The ultimate goal of such research is to be able to established whether or not watertight vats and amphorae were used in the manufacture and transportation of fish and/or fish sauce. These new forms of residue analysis are leading to a far greater understanding of the potential content of amphorae. It is, however possible that residue analysis alone will not be able to clarify amphora content in specific cases as it is possible to envisage a situation where amphorae were reused multiple times and were used to transport numerous different products which leave multiple signatures that could not be separated. As the practitioners of this new science sometimes work a considerable distance from the literary and consumption evidence, there is always the potential for false conclusions to be drawn when their complex results are interpreted in isolation. The research conducted under Smigra *et al.* (2010) with a contribution from Robert Curtis sought to identify the amino acids and minerals present in the residues from the fish sauce shop in Pompeii. They not surprisingly found that the profile was very similar to modern fish sauce samples with "umami dominating with free glutamate being the taste-dominant amino acid, followed by sweet-tasting glycine and alanine" (Smigra *et al.* 2010:442). Research by Garnier *et al.* (2018) found similar results in the residues from various *cetariae* at Troia, Nabeul and Bello Claudia. This research was able to "demonstrate the predominance of cholesterol" in all samples of fish bones that were fermented which has contributed greatly to the question of how to identify a fish sauce residue from residues of fish discarded fresh. The new tests have revealed an important conformity about the nature of fish sauce production across many regions and sites. Garnier (2018:322) has identified a signature for malic acid and tartaric acid derived from some sort of fermented fruit wines being added to fish sauce at Marsa in Morocco and at Kerlaz in Brittany, which has been corroborated by work by Pecci *et al.* (2018) who found a grape signature on the fish sauce amphora at the *Garum* shop in Pompeii. The addition of extra enzymes from tropical fruits occurs in modern fish sauce production in South East Asia to aid liquefaction and Garnier conjectured that the fruit wines were a particular ingredient added to these fish sauces made in the colder northern climate to aid the liquefaction process, because the colder climate is considered to reduce or slow it down. Garnier also suggested that the wine and fish sauce combination identified was, if not actually an *oenogarum,* then possibly a "drink" linked to the *garum* blended to the color of honey wine that was drunk, according to Pliny (*HN.* 31.44; Garnier 2018:322), though I think this is not credible as it is unlikely such mixtures were blended at the processing stage. From our look at the modern techniques to make fish sauce and the experiments conducted by the author and others it would appear that these additional fermented liquids should be better understood in terms of their principal role as a diluting agent to render what was a thick fish mash into a liquid sauce, and also to enhance taste, while the lower pH and reduced bacterial contamination were a secondary consequence, not understood in terms of the science in the past but clearly demonstrating a more stable sauce in ancient recipes. It is true that the higher temperatures associated with modern fish sauce, which are perceived to be lacking in Brittany (25–35°C) can be shown to increase and improve protein yield, which may result in better perceived taste and quality. It can be demonstrated from experiments conducted in the UK, when temperatures hovered at ±20°C in spring and autumn, that a sauce will indeed dissolve and hydrolyze, albeit relatively slowly compared to sauces that were made at the height of the summer and those made in the Mediterranean. It would therefore have been quite possible to create an everyday fish sauce for mass consumption quite easily in northern Gaul from small fish readily dissolved in the temperatures of their high summer ±20°C to 22°C.

Residue analysis by Morais *et al.* (2016) has revealed that a similar sweet addition to fish sauce was possible from samples taken from a number of Spanish and Portuguese amphora of Augustan date at Braga in Portugal and from Galicia and these results were corroborated by evidence from a *doliola* containing fish sauce residue at Boca do Rio in Lagos, though the latter is 4th/5th century AD. The results show evidence of exceptionally high carbohydrates as well as high malic and lactic acid, both of which were present in the residues found by Garnier above, and in that instance they have been interpreted as evidence of wine while here it is interpreted as the addition of honey or some sort of wort though it is not quite clear what is mean by this term. Adding beer to fish sauce would seem to be very unusual and wine, that is, must, would be the more logical addition. (Morais *et al.* 2016:107). Research into the degree to which wine residues are able to penetrate fine grained ceramics has been undertaken on replica vessel. The vessels were subject to prolonged heat as this is known to increase residue penetration and preservation. The results have revealed five wine biomarkers: succinic acid, malic acid, tartaric acid, citric acid and lactic acid. It was determined that malic acid could also act for white wine and lactic acid for red wine (Teodor *et al.* 2014:1022). These results would seem to suggest that when malic or lactic acid are found in fish processing containers they will probably refer to wine, rather than other forms of fruit juice. Morais also found evidence of aromatic herbs and the smoke derived from the burning of aromatic plants within all the amphora tested and the *doliola*. This appears to indicate the use of herbs in the process of making a fish sauce across a number of sites. The *Geoponica* quick recipe includes oregano which this may reflect.[9] The burning signature has been interpreted as evidence of subjecting the sauce to smoke. As noted by the authors, we may be able to link this to a recipe in *Apicius* which gives instruction to fumigate an amphora designed to hold a sauce that has begun to smell to strong (very likely due to oxidation)and in fact to add honey too, to mask the strong taste (*Apicius* 1.7; Morais *et al.* 2016:107). This action to stabilize a sauce may have been done to older and stale sauces at the point of retail and also by the consumer. There is always the possibility that the four signatures found here: smoke, aromatic herbs, carbohydrates indicating sweetener and cholesterol indicating fish were derived from separate fills of the amphora in question. The high levels of sweet liquids may also correspond to attempts to extend the life of the fish sauce as indicated in *Apicius,* rather than the addition of very sweet liquids in the initial manufacturer stage. A further alternative is that the sauces in question were forms of *confectio* that had been cooked with honey at later stage, like that found in the Gargilius recipe. Garnier has noted that at Baelo Claudia, the cholesterol found in the *cetariae* was associated with "abundant crinosterol, and with a 22-dehydrocholesterol" in half of the vats. This was also noted in an experimental reconstructed *garum* made with oysters and mackerel, and this form of cholesterol is also noted to be indicative of seafood generally. The team working with Garnier *et al.* (2018) determined that it was highly likely that this indicated that additional seafood such as oysters were added to make special mixed sauces (Garnier 2018:320). It is possible that oysters were purposefully added to a fish sauce. Pliny does tell us that oysters were used to make an elite *allec* paste, but crucially this did not generate a sauce first. The shellfish signature would suggest that molluscs were in the fish sauce, but it does not tell us that they were necessarily added intentionally as part of the recipe, or that oysters were added to a *liquamen*. This would seem to be a leap too far. One can easily imagine a situation where various kinds of sea mollusc

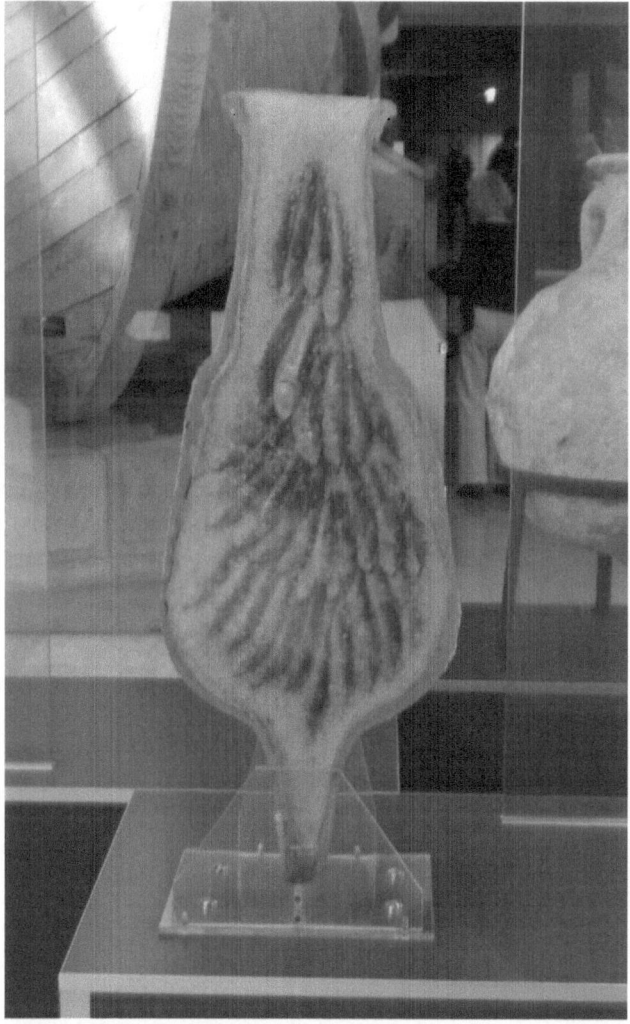

Figure 6.1 A Dre. 9 on display at the Musée d'Histoire in Marseille, filled with small mackerel, compacted in salt.

were simply there as they were part of the haul of fish and shells that were drawn from the sea on to the beach in the seine nets. These nets drag across the shoreline and could easily have gathered quantities of small, insignificant shellfish. One can only surmise that all non-kosher sauces were potentially contaminated with shellfish as a result of the vast quantities of fish that were processed without being sifted of such apparent contamination. For kosher sauces see page 33; 233; 239. Whether this level of contamination would result in this kind of shellfish signature is difficult to say. There is evidence of multiple shellfish species in the processing vats at Bello

Claudia but they do not appear to be in sufficient number to suggest that they were added intentionally. Bernal-Casasola (2018:340; table 3) sites the MNI for these shellfish finds, and the majority exist in very small numbers. The presence of this material has been interpreted as a purposeful addition to the recipe in line with the further idea that unusual "mixed sauces" were made at the site (Bernal, Exposito and Diaz 2018:347), though they acknowledge that these documented shellfish may have been incidentally captured, and not represent an intentional addition to the sauces. It is I think more probable that seafood pastes, as well as the idea of shellfish pickled in brine were made separately and independently from *liquamen*.[10] Garcia Vargas (and Bernal 2009) discuss the faunal evidence for sea food products and cite Cadiz amphora at Lixus (type T-7.4.3.3) at levels from the Mauritanian period (50 BC-10 AD) containing apparently mussels remains preserved in brine) and also a preparation based entirely on cockles found at Cerro da Villa (Garcia Vargas and Bernal 2009:146). Dario Bernal (and Garcia Vargas 2009:147) also note that oysters chipped open but otherwise discarded in quantity suggesting some kind of individual processing of the oysters to preserve them, otherwise the shells would be discarded randomly after consumption

The evidence for "mixed sauces" is not convincing. The single find is in a regional imitation of Greco/Italic amphorae and it is believed to contain a mixed sauce, composed of tuna and whole small fish, as well as what is assumed to be intentionally added meat from young ovicaprids and pork. This is the only example of meat and fish together in a vessel and while it is possible that such sauces existed, I would interpret this unique find as an accumulation of discarded rubbish inside the vessel and not a "regular" mixed product at all. For fish residue disposal see Chapter 11. My hypothesis is strengthened by the fact that the primary evidence at Bologna cited by Arévalo-González and Bernal-Casasola (2008:17) is mid. 2nd century BC, and therefore too early to expect any great diversity in fish sauce, which only began to emerge in the 1st century BC under Roman influence, according to the literary evidence. The idea of these "mixed sauces" is often repeated without questioning its validity (Marzarno 2018:440). It is my belief that the market simply did not exist (and probably never did) for a product made with meat and fish, though I am prepared to be dissuaded. We can be fairly certain that quadruped meat was salted in the *cetariae* when the fish was out of season, as large amounts of animal bones are regularly found at the fish sauce factories (Curtis 1991:75).

Synthesis of the literary review

The first reading of the *Geoponica* and the associated texts in Chapter 1 gave us a relatively clear and concise understanding of the two basic ancient fish sauces and their properties. One was a whole- fish sauce (of various species and size) with additional viscera for digestive purposes, and the other one a blood viscera sauce: there is a simple and straight forward separation of products. The review of the literary consumption texts in Chapter 2 alongside the archaeological chronology has not fundamentally altered this separation, in fact it has reinforced it. There were, it is true, many components that could potentially have gone into an ancient fish sauce: blood and also viscera from the larger species (tuna and mackerel); the meat from these same fish with and without extra blood and/or viscera; the small species (sardine, anchovy, sprat, herring, etc.) used whole with and without extra viscera and/or blood; wine or other fermented liquids, vinegar, honey and various aromatics herbs. There are numerous ways in which these components could have been incorporated into a sauce, and through the years of culinary development it is quite possible for the sauces to have evolved and changed immensely as they became more popular among the elite. We cannot be absolutely certain which of these components were used in which named sauce in the 1st century AD. As we have already seen (Chapter 1) we can be fairly sure which components went into ancient sauces in the mid- to late empire through the evidence in the *Geoponica*. If the clear distinction between the two sauces that is found in the *Geoponica* has to be rejected for the late republic and early empire, as many scholars believe, we are left with the option of trying to manipulate the remaining evidence in order to find what I would claim was an imaginary "single sauce" called *garum*. This single sauce continues to be understood as a combination of all the components listed above. However, by trying to bring the single sauce into existence, a great deal of the subtle differences in quality, perceived value and cost, consistencies, culinary uses (discussed Chapter 3), and residues (Chapter 11) are completely lost. My main aim in this study has been to separate the different strands of evidence for the distinct kinds of sauce, in order to counter this tendency within archaeology to unite the evidence. It is necessary to recognize that each sauce has characteristics determined by multiple criteria: species selection; the consistency of the resulting products; the residues (or lack of) from each type; the distinctive amphora shape used to transport each type; the perceived status of the consumer of each type; the various dining practices, the dining table etiquette and post-consumption discard practices. So long as the imaginary single sauce remains the focus in archaeology there is little chance that any of these aspects of ancient fish sauce and its trade and consumption can be really understood.

Notes

1 http://oxrep.classics.ox.ac.uk/bibliographies/fish_industries_garum_production_bibliography/
2 The Bountiful Sea: Fish Processing and Consumption in Mediterranean Antiquity Journal of Maritime Archaeology December 2018, Volume 13, Issue 3, (Eds) Angela Trentacoste, Rebecca Nicholson, Dimitra Mylona.

3 The useful websites on *Baetican* amphora derive from the University of Cadiz, *Amphorae ex Hispaniae* (http://amphorae.icac.cat/). Here up to date production, distribution and content analysis can be found for all the major amphorae types associated with fish products. Extensive use of this site will be made in Chapter 12.

4 The International Research Institute for Archaeology and Ethnology, based in Naples in collaboration with Alfredo Carannante.

5 It is apparent from experiments that those larger mackerel cut up to make the sauce do not need *extra* viscera to ensure auto-digestion as there is sufficient within each fish.

6 My experiments demonstrated that the medium-size fish, 10 to 20 cm, had not dissolved at a time when the smaller >10 cm sardines had fully disintegrated.

7 The confusion between the use of tuna or mackerel intestines for *garum* in this respect is entirely understandable. The *Geoponica* makes its *haimation* with tuna blood and viscera, and the only archaeological evidence we have for this kind of sauce comes in the form of tuna gill bones in a local baggy jar from Jordan from the late 1st century AD. This is interpreted quite accurately as the residue from a tuna blood and viscera sauce (Van Neer and Parker 2008). I will maintain that the fish used for this kind of sauce changed from mackerel (*sociorum*) to tuna (*haimation*) because the mackerel had been over-fished, while the obvious increase in volume that results from using tuna viscera was also a motivation for the change. Mackerel does not yield a great deal of either of these juices per fish. The two sauces were essentially the same nevertheless. See page 163.

8 The issue of fish sauce is not dealt with in depth, though she does provide valuable evidence of a 15th century BC Babylonian fish pickle called *siqqu* (Marzano 2013:90).

9 But see Chapter 3 on fish sauce in culinary sources, where it is also possible that much of the potential aromatic substances were added at a later stage by the cook in the process of making oenogarum.

10 However, a Mesopotamian recipe preserved on cuneiform tablets from the 17th century BC records that the fact that fish and shell fish were combined to make a fish sauce, suggesting that shell fish were intentionally added to sauces in early cultures (Dalby 1996:76; Theodoropoulou 2018:393).

7 Fermented fish sauce in Southeast Asia

Many of the fish sauces that are manufactured today in Southeast Asia would appear to be very similar to ancient fish sauces in terms of the finished product and the manufacturing process, and it is therefore necessary to consider modern techniques and scientific research into modern fish sauce production. The process of preserving large hauls of small fish to make a basic fish sauce occurs in tropical countries during the summer months in many areas across Southeast Asia. Fish sauces are variously named: *nuoc-nam* in Vietnam and Cambodia, *nam-pla* in Thailand, *petis* in Indonesia, *buda* in Malaysia, *patis* in Philippines, *yu-lu* in China, *shotssuru* and *Uo-joyu* in japan. A special Japanese sauce called *ishiru* and *ishiri* is also made in northern Japan, the latter resembling *garum haimation* in flavor and color. See page 141. Just as in the ancient world pastes may be prepared from fresh fish or the residues from fish sauce production. Shrimp paste is *belachan* in Malaysia, *trassi* in Indonesia and *ngapi* in Burma. The fish paste from *patis* is called *bagoong* in the Philippines and *kapi* in Thailand, *pra-hoc* in Cambodia, *mam-cho* in Laos and Vietnam, *yu-jiang* in China, and *gyomiso* in Japan (Ange 1999:264). There are a huge variety of different methods employed to make fish sauces, each subtle difference making small changes to what is a very simple commodity. In Vietnamese, nước mắm means literally "salted fish-water." Very simply, small whole fish, of an infinite variety but nonetheless dominated by sardine or anchovy, are mixed with varying percentages of salt and allowed to dissolve and generate sauces and pastes. Just as in the ancient world, other sea foods and even fish viscera are sometimes used alone, while in areas that are land-locked in Thailand fresh water fish that flourish in the rice paddies are used. The maceration of whole fish and salt initially produces a brine. The natural enzyme present in the viscera, begin to hydrolyze the flesh, and dissolve the protein into the brine in the form of free amino acids and peptides and ammonia. The best sauces would appear to have an optimal balance of enzyme/bacteria/acidity/sodium/water content that results in a stable fish sauce with high nitrogen levels and good keeping qualities. The modern sanitized crystal-clear sauces, of various shades of amber through to brown, known as *nuc nam* and *nam pla* from Thailand

and Vietnam dominate the world markets, and have become fashionable cooking ingredients around the world as they are recognized and valued for their umami-rich flavor enhancement qualities.[1] Often the fish sauce available in the west is of poor quality relative to what is considered valued in the east. All over Southeast Asia, there are two basic techniques: the small-scale anaerobic domestic approach in jars and bottles, which is probably quite rare today, and a large-scale industrial process. The modern large-scale bulk process is further divided into two techniques, one subject to the sun's heat in large concrete or metal tanks, which mirror ancient vats in many respects. These are covered for protection but not otherwise airtight while the sauce is frequently stirred, and the other technique involves using wooden barrels in which the fish is enclosed and compressed. The small-scale domestic approach uses a ceramic or glass vessel and sufficient freshly caught or bought anchovy or similar small pelagic fish. Varying levels of salt are used beginning at 25%. The Vietnamese sauce *nouc nam* is still made at home with minced fish and salt at a ratio of 3 parts fish to 1 part salt, sealed in small jars for 2 months. Mincing the fish is a relatively common practice as it allows the salt to penetrate to all areas and guarantees that pathogenic bacteria are eliminated and it also maximizes the brine yield. These small vessels are sometimes buried to keep out all light and left for a period of fermentation and maturation before the fish mash is strained and a first harvest of sauce is taken. The residue is always rebrined repeatedly. Each household would make their own sauce and as the process involved rebrining, one batch of fish could actually generate a considerable amount of sauce over time. Each day the amount of sauce required for each family could be taken and replaced with brine until all the fish was used up. Even though the volume of harvested sauce was relatively small each day, it was high over time and very efficient (Cuong Pham, owner Red Boat Fish Sauce Co., pers. comm.). Nevertheless, in Thailand and Vietnam today most people do not do this, but buy a clean filtered product from the supermarket. Today, the practice of making fish sauce in bulk open to the air is probably quite rare, even in regions where very traditional techniques are applied. I once had a memorable encounter with a US helicopter pilot who had flown in Vietnam during the war. When he realized that Roman fish sauce and the Vietnamese sauces, he had encountered, were so similar, he gave me a vivid account of the open-plan fish sauce tanks and the smell that rose into the helicopter cockpit as he flew over them. The pilot remembered the laborers stranding by the side of the tanks with long paddles to stir the sauce. These tanks were covered with rush or wicker mats, weighted down to prevent debris and vermin getting in, but these coverings were not airtight and had to be removed regularly. Evaporation was an inevitable consequence of this technique. The Thai traditional method for *nam pla* uses concrete tanks built above ground level. Whole anchovy is employed, mixed with salt in a 2:1 or 3:1 ratio. Often extra brine has to be added, and after 12 to 18 months of fermentation the sauce is taken, filtered, and allowed to ripen for a further 2 to 12 weeks. They make a second-quality sauce from the bone residue by rebrining in the same way as in Vietnam (Crisan and Sands 1975:106; Lopetcharat 2001:65–68).

The temperature at which fish sauce is manufactured varies considerably within a range of 25°C to 40°C, with a greater efficiency and yield of sauce the closer to the top range. On Phu Quoc island, off the coast of Vietnam, the daily temperature is in the 30°C to 35°C range, and this is also the region where the best fish sauces are made. There are limits to the level of heat that is ideal, however. A cooked sauce is never valued in Southeast Asia, and temperatures above 40°C are not desirable. Manufacturers take care to prevent the sauces overheating in the heat of summer (Anonymous 1982, New Zealand digital library). The majority of modern fish sauces are left to mature for at least 12 months and often 18 to 24 months are stipulated before filtering, discussed below. Even after this length of time the residues are rebrined to make weaker sauces and sometimes a hot brine is added to extract all the potential nutrition. The first and second yield is blended with the weaker to make the most sauce of an acceptable quality. A description of fish sauce manufacture in an application for an official Registration of Appellation of Origin for a fish sauce from Phan Thiet from the Ministry of Science and Technology of the Socialist Republic of Vietnam details the different techniques.[2] The open and continuously stirred technique is used in the Cat Hai and Hai Phong areas and an integral part of the process includes adding water and salt at least three times to maintain a brine of sufficient concentration throughout the process and to replace the water lost to evaporation. The Phan Thiet manufacturers however recognize that the enclosed and compressed technique is the most efficient at harvesting the maximum amount of nutrient in a given sauce (see below). It was also believed that stirring the sauce contributed to an increase in hydrolyzation from the agitation, but it was very labor intensive and inefficient in other ways. The official Registration of Appellation of Origin for the fish sauce from Phan Thiet states that an exposed sauce also loses more protein to increased decomposition than one that is enclosed and compressed. Owens and Mendoza have suggested that this loss is due to excessive conversion of the organic nitrogen into ammonia, which creates a less desirable smell and taste. The sauce becomes more pungent, with stronger fishy aromas, and might be considered by some as more "off," unstable and less valued commercially, though this is not in fact the case as fish sauce rarely goes bad, but rather becomes more pungent and over time it also darkens as it oxidizes, and becomes chemically altered by exposure to the air. The presence of small amounts of titratable ammonia is in fact normal in a fermented fish sauce (Anonymous 1982, New Zealand digital library). Owens and Mendoza (1985:276) suggest that the higher salt levels used in modern sauces today reduces this natural ongoing decomposition of the dissolved protein. They also note that the potential for increased ammonia may also be due to enzyme activity from microbial growth (see page 137).

The alternative bulk method which takes place in large enclosed wooden barrels involves the fish and salt being layered and compressed with weights. No extra liquid is added or needed and there is rarely agitation. As the fish are compressed, more space is created, and so more fish and salt are added every few days until the barrel is full. The first liquid is drawn off from beneath and poured back on a regular basis. This is a clean and efficient method with a high nutritional yield and makes for a stable product. A combination of the two techniques, whereby the sauce was initially stirred regularly and then compressed, was applied to the Phan Thiet sauce which they claimed made it exceptionally good. The finest fish sauce now available in the West is acknowledged to be Red Boat, an exceptional quality sauce manufactured from very small (<5 cm) black anchovy on Phu Quoc Island in Vietnam and made with the highest nutritional values and the lowest salt levels specifically for the American market. Its taste is quite different to that of the traditional Southeast Asian fish sauces, which are dominated by the taste of salt. Red Boat is made using barrels holding more than 50 tons of black anchovy layered with salt at a 25% to 30% w/w. Four kg. of fish generate 1 liter of finished sauce. The fish are not compressed and some stirring takes place initially. After 12 months the sauce is ready, while the residue, which we are told still contains up to 70% of its potential nutritional yield, is sold to fish sauce residue brokers, who manage the rebrining process. The salt levels during the manufacture of Red Boat are quite high but this is to secure a stable sauce, that is, to prevent protein decomposition and pathogenic bacteria in the early stages. The sodium levels of the finished sauce are subsequently reduced after the sauce is filtered by a form of vaporization. Red Boat is only marketed in America and as such it has to be pasteurized in order to enter the United States. The reduction in salt creates a situation where the fish sauce solution can take up and retain more nitrogen before it becomes once more a saturated solution of fish sauce. Red Boat uses the degrees N as an industry standard to measure the number of grams of nitrogen per liter of fish sauce which relates to the protein level. The highest quality fish sauces are greater than 30 N with the flavor becoming richer and more complex with larger N designation. Red Boat is marketed at 40 N and is also available at 50 N after ageing and the removal of more of the salt. The ordinary everyday fish sauce, whether local to Vietnam or the Western market, tends to have a nitrogen level of 10 to 15 N. We may wonder whether the Romans were able to make a sauce with similar levels of nutrition to Red Boat using their recipes? If time is a crucial factor then the answer might appear to be no, as ancient sauces were apparently made in as little as 3 months, but there are many other factors to consider. The lower salt might result in a sauce that was prone to protein conversion and as a result appear more pungent than Red Boat, while the traditional fish sauce makers of Phu Quoc Island, who market locally, state that a short fermentation time results in a fishier flavor while a long ferment results in a nuttier one. This might imply that ancient fish sauce was somewhat fishier, given that it was made in such a short time.

Modern fish pastes

There are numerous fish sauce products local to Southeast Asia of every kind of consistency and state of disintegration, made with all kinds of sea food and fish, which have some resemblance to the range of ancient products we have discussed, and particularly the production and use of *allec*. A sauce is made in the Philippines which involves the harvesting of the fish flesh in the form of a fish paste called *bagoong*. It resembles the *pissalat* of Southern France and the elite *allec* mentioned by Pliny the Elder (*HN*. 31.96; Curtis 1991:22; Crisan and Sands 1975; 107). *Bagoong* is either made from tiny anchovy or shrimp in a ratio of three parts fish to one of salt. The finished product is described as a thick suspension of undegraded fish flesh together with liquefied protein (Owens and Mendoza 1985:277). The fermentation time varies widely. In the Visayas region *bagoong* ferments for only a week. In Ilocos, in the northern provinces, *bagoong* is preferred to age for at least 6 months and up to 2 years. The environment of production is always enclosed and anaerobic. The resulting reddish paste is added to dishes near the end of the cooking process and is also used as a dipping condiment. The making of *bagoong* naturally also generates a clear sauce called *patis* which has remarkable parallels with *liquamen* and *allec*. A characteristic of *bagoong* would appear to be that, for some, the paste is viewed as the primary product, while the clear amber sauce a secondary by-product, while it would appear from some references in our literary sources that the reverse was the case with *allec* and *liquamen*. The presence of the bone in the finished product in a domestic setting is of interest. A recipe posted on a Filipino web site for a bagoong made with anchovy of 4 to 5 cm. in small preserving jars suggests that while the paste is forming in the brine – that will later be removed as the *patis* – the bones naturally sink to the bottom of the glass jar and can be avoided: "After at least 6 months, the *bagoong* is ready to be used and when taking some *bagoong* from the jar, remove any remaining fish heads or bones from the paste with a fork and spoon."[3] This would rather suggest that bone removal was relatively easy. It also suggests that the cranium bones and possible the vertebrae may remain sufficiently articulated to be removed in one piece.

Enzyme hydrolysis of fish sauce

The sauces that utilize indigenous enzymes from the viscera are generated through autolysis, which acts to dissolve the muscle tissues into water soluble free amino acids and peptides through chemical reaction with water ions (Curtis 2009:712; McIver *et al.* 1982:1017). Some modern fish sauce is generated with added enzymes from starches such as rice, though we have no indication that such practices occurred in the ancient world. Ancient sauces worked with added enzymes from additional viscera, which in theory would have allowed the digestive process to act on the outside of the small fish as well as that occurring from the naturally present enzymes. Klomklao *et al.* (2006) have experimented with the addition of tuna spleen, which is known to contain particularly high levels of proteolytic enzymes, added to a whole fish sauce, to determine how effective this additional viscera would be to an industrial fish sauce manufacturing process, an experiment of considerable interest as ancient manufacturers clearly used similar techniques to make their sauces. Their results, which conclude that between 10% and 20% spleen and 15% salt lead to the highest enzyme activity, demonstrate that extra viscera at the beginning of the process make a very significant difference to the amount of enzymatic activity and greatly speed up the entire process:

> "From the results it was noted that no proteolytic activity of sample without spleen addition could be observed after 10 days of fermentation in the presence of high salt (25%), nevertheless some proteinase activity of the samples without spleen was still found in the presence of less salt at every fermentation time."
>
> (Klomklao *et al.* 2006:444).

This demonstrates that too high a salt ratio actually impedes enzyme action and reduces protein uptake. However, though it appears that salt levels above 15% can seriously impede the hydrolysis of protein into a liquid form, salt concentration between 3.5% and 10% can be problematic. Under these low salt conditions, the growth of pathogenic bacteria such as C Botulinum is quite possible, as this bacterium thrives in these specific low sodium conditions. It is therefore necessary to ensure that the salt concentration remains within a band of at least 15% and above to guarantee the safety of the sauces (Essuman 1992:81). Klomklao *et al.* (2006:443) has also demonstrated that the enzyme activity in fish sauce production has a limited time span. As noted above, in adverse conditions (i.e., salt at 25%) this can be as short as 10 days, though this would seem to be particularly low, given the long process involved in making modern industrial fish sauce, which uses these high levels of salt. Klomklao (*et al.*) have concluded that in optimum conditions (i.e., lower salt and the ideal temperature) enzyme activity is detectable for up to 160 days of maceration, after which it is not clear what further effect there is on any muscle tissue that has not been hydrolyzed. This might suggest that the extended time period of production, 12 to 18 months in modern recipes, has a limited effect on actual nutrition uptake but would improve quality through maturation and ageing. A travel writer's description of a fish sauce factory visit in Thailand and personal contact with the manufacturers has helped to clarify these issues. It is apparent from these discussions that the extended aging process in the tanks changes the sauce considerably:

"The length of time that the fish/salt arrangement is left to sit has a great deal to do with the final quality. The family at this facility has been making fish sauce for 80 years. They originally left it for 24 months but the current owner was able to determine (by measuring the protein levels and observing the quality) that the extra 6 months was not necessary. Fish sauce that has been sitting for periods of time shorter than 18 months will be darker and have a more 'fishy flavor' .... Darker is not necessarily desirable in fish sauce – good fish sauces tend to be the color of whisky. After 18 months the fish sauce has become less dark and less 'fishy'."[4]

The last 6 months of ageing removes the pungency, replacing it with the meaty cheesy umami flavor we recognize, while the color is lightened by this ageing process (Michael Babcock (for) Kasma Loha-unchit Clark, pers. comm.). I have noted a slightly different process at work, as when sauces are stored for extended periods, at home, they darken due to oxidation. Cuong Pham has clarified this issue: higher protein sauces tend to oxidize faster therefore they get darker faster (pers. comm.). It possible to see that ancient sauces were also aged beyond the initial 3 months stipulated in the recipes, but this appears to have occurred away from the processing tanks in the transport amphorae, which, as it would naturally exclude light would also prevent oxidation (see page 238).

Bacterial fermentation of fish sauce

It is clear that all modern fish sauce products are produced using a unique combination of fermenting bacteria and hydrolyzing enzymes in the presence of a variety of salt percentages, which together liquefy, preserve and develop desirable umami tastes and textures (Owens and Mendoza 1985). There is some doubt however as to the degree of bacterial involvement, in ancient fish sauces. Some research suggests that modern fish sauces of the *nuc nam* and *nam pla* type hardly rely on bacterial fermentation at all. Other research suggests that bacteria is a fundamental part of the process even in high salt levels, and that this aids the liquefaction process (Crisan and Sands 1975). It is understood that the presence of bacteria, particularly lactic acid producing bacteria (LAB), result in a true fermentation, while the process of a true enzyme hydrolyzation typically functions with a low bacterial count. A high LAB count contributes to a higher acidity, which is particularly desirable as the acid prevents the pathogenic bacterial growth. Owens and Mendoza have attempted to classify sauces in terms of either bacteria or enzyme action, while also acknowledging that "any classification is of course arbitrary and it is likely that the processing of bacterially fermented products will involve some autolysis by indigenous enzymes and that microbes may make a contribution to the enzymically hydrolyzed products" (Owens and Mendoza 1985:274). The large-scale process in Vietnam today has been entirely modernized and sanitized so that the sauces can be recognized as safe for human consumption around the world. That of course is not to say that before this modernization fish sauce manufacture was unsafe; poisoning from tradition fish sauce manufacturing is very rare indeed, largely because the salt levels have always been higher than actually necessary (>25%) to ensure the perception of a safe sauce. There is a strong association between excessive salt consumption and cancer (Wang *et al.* 2009) which can militate against the health benefits of consuming fish sauce for its nutritional values, even when diets are largely vegetarian, as is the case in the developing world. One of the traditional sauces made in Thailand used to allow the whole anchovy to be held at ambient temperature for anything up to 48 hours before being salted for fish sauce (Crisan and Sands 1975:106). In these particular conditions a rapid increase in lactic acid-producing bacteria inside the gut causes substantial bacterial fermentation, which is known to destroy the pathogenic bacteria (Owens and Mendoza 1985:273). The fear of pathogenic bacteria is quite natural but is largely unfounded in relation to fish sauce production, while the dangers of toxins in solid fish products which are too lightly salted are often far more problematic (Paludan-Müller 1998:56). An example from an FDA (US Food and Drugs Administration) report illustrates the phenomenon and demonstrates how the size of the fish in sauce production is perceived as of the utmost importance for food safety. A factory making fish sauce in Thailand was inspected in order to apply for an import license to trade their anchovy sauce into the United States in June 2008. The inspection report was posted on line and it outlined why their license was rejected:

1 Evisceration or destruction of preformed toxins (clostridium botulinum) prior to processing into sauce. The following processing procedures are necessary to control clostridium botulinum toxin formation:

   a   Fish that are 5 inches (12 cm) in length or longer (head to tail) should be eviscerated. FDA is unaware of any other processing procedures, other than evisceration that can be applied to control clostridium botulinum toxin in fish that are five inches or greater length....
   b   Fish that are smaller than 5 inches (12 cm) in length and which are uneviscerated should have an operating step such as boiling prior to the salting process to destroy any toxins....[5]

These are highly restrictive requirements as it is generally understood that visceral enzymes are essential to a quality hydrolyzed fish sauce. Some modern sauces mince the fish in order to maximize the nutritional yield under these restrictive conditions. This sauce was not allowed into the

United States after this report as the makers were not prepared to alter their production process and lower the quality of their sauce to reach the American market. My own experiments have demonstrated that sardine and anchovy over 12 cm are slow to break open and expose the viscera to the salt, which could potentially allow botulism to develop inside the cavity. The addition of extra viscera effectively prevents this from happening. I noted that sardine and anchovy below the 12 cm size range were also more likely to dissolve fully into the brine. We will see in the discussion of fish size and the fish sauce residues identified from archaeological sites that sardine and anchovy of multiple sizes up to 20 cm. were regularly used for Roman fish sauce, but often in combination with fish as small as 4 cm (see table 11.1).

Harvesting and rebrining the sauce

Many of the traditional methods involve rebrining. The timing of this would seem to be the key to obtaining further sauces of reasonable quality. Some concrete tanks have a grid that is pushed down to about a foot from the bottom that keeps and holds the bones at the bottom of the tanks. My own experiments have demonstrated that some of the lighter bones can float free in the brine, which this clearly prevents. When the first sauce is removed, filtered and retained as a first-quality sauce, the remaining mash, of indeterminant appearance, but essentially bone and remaining paste, is covered with brine or hot brine and a second-quality sauce is taken. In order to produce enough sauce for local and foreign markets, the two qualities are often blended in Southeast Asia. A travel writer recounts a visit to a "Golden Boy" fish sauce factory in Thailand and gives an insight into the process of rebrining that is very useful:

> "After the liquid (sauce) is taken off and filtered, liquid is added to the remains of the fish and salt and over several days it is moved from tank to tank until a lower quality liquid suitable for dilution is available. They have to move it from tank to tank in order to keep the same rich color of the 100% fish sauce."[6]

This description of the brine added to what seem to be multiple tanks of residue and then moved from tank to tank to harvest the second sauce has many implications for our understanding of the ancient process. As we will see, many of the ancient factories contain numerous tanks side by side and often of different sizes, which we might consider had a similar function. We have been told that when the "Red Boat" fish sauce is filtered at 12 months, it leaves up to 70% of the potential nutrition still present in the residue. This finding is corroborated by modern scientific research into fish sauce. Beddows *et al.* (1979) note that in the commercial production of *budu*, a Malaysian fish sauce, it was shown that the maximum volume of liquid was produced after 140 days, and that proteolysis continued to occur until 200 days, when 56% of the insoluble fish protein had been hydrolyzed into soluble form, leaving 44% remaining to make a second sauce. These finding differ slightly from that of Klomklao *et al.* (2006:443), who found that enzyme activity only lasted 160 days in optimum conditions. It is not all clear precisely what this residue constitutes. It has been claimed that enzymes associated with fish sauce do not normally degrade collagen, and this might suggest that the bones remain articulated (Klomklau *et al.* 2006:16). What is the precise nature and consistency of the remaining liquid? Is it a thick or thin substance? Personal conversations with Cuong Pham, the owner of Red Boat fish sauce in San Francisco, suggest that it is not possible for more nitrogen to be held in the "first run" sauce, as it is in fact a saturated solution already. It is saturated with both sodium and nitrogen, the ratio of sodium to nitrogen being the key factor in the N classification. It seems from descriptions of the rebrining process utilized by the tradition fish sauce makers that, after a 12-month ferment, the residue is soaked in the new brine for a relatively brief time, at most a few days or weeks, before a second sauce can be taken. This might suggest that the remaining nutrition and flavor can be retrieved relatively quickly. It would appear that there is no need to referment in order to dissolve the remaining protein as the enzyme digestion has already transformed the solid fish protein and, as we have already concluded, there would be hardly any enzyme activity after 12 months. From this we can conclude that the nutrition in the form of the fish paste is in a more readily dissolvable state and once more liquid is introduced the protein can be harvested quickly. Modern sauces are pumped out from beneath already filtered efficiently into oil-tanker-sized trucks and delivered to the bottling plant. The travel writer's description of traditional fish sauce manufacture in Thailand aids us here too:

> "There were between 200 and 300 concrete sections. After a few days there is not very much liquid. By several weeks the amount of liquid has grown, although it looked as if it contained a fair amount of salt and scum. By 18 months the liquid is virtually clear: the owner put a glass in, swirled it around and it came out looking like fine whisky. No liquid had been added and yet the tank, save for a crusting of salt around one edge, was nearly completely liquid."[7]

The description is a little vague, but it looks as if the particles of fish flesh have fully dissolved as, if they were present, they would still be floating on the surface and this finished sauce is almost crystal clear. This was surprising as modern fish sauce experiments, described in the next chapter, have revealed that the fish flesh in the form of particles, float on the surface of the liquid for quite some time. Clearly over the 18 month this fish paste will fully dissolve.

Is there a modern blood viscera garum?

The concept of a sauce made just with fish viscera in modern cuisine is rare indeed but it does exist, though it is rather different to the ancient variety. It is of particular importance to acknowledge that the viscera of all mammals and fish that we normally consume are a valuable nutritional resource in term of protein. The livers of the larger fish species, including tuna but also the *Gadidae* species such as cod and skate, contain livers with oil with high levels of fatty acids and vitamin A, D, E, B1, thiamine, B2, riboflavin, niacin, B6, and B12. Fish liver oil is well known for its health benefits. There is a cultural aversion to all kinds of offal in the West today which appears to develop with increasing affluence. The internal organs of an animal (the digestive tract, liver, spleen, heart, lungs) are considered distasteful and unpalatable and are frequently seen as rubbish in the West, where most of this matter ends up in dog and cat food. In the developing world this sense of squeamish distaste for offal is a luxury few can afford if access to animal protein is limited. Many poorer people in regions of Southeast Asia rely on fish sauce for most if not all of their protein intake, and the inclusion of fresh fish offal into fish sauce production has been seen as a desirable way to improve the quality of fish sauces for mass consumption. In the developing world of Southeast Asia fish waste material from fish ports is often discarded inadequately and causes a major health problem once they decay. The nutritional value in this material even when it is fresh is still a difficult to concept to sell, but it has been recognized that if utilized invisibly in fish sauce it can increase the volume and quality of the end product without disturbing consumers. Sauces made just from fish viscera are rarer and it is really only through anecdotal and uncorroborated comments from food-based travel writers that they can be conjectured in modern-day China. It is quite possible that such sauces are made from all kinds of fish waste including heads and tails which will include some fish flesh. It is likely that once the process of liquefaction has taken place the resulting sauce would not look or taste very different from one made from fish flesh. Such sauces are likely to be used in exactly the same way, that is, utilizing the umami salty flavor to enhance the flavor of bland and largely vegetarian foods.

In Japan there is a unique fish sauce known as Ishiri that is made with squid viscera and salt. It originated in Noto-cho town, in Ishikawa Prefecture, and had until recently a very small market in the local area. The more familiar term for fish sauce in Japan is Shotssuru. It is virtually impossible to purchase Ishiri in the West, despite a growing interest among gourmet chefs in the United States. I was given a bottle through a colleague in the USA and after tasting it and looking at the recipes I realized that it had many similarities with the black and bloody viscera sauce that is the true *garum*. Ishiri is quite black, and allows little light to penetrate the mixture when it is held up to the light, an appearance that my experiments to make *garum* have replicated. I have been able to make contact with an

Ishiri manufacturer called Kaneishi-san, who provided me with the details of its manufacture.[8] Kaneishi-san uses squid intestines, largely the liver and the squid ink that is naturally present in the viscera, with 18% to 20% salt. This mixture is left to ferment in the heat of the summer and then left to mature for up 3 years. The process involves an exceptional fermentation process using naturally occurring *lactobacillus* bacteria which allow the sauce to seethe and boil as it gives off CO_2. It has a remarkable appearance similar to a very active sourdough at the height of the fermentation process. At the end of this long maturation process some manufacturers "cook" the sauce to give it a unique flavor, and there are other techniques to add fermentation agents to speed up the process, though this is not considered desirable in the finest Ishiri sauces.

During the experiments which will be described fully in the next chapter, many of the sauces were tested for nutritional content, and it was possible to compare modern Ishiri with a regular commercial fish sauce and my own experimental sauces. See table 8.1 page 168. The Ishiri sauce contained exceptional levels of protein, over four times the level of protein in a standard fish sauce known as "Squid Brand" (which it should be noted is not made with squid!) The testing procedures were not compatible with the N designation for nitrogen recognized by Red Boat and many other fish sauce facilities so it has not been possible to compare these findings with a Red Boat fish sauce. The web site at http://ishiri.jp/en provides evidence of the exceptionally high levels of free amino-acids and antioxidants in the commercial Ishiri. Ishiri and *garum* are made in quite different ways. The bacterial fermentation and long aging process did not compare to *garum*, which, according to the *Geoponica*, was ready in 2 months; and the obvious absence of squid ink, nevertheless the taste is remarkably similar. The only recipe for a true *garum* in the *Geoponica* does not imply long term fermentation as the time scale is relatively short, but this is not it itself enough to preclude this kind of lactobacillus fermentation in a true *garum* production, particularly as I suspect the sauce and it's residue may have been transported together for a longer period. *Garum* was aged, according to tituli picti, for up to 2 years and if it was retained in its unfiltered state, a state that I believe was indicated by the titulus *flos,* this fermentation may have continued inside the amphora (page 201).

The flavor of Ishiri is quite unique and hard to describe. It tastes salty and produces an umami effect but the background flavor is not at all fishy, and in fact one would not guess that it came from a marine resource. The flavor is somewhat metallic, and the taste is of iron. It is not unpleasant, but is certainly different! The experiments to make a true blood viscera garum will be described in full on page 163, but for now we can note that when this sauce was completed it had the same iron compounds in the smell and taste, which I associated with the inclusion of blood. Normally one would reject such a taste, especially in food, but this is not the case with Ishiri or *garum*. This sauce is unusual, and not for the culinarily faint-hearted but it clearly has its place in the complex cuisine of Japan, as *garum* surely did in Rome. Ishiri is used like many fish sauces as a final finishing ingredient, added close to the end of the cooking and often in combination with a standard fish sauce made with whole fish. It is used with great care; the instruction is to sprinkle "a few drops" to enhance the dish.

The case of fermented fish liver

The concept of fermenting fish viscera, particularly the liver, is obviously an issue that would seem to be closely allied to the idea of a blood viscera *garum*. The concept of fermented fish viscera has recently become part of a popular debate within the supplement market. The popular supplement that is cod liver oil is largely available as clear, attractive capsules of yellow oil. Few comprehend that until modern industrial techniques were introduced, the oil from cod liver was fermented to extract it (Butler 1948:5). In its natural state, cod liver oil is a much darker shade of yellow and sometimes brown, in contrast to modern cod liver oil, and the darker the shade, the greater are its nutritional benefits (Butler 1948:89). Today, nutritionists widely acknowledge that the health benefits of a fermented cod liver oil are greater than an unfermented oil (Peckel Möller 1895; David Wetzel, pers. comm.).[9] The liver of squid is exceptionally rich in nutrients and comparable to cod and skate liver oils in terms of human health benefit.[10] The livers of many fish species, and particular tuna, are valued for their oils too as they are reputed to have be particularly potent source of vitamin A and vitamin D, though it is not as easily extracted as cod liver oil (Barnes *et al.* 1944:159).[11] We may conclude that if *garum* was fermented it will have maximized the yield of nutrients during the process of manufacture. It is clear that fermentation has always been necessary to break apart the hepatic cell walls of fish liver so that the oil can flow. A *garum* made from tuna viscera including the liver will have naturally generated a substantial amount of oil, over and above what was naturally generated when fish sauce was made. We cannot be certain that the oil from *garum* was taken and consumed, as there is no documentary evidence to support a recognized market for the oil from *garum* production. It is nevertheless possible that a *garum* fish sauce made from viscera including tuna livers would have had considerable health benefits. See page 163 for further discussion of the ancient evidence for *garum haimation/sociorum*.

A supplement that is marketed as *Garum Armoricum* and also as Stabilium® 200 is a fish sauce derived from the processing of fish caught in Britany. The origin of this product is obscure. The web site states that "Garum Armoricum is an extract of a large blue fish which are over a meter long and which live off the coast of Brittany at depths of 1500 to 2000 m. The lack of oxygen and the high pressures have led to the development of a special metabolism in these fish, allowing them to produce substances which help them adapt to life in such an extreme habitat. The Garum

Armoricum is then obtained by auto fermentation (controlled enzymatic autolysis) involving the entire fish."[12] (The supplement is recommended for anxiety and stress.) The manufacturers claim that the recipe is based on sources from Galen but these are never sited and as we have noted, there are many later additions to the texts of Galen that are spurious and it has proved impossible to find the original source of this material. The difficulty in determining the original text of Galen is discussed on page 29. The marketing material for *Garum Armoricum* suggests that this special fish sauce from Britany was known about in Roman times and they claim that it was the Gaul's who taught the Romans how to make it, which rather down plays the contribution from Greece in the story of *garum*. It is also claimed that that Roman soldiers were able to endure the stress of military action from consuming it. There is even the spurious but entertaining suggestion that the magic potion in Asterix was this special sauce! Much of this is unaccountable and almost certainly invented but nevertheless an intriguing modern phenomenon related to the benefits of fermented fish sauce.

A modern and ancient comparison of methodology and some initial conclusions

It is clear that the production methods of the various kinds of modern fish sauce mirror the ancient techniques in many ways. The domestic process of salting and storing small quantities of fish in pottery or glass jar resembles the Gargilius Martialis recipe. This process would seem to generate relatively small quantities of both a fish paste, and a sauce at regular intervals just like *bagoong* and *patis*. The modern large-scale process using open cement lined tanks appears to mirror precisely the ancient techniques during the early stages, and while we cannot confirm that ancient fish sauce residues were rebrined, it would seem entirely likely and essential to maximize the yield of sauce from a given batch of fish. There is evidence for the use of wooden barrels to transport a small fish product and it is plausible that barrels were used to actually make the sauce, using the compression technique and with a drain at the bottom. It is more plausible to imagine a situation where the process of generating a sauce from a fish mash could occur in transit in such vessels. See page 243 for the wider issues of commodities traded in amphorae and barrels (Bowman and Wilson 2009:221). The most marked difference between ancient and modern fish sauces appears to be the time taken up in producing them and in their salt levels. The ancient recipes from the *Geoponica* stipulates a salt proportion of 15%, c. 7 or 8 parts fish to 1-part salt, compared to a regular occurrence of 3-1 in modern sauces. We may legitimately conjecture that the time stipulated in the *Geoponica* of 2 to 3 months would probably lead to far less of the fish being liquefied, which would result in an increase in the potential nutrition available to make a second and subsequent sauce. These considerations and conjectures governed how the experiments to make ancient fish sauces were designed and described in Chapter 8.

Notes

1 In Vietnam, the notion of *Nuoc mam* was first defined legally in the Order of Indochina Government (in French) on 21 December 1916: "[2]. Nuoc mam is the result of the maceration of fish in a concentrated marine salt solution, that is basically a saline solution of albuminoids (proteins) in a certain degree of disintegration" (Aldrin *et al.* 1969, Nghia *et al.* 2017).

2 Accessed on line in 2010, but now no longer available.

3 http://filipino.kitchen/article/how-to-make-bagoong. I am grateful to Sarahlynn Pablo from the filipino kitchen for her advice.

4 http://www.thaifoodandtravel.com/features/fishsauce3.html. I am grateful to Kasma Loha-unchit Clark and Michael Babcock for advice and clarifications on these points.

5 https://wayback.archiveit.org/7993/20170112200403/http://www.fda.gov/ICECI/EnforcementActions/WarningLetters/2008/ucm1048265.htm 05/10/2020

6 "A Visit to a Fish Sauce Factory in Thailand" by Kasma Loha-unchit http://www.thaifoodandtravel.com/features/fishsauce3.html

7 "A Visit to a Fish Sauce Factory in Thailand" by Kasma Loha-unchit http://www.thaifoodandtravel.com/features/fishsauce3.html

8 Grateful thanks are due to Tony McNicol, a photo journalist based in Tokyo who put me in touch with Kaneishi-san and also translated our conversation. There is a very informative Ishiri web site at http://ishiri.jp/en/.

9 A company in the United States called Green Pastures, developed by David Wetzel, has reintroduced the techniques of fermenting cod liver oil. https://www.greenpasture.org/blog/historical-perspectives-of-cod-liver-oil.

10 Fish oil products marketed under the name "Calamari Gold" are derived from squid liver oil and they have some of the highest levels of omega-3 fatty acids and are rich in form of essential fatty acid called Docosahexaenoic Acid (DHA). It is claimed that the capsules contain four times as much DHA as unfermented cod liver oil, even though these squid fish oil capsules are not fermented. https://bioglan.co.uk/calamari-gold

11 In China today, tuna (Katsuwonus pelamis) is caught in huge quantities, estimated at 4 to 6 million tons. The viscera, including the liver, from this catch is often discarded and recent research has concluded that a tuna liver oil can be extracted (Yizhou Fang 2018:2723).

12 http://www.clavisharmoniae.com/garum_sociorum.php

8   Modern fish sauce experiments

Experimental archaeology

Many of the problems that we face in understanding ancient fish sauce may be illuminated by a comprehensive study of the processes involved using experimental archaeology. It is very apparent from our previous discussions that being able to visual the nature of the residues of fish sauce and observe their formation would be invaluable in answering many of the current questions that we face. Experimental archaeology "replicates past phenomena" (Mathieu 2002:1), and seeks to reproduce the conditions under which the historical activity that is being studied can occur. For many there are difficulties with reproducing past conditions, as "one does not actually know what the past was like, so one cannot reconstruct it" (Outram 2008:2). The choice is whether to replicate the conditions entirely, that is, create a sauce in a concrete tank open to the sun's heat, as described in the *Geoponica*, or attempt to recreate a similar sauce under controlled laboratory conditions. An experiment under laboratory conditions with control over only one variable at a time has disadvantages: the conditions are sterile, tightly controlled and entirely unnatural. Experimental archaeology has been designed and requires that a single question is formulated and an experiment designed to test that particular question (Outram 2008:1). A negative or positive result to a given hypothesis is equally valuable and provides insights that can be used to design new hypotheses that can be subsequently tested. Much modern experimental work on fish sauce has been conducted this way. A team of archaeologist from the Universities of Cádiz and Seville conducted their first experiments in a sealed fermenter with a cooling jacket which maintained a constant day and night temperature that was envisaged for the process in dolia in full sun (see below; Vargas *et al.* 2014). This kind of rigorous approach can often be to the detriment of a good understanding of the process you are trying to understand as a whole. Current thinking in experimental archaeology research is that "experiential reports" where all that is recorded is what is observed are not valuable (Outram 2008:4), yet with a situation like fish sauce and the fact that there is still a basic ignorance of what actually happens to the fish, the formulation of a useful hypothesis is all but impossible without such primary knowledge.

An experiential experiment is entirely valid and scientific. The individual complex questions are not important in themselves and it is the accumulation of empirical knowledge that is important. An iterative process allows a vast amount of incidental detail to be assembled, which cumulatively leads to an invaluable body of empirical knowledge. This in turn produces a contextual framework from which to re-interpret raw archaeological data and literary evidence and ultimately to design complex and well-thought-out hypotheses to test. It also allows for the possibility of deductive leaps forward in comprehension: one must learn how to do something before one can do it successfully. Only with experience can legitimate hypotheses be constructed. An example of premature overdesign was an experiment to reproduce the brown residues that adhere to the sides of the *cetariae* from western Gaul, and particularly the ancient site of Plomarc'h at Douarnenez. It is apparent from the published report that there was a lack of a basic understanding of ancient fish sauces and their production process and a lack of information from modern research. The aim was to duplicate a sauce made open to the elements, yet the enclosed Gargilius recipe was followed, which was an error. The amount of salt used was excessive and a combination of these two errors contributed to a distorted result. A highly complex hypothesis was formulated based on the data gathered as to how the residue formation occurred, which as a result was potentially flawed but very valuable nonetheless (see below Driard 2012).

Many experiments in fish sauce have suffered from this lack of experience. One cannot acquire at speed the practical skills that ancient fish sauce manufacturers gained through day-to-day observation of the products and how they may have looked, smelled, tasted, and felt to the touch. One must imagine in the ancient world a long apprenticeship in which you learn from observation. In terms of fish sauce, we can never be sure what is normal, what smell is normal and what appearance is normal. It should be essential to acquire a degree of competence in the skills required before attempting a controlled rigorous experiment with a defined hypothesis to test, but with an experiential approach this is unnecessary, as what you are doing is simply watching and recording, so long as the initial design of the experiment (most importantly, the choice of recipe) is sound, and sufficient background knowledge from modern research has been acquire. I discuss below the previously-published fish sauce experiments and highlight the advantages and disadvantages of each.

Previous experiments with fish sauce

There are currently numerous reports of experiments to make fish sauce available in the academic press. Comis and Re manufactured only one sauce on a very small scale using the Gargilius recipe but, because of a number of errors of interpretation which included evisceration of the fish and excessive salt, the sauce failed (Comis and Re 2009:36). Comis and Re reported that the yield of water that forms the brine in the Gargilius technique was very small, and this was due in part to the dry environment created by the high salt content and the dry herbs, though the reduction in viscera may have contributed to this. It was noted by Commis and Re that much of the salt actually remained in crystal form, as there was insufficient natural water from the fish for the salt to dissolve into. The experiment conducted by Driard (2012; see below) also noted that much of the salt remained in crystal form in his initial experiment where the ratio of fish to salt was 2:1 and it was not until he started adding water regularly to counter evaporation that the experiment began to achieve results (Driard 2012:54).

The experiments conducted jointly by the Universities of Cadiz and Seville have instigated a research project into the fish sauces found at the Officina del *Garum* (Regio 1, insula 12, doorway 8), known as the Bottega del Garum (*Garum* shop) in Pompeii (Garcia Vargas 2014:66; Peña 2007:83). They have sought a recipe to replicate the kind of fish sauces they found in dolia from the shop, and chose to use the Gargilius recipe as they concluded that as the recipe was in Latin, it was closer to the heart of the Imperial Roman period (but see a different conclusion in Chapter 1). It was also concluded that the small scale and enclosed nature of this recipe was in keeping with the small dolia found in the *Garum* shop (Palacios-Macias *et al.* 2016:97). The experiments conducted by the Cadiz team have been published widely (Vargas *et al.* 2014; Palacios *et al.* 2016). They layered herbs with anchovy of indeterminate size at a ratio of between 2:1 and 3:1 fish to salt. The temperature was controlled throughout the day and night, with a day temperature of 50°C, which is considered too high in modern fish sauce production.[1] The experiment was not fully published, so they we have no data on the actual weight of fish and salt, duration of the experiment, yield of sauce and also no acknowledgement of the potential low yield of sauce given the high salt levels. The various reports in press do not state whether they rebrine the residue, or discuss the necessity of this. As we have discussed in Chapter 1, it seems doubtful that the Gargilius recipe was suitable to replicate the potential production process envisaged in the dolia as these were simply covered with roof tiles which did not fit tightly and could not prevent evaporation, while the Gargilius recipe requires a sealed vessel. It will become apparent that the initial brine which forms is subject to evaporation even when made in dolia if the covering is not airtight, and it is clear that this loss needs to be replaced on a regular basis to maintain a desirable volume of fish sauce (Driard 2012:54; see below). The Gargilius recipe also creates an exceptionally dry environment, which is not conducive to maximizing the yield of sauce.[2]

An initial experiment carried out to manufacture a fish sauce by Cyril Driard (2014:54, see below) that left the fish and salt open to the sun according to the *Geoponica* method produced a brine, but it rapidly evaporated and eventually completely dried out. This occurs even if a loose lid is used to prevent contaminants (Grainger 2010:55). The fluid that is generated has to either be entirely prevented from evaporating with an airtight lid, as in the Gargilius recipe, or the loss has to be constantly replaced to maintain the desired volume of fish sauce. We have seen that in modern fish sauce production, when processing occurs in the open, there is a technique whereby the loss of volume is replaced on a regular basis. Driard's experiments were designed to replicate the residues identified on the fish processing tanks at Plomarc'h in Douarnenez, Brittany. The initial analysis identified various different types of residue that were classified by color, opacity, texture and depth, as well as the position of the residue on the surface of the vat (two thirds of the way up) and the roughness of the surface beneath them. These distinct residues had only been noticed in Brittany before but they have recently been discovered on the walls of the vats from Baelo Claudia (Bernal-Casasola *et al.* 2016), and would seem to be characteristic of a long-term use of the vats for the production of fish sauce without intensive cleaning. Driard's experimental sauces were made with sardine in plastic buckets with a lining of plaster so that the residues could be identified on the inner surface of the bucket. Driard makes no mention of using extra viscera. Sauces were also made in fish tanks. Driard used a ratio that was very high, 2:1 fish to salt, which resulted in a large percentage of the salt remaining in crystal form, and it produced a thick layer of salt at the bottom and a solid raft on the surface (Driard 2102:54). Subsequently, Driard also began to add extra water, as he had noted that evaporation was constant. Even allowing for this extra fluid (0.5 liters per week), the undissolved salt remained until the end of the experiment, estimated at 133 days. I do not think that the salt at the bottom and the crust on the top would have been present in an ancient fish tank, as the salt level would have been considerably lower and there would have been sufficient water to dissolve it all. The primary aim of the experiment was to encourage the formation of the different kinds of residues that adhere to the sides of the vats, and the hypotheses which Driard finally puts forward are very interesting. However, he found that the salt raft actually encouraged the formation of residues, which though an interesting point, is less significant, given that the raft of salt was unlikely to have been present. The sauces that Driard eventually produced were quite dense, with a high specific gravity, as not only did the salt raft float but many of the bones also floated on the surface when the raft was removed. This density is a phenomenon that I have noticed with my own experiments and will discuss more fully on page 161. Once the raft and bones had been removed, using a basic sieve, the sauce was transferred into another vessel to settle out. Driard noted that the fish paste residue, which he correctly identified as *allec* and which is formed of undissolved particles of flesh, also floated, and it allowed the particular type A residue (light brown, rough, thick and opaque) to be deposited close to the top of the bucket. The clear brown sauce beneath the layer of fish paste was also able to form a residue, identified as type B (dark brown translucent thin and smooth) on the lower area of the tank. In the archaeology the combination of residue B deposited on the top of residue A has also been noted, where it was assumed that subsequently sauces were transferred from one vat to another. In concluding, Driard considered that the excessive salt he used, far in excess of what was normal in modern recipe and ancient recipes, was necessary because sauces in Plomarc'h were made so much further north in a colder climate, and that this would have increased the time required to dissolve the fish. This is a false argument as it is not the salt that contributes to the dissolving process but the digestive enzymes (which are actually impeded by too much salt) along with temperature. There is also a belief that sauces made in Brittany would have required some kind of extra help, in the form of fruit wine additives (Garnier 2018:322; discussed page 125) to become hydrolyzed because the temperature does not reach the levels found in the southern Mediterranean where the majority of fish sauce was made. I have found that though slower, it was possible to make a fish sauce in the average summer temperatures of Britany of 20°C to 22°C.

Roman fish sauce: an experiment in archaeology[3]

My own experiments aimed to reproduce the bulk and exposed *Geoponica* recipes and observe the liquefaction and residue formation. My subsequent aim was to compare the generated fish bone residues from these experiments with those found associated with amphora and ship wrecks and those from processing sites, in order to contribute to the debate as to which product, solid salted fish or liquefied sauce, these bones represent. Fish sauce is a very simple thing at heart and the most effective approach was simply to buy various kinds of fish some small and some larger, some salt and some large fish tanks (to make the process visible!) and with the *Geoponica* to hand, begin to make fish sauce and observe the liquefaction process (Grainger 2012, 2013).

Seven basic experiments were conducted in order to understand what happens to fish under a number of different conditions. Salt levels were calculated initially using a conservative 17.5% w/w. while later in the experiment I began to add brine made up at 15% v/v.[4]

The experiments were as follows:

1 Small sardine (*Sardine pilchardus*) without the additional viscera.
   The mature sardines (10 kg = 108 individual fish) were between 8 and 24 cm, the majority were over 10 cm and a third were over 15 cm in length. They were layered in the salt at 17.5% w/w in two tanks.

2 Whitebait (*Sprattus sprattus, Clupea harengus* sprat and herring) with no additional viscera. The fish were 5 to 10 cm in length with 2 cm body depth and with an average weight of 4 to 5 g. A total of 8 kg (c. 1,940 individual fish) were salted at 17.5% w/w in two tanks.

3 Sardine with the additional viscera. 3 kg of un-eviscerated fish and 17.5% w/w with 170 gm of mackerel viscera.

4 Mackerel of 15 to 24 cm (*Scomber Japonicus*), freshly caught in the Solent, individually cut open to ensure the viscera were exposed to the salt.

5 A salted mackerel layered in 2 fingers of salt to monitor brine yield.

6 A blood viscera garum using viscera and blood from c. 50 large mackerel.

7 An artificially heated mackerel sauce. Temperature between 35°C and the minimum of 22°C.

Temperature data

This was monitored and corresponded to levels of direct sun from 9 am to 3 pm and a max air temperature of above 40°C+ at midday. On cloudy days the greenhouse maintained an air temperature of at least 25°C at midday. There were considerable levels of naturally fluctuating heat, reaching an absolute maximum of 58°C with a mean high of 31°C and a mean low of 10.8°C overnight. These levels do not quite match the heat levels found in Southern Spain, particularly the night-time temperatures, which rarely drop below 15°C overnight from June to September, though the daily highs do not go above 35°C, a figure readily matched by the greenhouse temperatures.[5] The temperature of the sauces was taken daily during all the experiments and they rarely rose above 25°C even on the hottest days. The air temperature for traditional fish sauce production in Southeast Asia is between 20°C to 35°C, while modern techniques use up to 40°C for optimum solubilization (Klomklao *et al.* 2006b:441; Owens and Mendoza 1985:274).

The results and discussion

From the information gained from modern fish sauce production and the initial observations made during these experiments it is clear that there are a number of key issues which are fundamental to determining the processes that occurred in the ancient world. The salinity has been dealt with sufficiently already see above. Discussion on experiment 5, the blood viscera sauce, will be postponed until the essential substance that is *liquamen* has been fully elaborated. The key issues are these:

1 Initial brine yield;
2 What actually happens when the fish turn into sauce;
3 Presence and absence of additional liquid;

 4   The speed of initial breakdown;
 5   Presence and absence of additional viscera;
 6   The size of the fish and the effect on liquefaction;
 7   The formation, separation and removal of the emulsion;
 8   *Allec:* as it appears in experimentation.
 9   The bone residue itself
10   Experiments with the blood viscera *garum.*
11   The nature of the finished sauces: protein analysis.

1. *Initial brine yield*

The initial brine yield generated from all the experiments was as low as predicted from previous studies. None of the experiments generated enough brine to cover the fish without some form of compression with a weighted mat, and in most cases extra brine was necessary, either initially or after the first evaporation. Experiment 5, using salted cleaned mackerel, yielded the lowest levels of brine: the majority of the salt stayed in crystal form, as there was insufficient liquid from the flesh for the salt to dissolve. Subsequently I learned that had I scaled the fish it would have contributed to salt penetration and increased brine yield. The sprat and sardine sauces (1, 2, and 3) without extra viscera generated sufficient brine to cover the fish when compressed, and this may have been because many of the sardines had already burst open, allowing the salt to penetrate. This may be due to the seasonal nature of the catch: just before spawning, the fish are particularly fragile. The sprat generated a surprising amount of brine, particularly as modern whitebait is normally lightly salted on board ship before being frozen. This might well result in loss of some natural water content.

2. *What actually happens when the fish turn into sauce*

The ideal conditions to ensure efficient liquefaction require sufficient enzyme activity from indigenous and additional viscera, an air temperature of c. 30°C to 40°C, salinity over 15% w/w, and a maintenance of the initial volume of brine. In these conditions, the fish begin to soften in the brine. The skin sloughs off in small patches, and the flesh beneath whitens and beings to shed itself in very small particles into the brine. As more of the flesh is shed into the liquid, the brine becomes thicker and thicker with this grey/brown fish paste material. The sauces form in slightly different ways, depending on the size of fish and the density of the flesh (for discussion of the individual sauces see below). The bones as they shed their flesh sink to near the bottom and largely remain in articulation. As the sauce progresses and more flesh is shed there is a separation of liquid and particles into layers (illustrated in Fig 8.1).

Figure 8.1 The brine that formed within a tanks of whitebate *liquamen* with the fish paste forming around it. Copyright: Sally Grainger

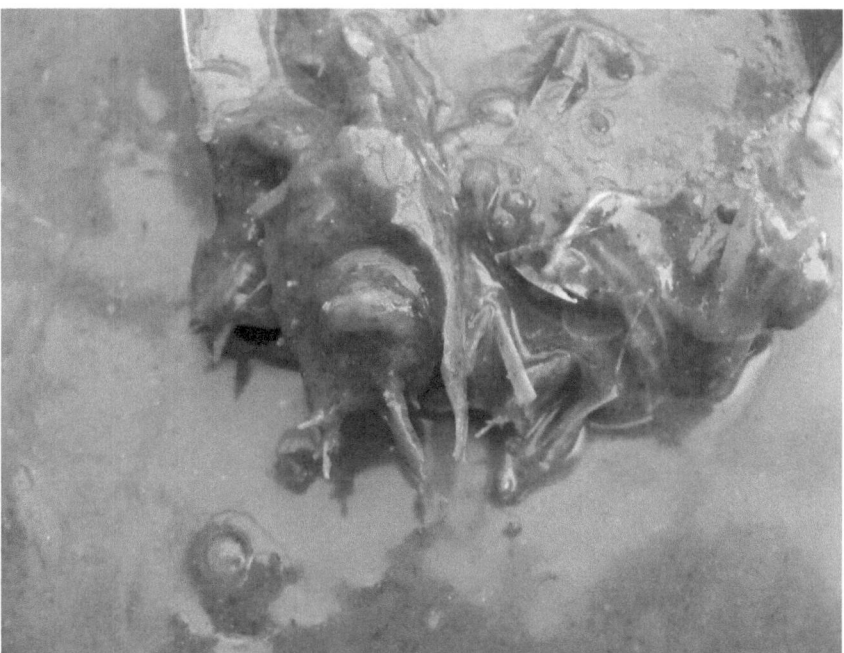

Figure 8.2 Sardine bones cleaned of flesh through enzyme hydrolyzation Copyright: Sally Grainger

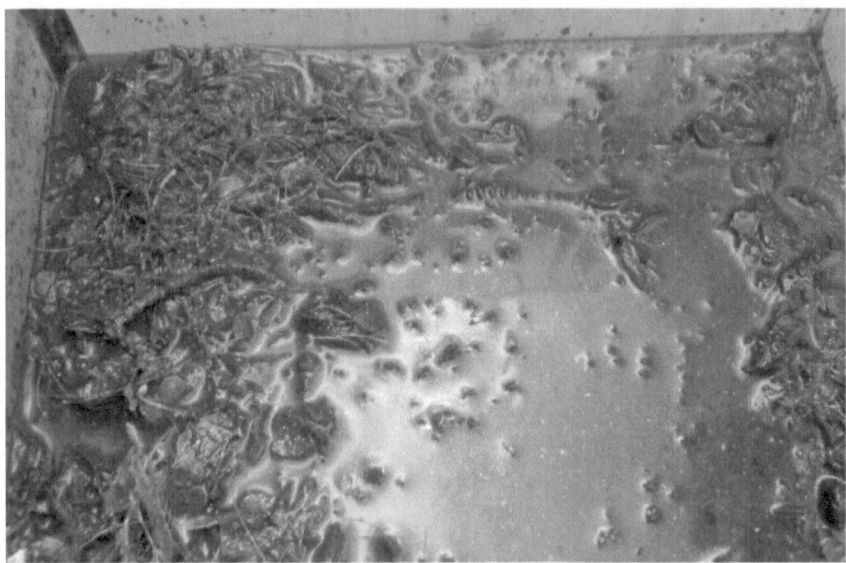

Figure 8.3   A failed sardine sauce demonstrating disarticulated bone and a thin pungent fish paste. Copyright: Sally Grainger

Figure 8.4   The thick creamy sauce from mackerel ready to take. Copyright: Sally Grainger

Figure 8.5   The basket used to filter the sauce from the *allec*. Copyright: Sally Grainger

Figure 8.6   A *liquamen* sauce demonstrating the separation of liquid from the *allec* paste. Copyright: Sally Grainger

Figure 8.7   Filtering a mackerel sauce to remove the *allec*

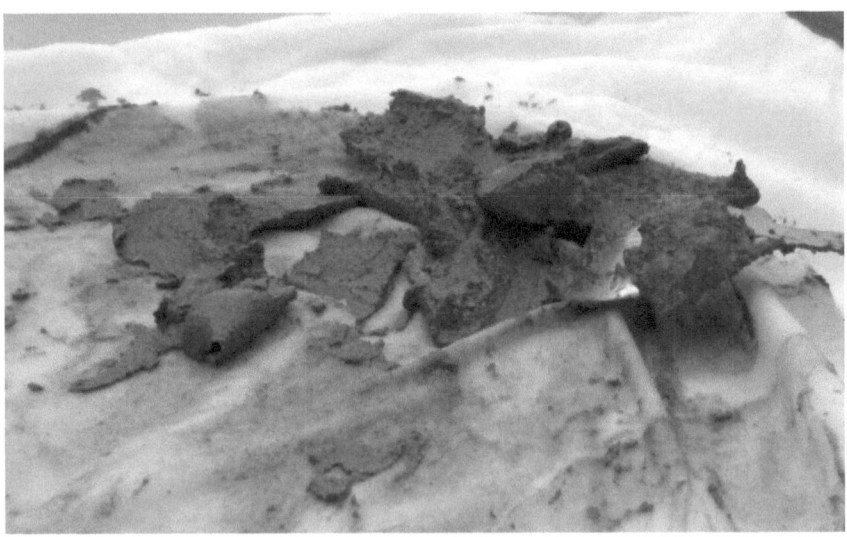

Figure 8.8   The fully filtered *allec* fish paste

3. *Presence and absence of additional liquid*

The consequences of the lack of natural fluid are quite considerable. Without sufficient fluid, the sauce simply cannot form. A limited volume of fluid would also make stirring the product very difficult in bulk production. An additional influence on the volume of sauce generated is the effect of evaporation, as noted. The loss of volume due to evaporation is varied and probably dependent on fluctuating temperature. In my experiments, the sauces lost anything from 10% to 50% of their volume (Grainger 2010:57). This loss of volume resulted in the sauces becoming more concentrated, which slowed down the liquefaction process, and in some cases this ceased entirely. The sardine sauce (3) was very thick and viscous after losing 30% of its brine and there were many pieces of soft fleshed fish within the tanks, and over a period of 1 week these pieces remained the same size. This was after a period of very active disintegration. The suspension of activity seemed unusual and it was conjectured that the water content of the sauce was effectively saturated on a macro scale with the particles of fish flesh, and also that the brine itself was saturated with salt and dissolved protein. The liquid simply could not encompass any more free-floating particles or dissolved protein, and so the fish pieces remained solid, effectively in stasis. Of the two detailed *Geoponica* recipes, the one for the Bithynian sauce states that for every *sextarius* of fish, two *sextarii* of wine are added. Garnier *et al.* has confirmed this with the detection of "malic and tartaric acids characteristic of black grape" (2018:322), which have also been detected alongside fish markers in amphorae from the *Garum* Shop at Pompeii (Pecci *et al.* 2018). Given the amount of volume lost to evaporation, the wine might have two roles, to replace the lost volume and also to add flavor and acidity, which would act just as effectively in reducing unwanted bacterial action as a brine, and would ultimately dilute the sauce further to the final desired consistency. This ratio may in fact be a good indicator of the amount of fish sauce that can be expected from a given amount of very thick fish mash: that is, when we come to relate this to amphorae, there is a very high possibility that one could place a 1:2 mixture of a thick fish mash and liquid (brine or wine) in a vessel, and the result would be the optimal utilization of the mash, to generate a standard fish sauce of the *liquamen* type. It was subsequently decided that only the lost volume should be restored by adding wine in the case of the sardine sauce (1), and in the case of the other recipes, brine of 15% v/v, added when required in an ongoing process.

4. *The speed of initial breakdown*

The speed at which the fish liquefy proved to be the most important factor in determining the form and quality of ancient fish sauces. There were considerable differences between the various types of sauce in this respect. The sprat sauce (2) turned into a mash of indistinct fish particles in a matter of days. Overnight, the vast majority of the bodies had burst and after 48 hours, the majority had lost their integrity and merged into a paste. The repeated stirring, which occurred twice before the fish were considered completely broken up, was instrumental is allowing this to happen. These smaller sardines also shed their meat from the bone readily and many of the bones became visibly cleaned of their flesh within 7 days (see Fig. 8.2). The larger sardine and mackerel (1, 4) took considerably longer to disintegrate, and the rate of disintegration was largely based on size and the rate of enzyme activity, which was in turn dependent on the presence of extra viscera and/or exposed viscera (see below).

5. Presence and absence of additional viscera

Experiment 3 used sardines with extra viscera and demonstrated a rapid and effective liquefication of the flesh. The viscera of four mackerel 30 cm in length (270 gm) were placed in the tank along with the whole fish, and the sardine had begun to whiten, soften and shed its flesh quite rapidly compared to smaller fish without additional viscera. Many of the smallest in use (i.e., 12 cm and below) had effectively disappeared within 2 weeks. This was maintained as long as the volume due to evaporation was estimated and replaced regularly. It appears that the extra viscera (only 10% of the volume) gave the dissolving fish a boost of enzyme activity at the beginning, which caused liquefaction to accelerate. This conforms to the modern fish sauce analysis of Klomklao *et al.* (2006:444). The speed in which this batch of sardine liquefied was remarkable, and the additional viscera was therefore identified as very significant. However, it was also apparent that the speed and efficiency of the liquefaction was as a result of both the extra viscera and a balance of water content sufficient to allow full liquefaction. The 10 mackerel used for experiment 4 (viscera exposed and freshly caught at sea) were covered with a quantity of sea water (c. 4.5 liters: the minimum to cover the fish) and an estimated volume of salt was also added at sea. This proved to be an inadequate estimation of salt levels as the sauce subsequently failed. A powerfully unpleasant smell entirely different to the other sauces, suggested that the sauce was contaminated with too many of the wrong kind of bacteria. It demonstrated that adding too much liquid of an unpredictable salinity, that is, sea water, rather than a concentrated brine, was potentially disadvantageous to a successful sauce. It was noted that with this level of bacterial contamination, the collagen was also digested and this resulted in fully disarticulated bone residue, illustrated fig 8.3.

6. The size of fish and the rate of liquefaction

When the residues from a sauce that was made with a sardine of a wide size range and with extra viscera, was assessed (3), it was found that the liquefaction rate was quite varied. The sardines ranged in size from between 8 and 24 cm, the majority were over 10 cm and a third were over 15 cm in length. After 2.5 months, when this sardine sauce was considered ready to take (i.e., there was an abundance of creamy liquid), it was apparent that a moderate amount of the fish flesh was still intact. It became apparent from later analysis that the largest fish, c. 18 to 24 cm were the ones that had the most flesh still adhering to the spine, while those under 10 cm were all fully liquefied. Those in the middle range were in various stages of disintegration: the larger the fish, the more flesh was still present. It is apparent from the *cetaria* evidence that fish bones from multiple species with a wide size range of 3 to 20 cm were all mixed up in the salting tanks associated with the Portuguese fish sauce processing sites and at many other facilities around the Mediterranean (Desse-Berset and Desse 2000:86;90). The volume of fish envisaged (600 m³) in the case of the Troia vats, precludes the idea of the upper size range being individually cut to expose the viscera, and the implication is that in order to ensure rapid and efficient hydrolyzation of this diverse size range of fish, both extra viscera, and or extra *allec* and extra brine was essential. Fish of larger size used to manufacture a bulk sauce such as the mackerel envisaged in the Gargilius recipe would have had to be cut up to expose the viscera to the salt, and this is confirmed by the fish processing description in Manilius (see page 37). It is my contention that the reference to the exposed viscera in this recipe means that additional viscera is not necessary, but that additional fluid is essential for every form of bulk production to ensure all the fish liquefy. The fact that the larger fish did not fully dissolve while at the same stage there was more than sufficient sauce that could have been removed demonstrates the utility of rebrining these residues.

7. The formation, separation, and removal of the emulsion

It was apparent with each sauce that as the flesh of the fish was loosened and shed into the liquid by the action of the enzymes, it formed a creamy emulsion of small particles of fish meat in suspension. The way in which the particles behaved was largely dependent on the volume and density of the liquid that they were suspended in. After the additional liquid was added to the sprat mixture, the sauce began to form in distinct layers, and it was possible to see a small amount of clear brine on the surface; layers were also distinguishable in the middle and at the bottom. The bones and small pieces of fish in the sprat sauce, being very light in weight, also remained in suspension for a considerable time illustrated Fig. 8.1). After

60 days the majority of the fish particles had settled on the bottom with the bone, while much of the fluid had risen to the top. The sardine sauce behaved in a similar fashion, resulting in a distinct separation of liquid from the particles of flesh and bone. An attempt was made to extract the pure clear liquid but it proved impossible to extract only the clear liquid by any process, as the very act of attempting to take it disturbed the particles and they remerged with the liquid. It would be impossible on a large scale to remove the surface brine and (as conjectured and it later transpired) detrimental to a sauce of quality. Over time, as the sardine sauce was continuously stirred, a strange reversal of these layers occurred. The embryonic clear pale brown sauce suddenly did not return to the surface, but remained below a layer of fish paste. It was noted that, despite the fact that it was only 2.5 months in production, there was a substantial amount of sauce available to harvest, and I decided to take this and review the residues, illustrated Fig. 8.4. In the *Geoponica* recipe, the sauce is taken using a basket and I was able to place a small plastic bread basket in the tank which allowed the emulsion to flow inside and a ladle was used to extract the sauce. The sauce settled in a fresh tank with an even more pronounced layer of fish paste on the surface and the brown clear sauce beneath, illustrated Fig. 8.6. A useful advantage at this point was that the fish paste layer reduced evaporation considerably.

The sprat sauce which had continued to settle out with the brine on the surface after stirring was fully dissolved at 80 days and was taken from the bone residue and similarly placed in a fish tank and allowed to stand. As this sauce settled out in the tank, the fish paste rose to the surface and the brown sauce settled beneath. It was not possible to determine nutrition levels before and after the removal; however, the process of removal and the agitation that was necessary encouraged and promoted liquefaction of the muscle tissue and increased the level of dissolved protein in the sauce; as a result, the density of the liquid increased to a level whereby the fish paste was able to be suspended above the liquid. Thus, the density of the clear sauce is fundamental to understanding what happens to the fish paste residues, where it forms within the liquid, and how accessible it is. Once this unfiltered sauce is free of its bone as a suspension of fish pastes and liquid, it would seem that the sauce can continue to mature. This is fundamental to an understanding of how fish sauce was made. The unfiltered sauce does not need to be left in the processing tank to ferment for the full 3 months. It is clear that once sufficient sauce has formed, it can be removed from the bone residue and transferred to an amphora, where the *allec* can continue to enrich the liquid over the period of its trade and commercial distribution. We should note that modern fish sauces are left for a period of a least a year and up to 18 months, at which time we are told there is no fish paste residue left floating on the surface, as it has presumably all dissolved. Ancient fish sauces were clearly made in far less time, and as a result a considerable amount of fish paste was available. This form of *allec* is naturally bone free and as it floats it was easily harvested as a paste.

Experiments 3 and 4, using sardine with extra viscera and mackerel that had been cut open, had considerably more extra fluid added and consequently the sauces behaved a little differently. As the sauces formed, they settled out with a distinct layer of particles on the surface and another layer on the bottom that also included the bones with the brown sauce forming in the middle. This was difficult to interpret. The sauces from experiments 1 and 2, manufactured initially without sufficient extra fluid or extra viscera, settled out after 10 weeks with all the fish residues on the bottom, and these subsequently moved to the surface, which implies that the enzyme action was both slow and inefficient initially, but later improved. It may be conjectured that the floating particles, having been digested more efficiently, are lighter relative to the density of the sauce and so they float, while those at the bottom retain more solid protein, are heavier relative to the liquid and so remain at the bottom. Stirring would logically promote continuing solubilization and result in the sauce becoming enriched, increase density and allow the remaining *allec* to rise to the surface. This implies that an efficiently made sauce, one that has sufficient liquid and extra viscera, will also settle out with more of the particles of fish flesh, the residue or *allec*, on the surface of the sauce and not at the bottom with the bones. From this we can see that there is a distinct separation of the three components: bones, paste and liquid which would clearly facilitate harvesting.

8. *Allec: as it appears in experimentation*

The first sardine sauce, using 10 kg of fish, was taken after 80 days and yielded 7.5 liters of sauce. It stood for a further 30 days, at which point it was filtered through a couple of sheets of muslin, and the result was 25% semi-solid fish paste (= *allec*) and 75% liquid sauce (= *liquamen*). The resulting fish paste amounted to 1.8 kg of a reddish-brown thick smooth paste. It is important to note at this stage that this form of *allec,* in these circumstances, that is, having been removed together with the liquid sauce, appears to be inherently "bone-free" when derived from the fish sauce process. The filtering process and the resulting fish paste are illustrated Figs. 8.7 and 8.8. It tasted pleasantly fishy, like an anchovy paste, and spread thinly on toast was quite appetizing. The results of these experiments have demonstrated unequivocally that *allec* was readily able to be harvested and retailed as a bone-free fish paste, and it almost certainly had to be so to be of any value commercially as a food product. These considerations inform the debate about the nature of *allec* and will be consolidated into a final discussion in Chapter 11 on fish bones.

In my experiments, the fully filtered solid fish paste harvested from the sardine fish sauce was subsequently rebrined with an equal quantity of 15% brine by v/v and allowed to settle and generate a second sauce over a further 2 months. The concept of rebrining is well documented in Southeast Asia, and while we cannot be sure that the ancient did it, it is very likely, given the amount of wasted fish protein that would have ensued. Pliny's source implies that the residue was valuable when it was not "used up" (*Pliny, HN. 31.95*.1), and it may be possible to see this reflected in the *allec* rich in undissolved protein.[6] All the sauces manufactured in the experiments were tested for nutritional content. Data is cited below. It was quite startling to note that this second sardine sauce actually exceeded the protein levels of the first sauce from which it was taken. This result has some quite important consequences. Firstly, had I allowed the initial yield of sauce to remain unfiltered for an extended time, the first sauce would have increased in protein and quality. Clearly a sauce allowed to stay unfiltered for an extended time would have been superior to one filtered early. The process of enzyme hydrolyzation is understood as a single process, but from these observations it may be better, empirically, to understood it as a two-stage process. Under the enzyme action, the fish muscle tissue "disintegrates" into particles, which can more readily "dissolve." The application of heat (see below) and agitation contribute to an increase in this process. The enzyme action is necessary to break up the fish into particles, but the heat and agitation increases protein yield. These secondary processors can clearly continue for considerably longer then the estimated time that active enzymes are detectable in fish sauce production, that is, 160 days (Klomklao *et al.* 2006). We have already noted that from 30% and up to 50% of the potential nutrition is left in the residue from the finest fish sauces of South East Asia, and this is the case even after an extended fermentation of 12 months in the case of Red Boat fish sauce (Quam Pham pers. comm.)

9. *The bone residues itself*

The question therefore remains as to what happened to the bone and remaining fish paste residue at the bottom of the tank or vessel. The remaining bone residue of the sardine sauce consisted of a considerable number of solid pieces of fish with the flesh very soft but still intact (see Fig 8.5). It therefore represented a substantial quantity of sauce still to be generated. The pieces of fish and bones were surrounded by a thick viscous sauce. In the experiment using sardine and extra viscera these fish bone residues were rebrined and a further second sauce generated over a period of 2 months. After this second fermentation had produced a sauce the remaining bones were largely disarticulated, clean of flesh and surrounded by a relatively weak liquid which I would conjecture was unable to generate any further sauce. The *allec* was effectively used up and exhausted and suitable only for disposal. The final tally was three batches of fish sauce from one batch of fish if that batch was of multiple sizes, that is, 3 to 20 cm. One initial harvest, of sauce which contained a thick layer of *allec* on the surface which could either be left to enrich the liquid until saturated and/ or be removed and rebrined to generate a food or a second sauce and one further sauce when the remaining *allec* is rebrined. One may conjecture at this point that when small vessels such as *doliola* were used, the rebrining took place continuously and one can imagine a continuous harvesting of sauce not unlike the traditional South East Asian types made in small ceramic jars, (see page 132). Here the housewife took the liquid generated and replaced the lost volume so that the rest of the fish could dissolve. We note that without the presence of a water-based liquid the process of hydrolyzation cannot occur. In these circumstances, rebrining would not create a *second quality* sauce but many, *first* quality sauces! For the repercussions of these conclusions in terms of the residues of fish sauce in the archaeology see Chapter 10 for the infrastructure of fish sauce production, Chapter 11 for fish bones in the archaeology and Chapter 12 for amphora use.

Experiments with the blood viscera garum

The processes involved in trying to make a blood viscera sauce were complex, as there were many logistical difficulties in acquiring such material. I approached and was subsequently able to accompany local amateur fishermen who traversed the Solent between the mainland of Portsmouth and the Isle of Wight. They caught mackerel regularly and I was able to observe the techniques they used for bleeding. The gills were partly pulled out and the fish was then placed in a bucket of sea water to drain which prevents coagulation. Only later were they gutted, at the end of the day's fishing. The separation of bleeding and gutting was interesting, and I wondered whether a similar situation may have occurred in the past. A special butchering technique has been detected in ancient residues employed to facilitate bleeding, which is also commonly used today in Spain and is discussed on page 174 (Desse-Berset and Desse 1993, 2000) In this technique, the front of the cranium of mackerel (the nose in effect) is cut off, which allows rapid bleeding, and this happens directly they are caught in the modern examples, so that the blood can flow from the head. Mackerel is nevertheless a particularly bloody fish; it contains considerably more blood then most fish of a similar size. Despite this, when I bled my first fish, a 30 cm mackerel, it yielded no more than a few table spoons of blood. As for the viscera during experiments to salt mackerel caught in the Solent, 12 fish of a size range of 35 to 38 cm and a weight range of 400 to 450 g (total weight 5 kg) yielded 610 g of viscera, giving approximately 120 g per fish, about 20% of the total weight of each. The volume of blood available from these fish was harder to estimate as the harvesting process was inadequate, but it did not exceed 10 ml per fish (Grainger 2010:121).

My now friendly amateur fishermen eventually agreed to bleed and eviscerate all the mackerel they caught over an entire day and keep it for me. This worked out at about 50 fish of varying sizes between 20 and 38 cm. I did not feel able to ask them to cut the cranium of all the mackerel while alive, as they were reluctant, as was I. I eventually acquired 1.5 liters of blood and viscera which I salted that same day at 15% and left in an airtight Kilner jar in the greenhouse alongside the other fish sauces for the allotted 2 months. Overnight the mixture settled out with the blood towards the bottom of the jar, while all the viscera floated at the top. After shaking the jar, the same layered structure returned within a few hours. The jar is illustrated Fig. 8.9. I took this to indicate that the bloody fluid was quite dense, a phenomenon also noted in the experiments producing *liquamen*. The blood of fishes is similar to that of any other vertebrate. It consists of plasma and cellular components such as red and white blood cells and thrombocytes. Fish blood contains proteins (fibrinogen, globulin, albumin). Other nutrients include glucose, lipids, and amino acids. Just as quadrupeds' blood is valued in such foods as sausages, fish blood is essentially pure nutrition and already liquefied, provided that coagulation has not taken place. This would explain the high density and the fact that the residues float right from the beginning, while the liquamen-type sauce initially does not display this tendency until enough protein has been dissolved and absorbed into the liquid. Over the 2 months I regularly shook the jar, but in each case the sauce returned to the same layered effect, with bands of clear, dark sauce and opaque fluid at the bottom and the increasingly liquefied viscera on the surface. This material was transformed over the 2 months into a thick red/grey sludge. On the surface a layer of dark oil also formed. The process is therefore initially similar to *liquamen*, in that the enzyme action dissolves the viscera. Thereafter, the harvesting techniques appear to be different. From the recipes we learn that it has to be harvested when the vessel is pierced so that the sauce can flow out, from beneath. This seems a particularly efficient means of extracting the desirable bloody liquid while cleanly containing the unwanted material. As my vessel was glass, I was unable to piece the fabric and drain out the liquid. I poured the mash through a metal sieve and the resulting sauce was dark and bloody but also very cloudy and viscous. This fluid continued to separate out into a layered structure, though now with very fine grey particles settled on the surface in a layer representing a third of the volume. It was eventually possible to produce a sparkling clear black sauce after passing it repeatedly through many layers of fine muslin, but this was a slow process as any attempt to speed it up resulted in a cloudy sauce. The ancient method of piercing the vessel with a fine hole may have allowed the sauce to emerge free of most of the cloudiness. There are many questions that do remain unanswered. When the fermentation was complete, was the sauce taken from the tank without the visceral mash and transported as a cloudy sauce, or was the

Figure 8.9 Kilner jar of *garum* in production with the thick layer of paste substance on the top, a thin layer of oil and the dark bloody liquid forming beneath

visceral mash placed in the amphora and the fermentation continued in situ. Some fish amphorae have holes drilled in the spike and this may correspond to a process of filtering from the amphora (page 201). We will discuss the evidence of special *cetariae* with lead piping and drainage channels at the base which would seem particularly suitable for the manufacture of this kind of sauce in Chapter 10. We can be sure that the *garum* residue resembled something quite disgusting, thick and slimy, and included pieces of visceral material which, though they had some structural integrity, could not be identified. This was not used a second time, but discarded. A fellow fish sauce enthusiasts and colleague Alastair Bland has recently manufactured a inauthentic form of "*garum*' using mackerel viscera and the defilleted carcass, (with residual but no harvested blood and minimal flesh). The *allec*, after 6 months was bitter and unpleasant and did not generate a paste of culinary value (Alastair Bland pers. comm.). The idea of a second quality *garum*, if the term refers to rebrining, would appear to be unlikely. The blood was liquid when harvested and formed the bulk of the liquid which subsequently became sauce, taken as a first yield: there was nothing left to rebrine. It is self-evident that attempting to make a true *garum* today using tuna blood and viscera would be impossible in northern Europe. Even in southern Europe the welfare issues would surely preclude it. However, there is a much greater volume of blood and viscera obtainable from tuna and consequently this form of blood sauce would be far easier to make in bulk, and one can envisage that tuna began to be used to make the blood viscera sauce precisely because of this higher volume.

10. *The nature of the finished sauce: protein analysis of sauces*[7]

The analysis of the sauces manufactured in the experiments revealed some surprising results. It was possible to test and compare many different modern and experimental sauces to access the nutritional quality and therefore the potential value that the purchaser and consumer would place on these sauces. Protein levels are directly connected to taste, and therefore they are a direct indicator of quality. The results are shown below in graph 8.1.

Graph 8.1 Protein concentration (mg/ml) for sauces made in 2009.

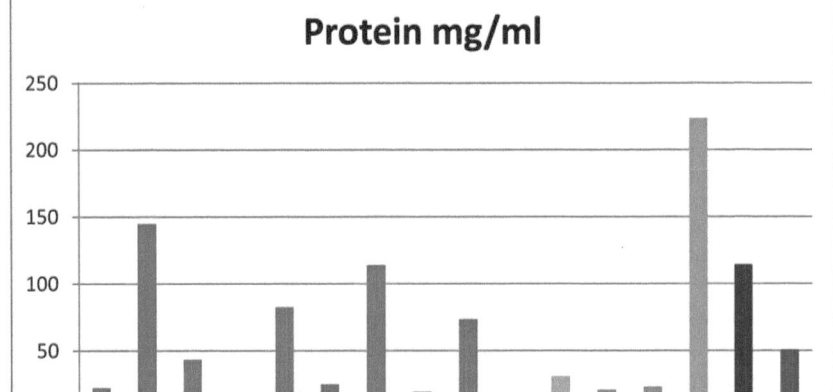

A1: 1st sardine; A1ℓ: 1st sardine residue; A1ℓ+: 1st sardine residue with 15% brine for 2 months; A2: 2nd sardine; A2ℓ: 2nd sardine residue; B1: 1st sprat; B1ℓ: 1st sprat residue; B2; 2nd sprat; B2ℓ: 2nd sprat residue; C: salted mackerel brine (*muria*); D: eviscerated mackerel; E: 3yr old garum (made in 2006); F: new garum; G: squid blood sauce; H: artificially heated mackerel sauce; I: modern fish sauce ("Squid Brand").

The results reveal an extremely wide range of protein concentrations which are specifically related to particular modes of manufacture. The "Squid Brand" fish sauce (I), which is readily available in the UK, has 50 mg/ml protein, which is an indication of the quality now expected from modern fish sauce, but it is not clear how readily such levels of protein can be detected through taste, especially as modern fish sauce is so salty.

Samples A1ℓ, A2ℓ, and B1ℓ; B2ℓ were the filtered residues from the sardine and sprat experiments 1 and 2. The protein levels are naturally high in this paste as it is mainly muscle tissue. This suggests that there was substantial nutritional value to these residues, which we can see were consumed by many different groups in ancient times, as well as just slaves and less elite groups. The high protein levels in the residue and the relatively low levels found in most of the experimental sauces are very likely due to the fact that these sauces were fully filtered to early, that is, the *allec* fish paste was removed shortly after extraction. A sample of this residue filtered from the first sardine sauce was blended with 15% brine at a ratio of 1:3 and left to stand for 2 months, and then this sauce was also tested. The resulting level of protein, 43 mg/ml, is considerably higher than that of the original filtered sardine sauce from which the *allec* was derived and which was, in comparison, quite weak at 22 mg/ml. These results suggest that a stored emulsion increases its protein levels considerably. This result (sample A1ℓ+) reveals that enzyme activity and protein hydrolyzation continued (albeit at a slower pace) for a considerable time and in fact, according to Klomklao *et al.* (2006:443), there is still enzyme activity after 180 days. These results do confirm that the longer a sauce was left unfiltered the richer it became. It may also be conjectured that there would be a protein saturation point, given that modern South East Asian sauces are only able to extract up to c. 50% of the potential nutrition for a batch of fish. This might suggest that the remaining *allec* would still retain within it the potential for another good quality *liquamen* sauce. The consequences of this phenomenon are immense in terms of the nature of *allec* and its role in the economy and in trade. It can be used as a concentrated fish sauce in paste form than the cook could have rebrined at a time that is convenient in the kitchen. If this was the case, then the role that *allec* had as a food item simply for consumption as paste may be diminished.

A subsequent experiment (H) was devised to attempt to dissolve a small amount of mackerel *liquamen* using gentle artificial heat. This was applied underneath the vessel, maintaining the temperature of the liquid to a minimum of 20°C, and reaching a maximum of 30°C to 35°C over brief periods during the day. This resulted in exceptional nutritional levels at 114 mg/ml of protein. This result indicated that gentle heat levels applied to sauces while in production increases nutrition, but it may also suggest that heat could have been used to extract as much nutrient as possible when reconstituting the *allec*. Modern fish sauce residues have hot brine added to ensure everything is extracted, but they are still not brought to the boil or cooked, as this is considered undesirable (noted above).

I was able to test the theory that gentle heat could be used to reconstitute an *allec* and make new *liquamen* in a more recent experiment using sardines to make a sauce in 2019. A batch of fully filtered *allec* weighing 500 g was mixed with 250 ml of hot brine at 15% and held at c. 30°C to 35°C for about an hour. It was then filtered again, and the resulting *allec* reweighed. It had lost 70% of its weight and volume. The *allec* had clearly dissolved into the brine, making a perfectly acceptable *liquamen* sauce. The processing sites included simple furnaces structures and perhaps also hypocausts which are normally used for underfloor heating systems, could have been used to raise the air temperature of the storage rooms where the filled amphorae were placed so that the sauces would continue to be enriched. However, none of the structures suggest that cooking per se took place at the same time as the fish sauce was fermented in the *cetariae*.

The sauce with the highest protein levels was a modern fermented squid blood sauce (G, *Ishiri*, page 141). Though blood is very high in protein, the difference is quality between this and the experimental *garum* sauces was considerable (samples E and F). However, it is possible, with hindsight, to see that the experimental blood/viscera sauces which I made did not harvest enough blood from the fresh mackerel.

One might assume that ancient sauces were of lesser quality compared to the modern varieties given the time scale for producing them was so different. Ancient sauces were taken after just 3 months which is considerably shorter than the modern method which is at least 12 and often 18 months. The nutritional quality however depends on the consistency of the sauce at 3 months, whether there was undegraded fish flesh still to be dissolved, whether there was any fish paste still undissolved and whether the sauce was stored unfiltered in the amphora before sale. A *liquamen* sauce taken at 3 months but stored unfiltered and aged in warm conditions for even a few more months would have resulted in a sauce of equal if not greater quality to a modern fish sauce and particularly as the salt levels were so much lower.

A synthesis of the experimental approach

We have seen from our review of the modern and experimental approaches to fish sauce that there are a number of key issues that we can carry forward when interpreting the fish sauce production and residue formation evidence in the archaeology. These issues involve:

- The duration of initial fermentation.
  Modern recipes ferment for ±18 months. Ancient sauces appear to have been fermented for 3 months. The shorter time will result in considerable differences in disintegration and liquefaction but not necessarily quality;

- The amount of additional brine required.
  We have seen from more than two separate experiments that evaporation in the bulk process is constant and in order to maintain if not maximize the yield, additional fluid is required throughout the process. When the fish mash becomes too thick disintegration and liquefaction can slow and potentially stop.
- There certainly appears to be a time lag between "disintegration" of the fish and the subsequent "liquefaction" and solubilization of the muscle tissue.
  Given the shorter fermentation time it is more than likely that up to half or more of the fish utilized will not have disintegrated or yielded their nutrients into the liquid after 3 months. The degree to which this will occur depends on factors such as the size of species used and whether there is a mixture of multiple sizes all blended together or a distinct size range. The smaller the fish and the more consistent the size range, the more likely it is that they will disintegrate and liquefy simultaneously and leave a residue of lower quality for rebrining. The larger the fish or more diverse the size range, the more likely it is that the residue will contain fish with structural integrity, and of sufficient high quality that it will generate a second stage sauce of similar quality to the first.
- The first harvest of sauce and its consistency and quality.
  The sauce is taken as a creamy liquid, a suspension of particles of muscle tissue and brine. Over the period of transport and storage the sauce continues to enrich the liquid. When the particles of fish flesh are filtered from the liquid they produce a good quality bone-free fish paste, and this is potentially what we should consider *allec* to be when it is mentioned in consumption texts. The bone and undigested fish mash was a true *faex* or "residue," as Pliny states (*HN.* 31.94), and not a food item in itself.
- The number and quality of subsequent sauces taken from the residues.
  The number of sauces that can be taken from one batch of fish will largely depends on the range of sizes of fish and the range of species. This corresponds to the amount of flesh on each individual fish. It is also dependent on whether there is sufficient liquid in the form of brine present, as we have seen that the nutrition cannot be converted and absorbed into the liquid unless there is sufficient liquid in the first place. It is possible to envisage three sauces of equal quality being produced from one batch of fish, especially if the size range is diverse and the manufacturer is able to add the correct amount of liquid at the right time and judge the extraction process so that the three sauces have an equal percentage of the potential nutrition.
- The bone residue.
  The dried out "residue" of fish sauce in the form of disarticulated bones left inside the processing tanks or beside or inside an amphora

may represent the waste matter after the sauce has been poured off. It is unlikely to represent a product with any commercial value. Given these conclusions it is less likely that the residues reflect a products that has been subject to organoleptic degradation.

Notes

1 This is considered too high in modern fish sauce production as it gives a cooked taste. My own experiments measured both the air temperature and the liquid temperature, and while the air inside the greenhouse could be 40C+, the liquid inside the fish tanks did not reach above 25°C. This would suggest that in the processing tanks, particularly those below ground level, the temperature of the sauce were not likely to exceed 25°C.

2 As noted in chapter 6 the surface brine that appears when the fish are first salted to make sauces is an issue as it has been suggested that it could be removed as a form of *muria* (Varges *et al.* 204:71; Palacios *et al.* 2016:92). However, in their reported experiments they did not remove the brine, as they recognized that it would have reduced the yield

3 This section is based on my unpublished MA dissertation (Grainger 2010) with some additional material from later experiments.

4 The recipe indicates 15% but this is a minimum for safety. It later proved impossible to monitor the salt levels in the ensuing sauce in any precise way as the conventional means of measuring salinity such as the hydrometer are ineffective, as the dissolved protein effects density, and digital measuring devices are rarely calibrated to measure at this level of density. Access to laboratory analysis was always limited through my experiments.

5 The temperature data was compared with the daily temperature levels found in Gibraltar during 2009, which were accessed on line http://www.cpc. ncep.noaa.gov/products/global_monitoring/temperature/tn08495_30.gif' 04/07/10.

6 The concept of a second-quality sauce, which may or may not be connected to re- brining, is found in the Diocletian price edict where a *liquamen primum* and *secundum* are recorded (Lauffer 1971:3.6-7).

7 Protein levels and bacterial count (not discussed here) of the 2009 sauces were assayed in the laboratories of the Department of Food and Nutritional Sciences, University of Reading. The analyses were kindly conducted by Dr Kieran Tuohy and Dr Duncan Brown. The Lowry assay test involves reaction between copper ions and peptide bonds which turn the solution blue green, and this color spectrum can be compared with known protein levels (Duncan Brown, pers. com.). The Soil Sciences Department of the University of Reading established the pH readings.

9 Fishing in the Mediterranean

The origin of the techniques involved in the fermentation and or salting of fish to preserve it for later consumption, whether as sauce or solid product, are difficult to discern. Many claim that the origins were to be found in one particular culture, either that of Greeks in the Black Sea or Phoenicians in Spain, and that the skills spread around the Mediterranean as a new idea by culture transfer, from one to another.

> The indigenous populations of the western Mediterranean region undoubtedly practised fishing as a means of sustenance, but the techniques of processing fish into *salsamentum* and sauce for later consumption were most probably introduced by peoples from elsewhere in the Mediterranean basin.
>
> (Trakadas 2005:71)

It is possible to see that the industrial scale production process with all its inherent technical details will have involved culture transfer and could potentially be attributed to one particular culture. However, the various way in which fish can be processed are not binary: either industrial or subsistence, as there will have been many different systems where the preserving of fish was well organized, involving bulk catches and infrastructure but it was not archaeologically visible and is therefore not well-illustrated. The issue resolves around the scale of production and concerned not with the act of salting but with the technological aspects required to trade a bulk product for long term profit, which involved the provision of ceramics, processing tanks, shipping and labelling. Salting fish can be seen as an intuitive practice, which undoubtedly happened spontaneously on many a beach, where early man fished with primitive equipment. Recent research has identified a hunter-gatherer community which dates to the 7th millennium BC salting catfish on the Nile in Sudan, while it has been recently shown that hunter-gatherer communities used rotting techniques with low salt to preserve meat and fish in the colder Northern Hemisphere (Speth 2017:44; Maritan *et al.* 2018). Evidence suggesting the fermentation of fish has been identified in Sweden at an early Mesolithic site (Boethius 2016).

There is also evidence of intensive salting and curing on Cyprus and Crete in the Mesolithic. At the Cave of Cyclops on Yioura, small *Scombridae*, tunnies, were processed in a way that suggests salting (Mylona 2008:62–63; Mylona 2014) and mackerel bones at the Mesolithic sit at Maroulas on Kythnos, Cyclades, suggest that fish processing and curing took place (Theodoropoulou (2018:393). It would seem likely that the specific ceramic, zoo-archaeological and literary evidence that might demonstrate to us who was salting fish efficiently is no indicator of the beginning of the process (Theodoropoulou 2018:390). The techniques that early fishermen used to preserve their small hauls of precious fish were basic, requiring nothing more than a vessel and either salt or brine. For a discussion on the earliest literary evidence for fish processing see Chapter 2.

The potential fishery of the Mediterranean

The range of species found in the Mediterranean and their migratory patterns make the practice of fishing diverse in nature. As an enclosed sea, the nutrient levels (and therefore fish resources) are not particularly high overall, but there are areas in the north that are richer in nutrients, and this is due to the number of fresh water rivers that flow in from the continent (Mylona 2008:33). There are many species of what is known as "table fish" caught and consumed in the Mediterranean, but they are not generally salted whole, and so will not figure greatly in our discussion. Fish species are either migratory or sedentary and many in the sedentary group were consumed at table when mature and these include *Sparidae* such as gilthead (*Sparus aurata*), seabream (*Pagellus)* white seabream (Diplodus sargus), red and grey mullets (Mugilidae). These are caught all year round and are frequently seen when immature in fish sauce residues of a local small-scale variety (see Chapter 11; Morales-Muniz *et al.* 2017:180). The dominant species in terms of the archaeological finds are the pelagic (swimming neither at the bottom nor regularly near the shore) and migratory species, such as scombrids, particularly tuna and mackerel, and the smaller species such as anchovy and sardine, sparids, and clupeids. These species are exceptionally abundant in the six regions where fish processing took place: the southern Spanish coast, the Atlantic coast of Portugal, Morocco opposite the Straits, Sicily, Brittany, and the Black Sea. These are precisely the locations of the large processing factories with high-capacity *cetariae* fish tanks designed to make a bulk product. It has been noted by Morales-Muñiz that chronologically, the archaeology is initially dominated by tuna in the 2nd century BC to the 1st century AD, followed by mackerel in 1st to the 3rd centuries AD, and as the Western empire reached its zenith in the 4th century AD, there was a preponderance of small fry species such as clupeids and sparids (Morales-Muñiz and Rosello-Izquierdo 2016:38). All species were caught at all times, but in each period one species seems, from the vagaries of the shipwreck evidence, to dominate in trade. This is a phenomenon which may have a direct correlation with the way in which the fashion in fish sauce evolved and changed over the same period. We can see from the literary evidence that in terms of fish sauces, tuna and the use of its *muria* dominated in Punic Spain and up to the Hellenistic period, and was discernible in the archaeology from the abundant amphorae with the large wide-open mouths that transported it (page 245). Occasional tuna bones also corroborate this trade, as in the 4th century BC Corinthian amphora building (page 50), the tuna bones from 2nd century BC Baelo Claudia, where tuna dominates in the evidence at this site (Bernal-Casasola 2018:340; Bernal-Casasola *et al.* 2007:364), and the 1st century AD wreck at Chiessi Elba I (Table 11.3). Any use of the small fry to make the simple early *garos* at this time is very hard to see in the texts and equally almost impossible to see in the archaeology. A site at Olbia in Sardinia from the 4th/3rd centuries BC provides potential evidence for this as the local undefined and fragmentary amphorae (spike and lower portion only) contained very small picarel (*Maena Smarnis*), which we may conclude were a fish paste or sauce residue (Delussi and Wilkins 2000:53). The practice of removing the neck and upper body of an amphora at Olbia is subsequently mirrored quite frequently in later examples where a fish sauce residue is assured (see below). Mackerel dominated in the late Republic and early empire, and this is reflected in the abundance of shipwrecks which carried a mackerel product of disputed character (Table 11.3). We can also see that mackerel "*garum,*" that is, in reality a dissolved *liquamen*, was reflected in the satires of Horace (Horace, *Sat.* ii.8.42). Later, with the introduction of the blood/viscera sauce, *garum* and *liquamen* from mackerel are doubly visible in terms of *tituli picti*. There is every possibility of an on-going trade in a mackerel *salsamenta* generating a *muria,* which we can see in Martial's satires (Martial, *Ep.* xiii. 103) and also in a rare *titulus* on a amphora from Cologne for a *m(uria) scombri exel* (Kat.-Nr.258; Ehmig 2009:395).

In the later period, the smaller species appear to take over as the dominant fish species for *liquamen*, and this is seen in the ubiquity of *liquamen flos* or *optimum* without a species selection on *tituli* and from shipwrecks such as Grado and Randello, where sardine and anchovy dominate. It is apparent however that the ordinary *liquamen* type sauce was always made with the smaller and smallest fish from the beginning of the industry in Archaic Greece and throughout the Hellenistic and Roman period, but we are often simply unable to find the residues as they are subject to too many adverse conditions for survival and identification, discussed in Chapter 11.

The tuna fishery

Blue fin tuna from the Atlantic take a very predictable course through the straits of Gibraltar in order to spawn in the Black Sea and other regions of the Mediterranean and return to the Atlantic waters. The capture and processing of the larger scombrid species requires a complex and sophisticated fishing technique. In many regions of the Mediterranean tuna swim close to shore, and when "look outs" or *thynnoskopia* are deployed, great numbers can be caught using fixed and deployed traps. In Oppian's description of tuna fishing, "the nets are set forward in the swell like a city, the net has its gate keepers and city gates and internal canals" (Oppian, *Halieutica* 3.637–644). In the description of tuna fishing in Manilius' *Astronomica*, a purse net has been employed and the haul is pulled close to the beach: 'the fish struggle against their bonds, await new assaults, and suffer death by the knife; and the sea is dyed mixed with blood' (Manilius, *Astronomica* 5.664–666). From Manilius' description, dated to the early 1st century AD, we have seen that the tuna was killed in the sea or close by and they bled into the sea. This mirrors modern bleeding practices where often the fish are left in the water tied to the boat to bleed out. There is considerable blood that travels within the muscle tissue of a tuna and large blood vessels run near the pectoral fins and serve as the main bleeding vein, rather than gills, in the largest species. Modern tuna fishing requires that the blood is removed very quickly while the heart beats, in order to drain the blood from the muscle tissue. If the blood is left to coagulate within the flesh it can stain and taint the tuna meat.[1] In fact all but the smallest fish of any species, <20 cm, may have needed to be bled into the sea or in a vessel of sea water to preserve their flesh regardless as to whether the blood had a use. This may have implications for our understanding of the blood/viscera *garum* as harvesting blood is a complex procedure and allowing the fish to naturally bleed out into the sea would have been the most common practice before fish blood became an ingredient in some fish sauces. There are many examples of a particular cut to the cranium of large species of fish, mainly mackerel, which was utilized to facilitate bleeding and a similar cut has been found on 6 young tuna of 40 to 80 cm length in an amphora at the Hellenistic-Roman (1st BC) town of Dion in an elite town house (Theodoropoulou 2018: 396; Desse-Berset and Desse 2000:79; Desse-Berset 1993:345). I have argued in Chapter 1 that the species of fish in Manilius' poem were not tuna but mackerel though we note others believe that tuna remains the topic of the entire section. Regardless, the haul of unnamed fish in the poem are then brought onto the beach and butchered in situ. From Renaissance sources in Spain it is clear that the tuna was butchered on the beach or very nearby and the bones often burnt to make a fertilizer and we can now conclude that they were also used as a source of fuel for the kilns that were making the amphorae (Romero 2019:16). We also find evidence of the accumulation of tuna bones at the foot of a pier at Nabeul, which suggest that much of this material could also be "rejected at sea" (Sternberg 2008:375). Further evidence of large accumulation of fish and animal bones have also been found in the sea beyond a pier at Istanbul from the Byzantine period (Onar Vedat 2021).

The incidence of tuna bones surviving in ancient archaeology is quite rare, and this would seem to be the explanation for much of the absence of evidence (Morales-Muñiz 1993). It is interesting to note that modern fishermen and chefs in Japan value the head and cheek meat almost more than the loin and belly of the tuna, and it is clear that this has been known for centuries. *Athenaeus* quotes Alexis' *Odysseus Weaving*, in which a tuna head is served, and also a play by Aristophon in which tuna "keys," *kleides*, are praised as "fancy food" (*Athenaeus* vii.303a/b). The *kleides* are believed to be meat from the cheek and possibly the roof of the mouth, which is also prized today. We do not appear to be able to detect practices associated with the processing of the head in ancient archaeology.[2] We do know that vast quantities of tuna meat were traded across the ancient Mediterranean and, though it is largely archaeologically invisible in terms of bones, there is extensive evidence of this trade from Spain into the Mediterranean from as early as the 5th century BC through literature and amphorae that we are fairly sure were designed to trade this salted fish. These issues and the *tituli picti* for these products are discussed in Chapter 12.

The size of the various forms of tuna at the different stages of development in the ancient Mediterranean are not well understood. There are strict restrictions as to the size and weight of the tuna that can be commercially caught today. In the Mediterranean fishermen are not allowed to catch anything smaller than 50 cm, to ensure that supplies are maintained. The blue fin tuna found in modern waters range from 80 to 200 cm and can weight up to 500 kg. We might estimate that a single 500 kg tuna would generate a substantial amount of meat which simply could not have been monopolized by the elite given, that fresh table fish and sea foods were by far the preferred dining option at elite banquets. In ancient sources, tuna over one year old were called *thynnus*, and when even larger and older they were called *ketos* (Athenaeus. vii.301). When under one year they have the name *pelamys*, and the very young tuna fish were called *cordyla/cordula* (Pliny the Elder, *HN*. xxxii.146; ix.15,18). These juvenile fish are never caught today, but appear to have been heavily exploited in ancient time which, as we imagine put great stress on fish stocks. Modern analysis tells us that tuna fry at 30 days measure just 5 cm, and they swim in vast shoals where cannibalism and predation take up to 50%. By 60 days they have reached over 20 cm and by 120 days they measure 35 cm, while at one year the average tuna measures about 60 to 75 cm (Jusup *et al.* 2011).[3] *Cordula* is *pusilla*, "very little, insignificant" (Pliny the Elder, *HN*. xxxii.146), and this suggests the easy-to-handle fish that, according to Martial, could be wrapped up in the sheets of his scroll of verse (*Ep.* xiii.1), though this was, as noted by Djaoui (2016:119) destined to become *emporitic* wrapping paper (Pliny the Elder *HN*. 13.74–82) used by shopkeepers to wrap food, particularly fish. A *cordyla* deemed *vetus*, "elderly," is potentially larger than what was deemed a small *lacertos* or "mackerel" according to Martial (*Ep.* Xi.52) when it was served at table garnished with egg and rue, though this really tells us very little. All these references suggest a *cordyla* tuna of perhaps 20 cm. There are few examples of residues of this species to date. The remains of small tuna which we can probably define as *pelamys*, of 40 to 80 cm length, were found in an unidentified amphora at the Hellenistic-Roman town of Dion in an elite town house (Theodoropoulou 2018:396). One possible example of tuna was found in the wreck Titan A dated to between 80 and 50 BC off the coast of Île du Lavant (Benoit 1958:5–8). The wreck carried Dressel 12 amphorae containing tuna and bonito vertebrae of unknown size, but given the neck width of these amphora the pieces must have been cut very small to get them inside and one must ask how they were removed? It does not seem probable that the meat from a tuna fish of the size of *cordyla* would have been deboned before sale, though we cannot be sure. It is not clear how this kind of tuna was processed, whether potentially served fresh or generally always salted for the table, though the later *tituli* for tuna suggest ageing and storage, and so salting whole and thus generating a *muria*. This is Curtis' assumption and it is mine, that the fish themselves are secondary and not the principal reason for the salting process (Curtis 1991:58). The brine is the raison d'être, taking on the flavor and goodness from the fish meat which may not even have been consumed. Crucially, this way of preserving tuna does not utilize enzyme fermentation to dissolve the flesh, but is a 'clean' process whereby the flesh simply sits in the brine and the flavors are transferred, over many years it seems, rather than by fermentation in a few months. My own experiments have demonstrated that a fish brine of c. 2 months duration is delicate and fresh, without the pungent umami from *garos/liquamen,* but we may conjecture that one left with the fish for up to 4 years (as the *tituli picti* suggest) would be considerably stronger with more umami and the fish itself uneatable.

Archestratus tells us that "the slice of Sicilian tuna" was "cut up when it is about to be pickled in jars" (*Frag.* 39; Athenaeus 117a). One might gather from this that the pieces of tuna were first salted in larger pieces and only cut into slices and other shapes after the initial salting. There are a number of different shapes associated with salted tuna (Curtis 1991:7). These include cubes triangles and squares. The simplest and presumably the cheapest would seem to be the cube (κύβιον), yet Athenaeus says that "cube," *horaian* was the very best kind of salt fish (Athenaeus iii.120). One may imagine that cube means roughly cubed-shaped pieces hacked off the carcass after the best meat from the loin and belly had been removed in a larger piece. This small size was ideally suited to a narrower necked transport amphora as the pieces would be easier to place inside and remove. The small pieces would have provided a cheap food resource, while also certainly generating a *muria*. Like all fish when salted, tuna meat shrinks by upwards of 50% in weight and size, reducing a moderate size cube into a small, firm and very unappealing lump. Bernal concludes that all tuna was cut into relatively small pieces, thus possibly allowing it to be accommodated in these vessels though this does not seem to accord well with diversity of terminology used or the vast quantities of tuna meat that would have been available. (Bernal-Casasola 2015:64). The trade in the lean flesh of tuna is largely invisible. Alternatively, large steaks from the belly and loin could have been dry salted and then fully dried and thus traded without a brine or amphora, or potentially wet salted in wooden barrels and, when skin and scales are removed, equally invisible. We may be able to see the dried tuna in the description of *melandrya* in Pliny (*HN*. ix.48), which is a term in common usage (*vulga*) and refer to a shape like 'oak wood' and described as looking like a "dark roots" (Curtis 1991:10, note 16). It has to be admitted that it is not impossible for the tuna product in the Titan A to have been a type of *liquamen*, as suggested by Djaoui (2016:119) given the amphora shape. The date of this find is 80 to 50 BC and by the mid-century we are hearing of fine sauces being made with mackerel and there must have been a growing interest in different types of sauce and a dissolved tuna may have been a novelty. The majority of *garos/garum/liquamen* (before the *garum* made with blood was developed) was still likely to be a mixed sauce using small fish. I fully acknowledge that the rarity of faunal evidence alone is not sufficient to deny a tuna *liquamen*. However, if tuna were used specifically to make *liquamen*, one would expect some form of textual or epigraphic evidence to support it (but note Djaoui 2016:119–123 for a different view; see also page 36; 236).

The initial salting process involves using dry salt, and the normal process is to stack the fish between layers and then compress them to force the fluid out. There is a description in Columella of how to salt pork and this is understood, for want of other evidence, to be the same technique to salt fish. The layer of salt is two fingers high between the meat pieces and the pieces of meat must not touch (Columella xii.55.4). It is very likely that meat would require a considerably higher salt to meat ratio to remain free of bacterial decay. Salting techniques are diverse around the world, but a typical salt to fish ratio of between 10% and 20% by weight and mass is recommended for mackerel (Sikorski 1990:152). This would not provide enough salt to create separate layers between the fish pieces. The salt concentration of fish salting brine is typically in the range of 17% to 24% when fully saturated (Hall 2012:55). It is difficult to estimate the salt to meat ratio reflected in the Columella description, but it was almost certainly considerably higher than the ideal for fish. The experimental approach has also demonstrated that when salting cleaned fish, the amount of fluid that a mackerel generates is not enough to dissolve the salt (even when 20% salt by weight is used) or to cover the fish entirely, and additional brine is needed. The time between salting and brining is quite short according to ancient sources. There is a description of how the ancients salted their tuna and mackerel in Archestratus. "But get a mackerel on the third day before it goes into the salt water (i.e., brine or *muria*) within the transport jar as a piece of recently cured half salted fish" (*Frag*. 39; Athenaeus 117a). We can therefore imagine a situation where the fish is left for three days in dry salt, merely sprinkled between the layers, in a container such as a small *cetariae,* at which time it is taken out and put inside another *cetaria or* amphora containing brine – likely evaporated sea water – along with any brine that the fish had produced. At this point it was considered fresh salted and desirable. Galen, in the 2nd century AD, reported that the best salt fish (*tarichos*), understood to be tuna, was from Cadiz and also Sardinia. Martial seems to imply that tuna *muria* (its brine) was less desirable than that from mackerel, which we might take to mean that the former was more readily available (Martial *Ep.* Xiii.103). One would need to know which form of tuna generated the better *muria*. The prevalence *tituli picti* for this product might suggest that *cordyla* was the more appreciated form but we cannot be certain. I have a sense that *cordyla* may have been a somewhat anomalous product reflecting a quite a range of tuna cuts, sizes and ages rather than specifically one of 20± cm, but this is admittedly just a feeling. For a wider discussion of *tituli picti* for *salsamenta* (see Chapter 12). We should also note that tuna that has swam a great distance has lost all its stored-up fat and is less desirable to one caught on its home territory before it leaves to spawn according to *Athenaeus (Athenaeus* vii.315d). This is also suggested of the moray eel that was served at a banquet in a Horace satire (Horace, *Sat.* ii.8.42). We might therefore consider that tuna caught in the Black Sea before it has left the Sea of Marmara, or one caught in the Straits as it enters the Mediterranean and has been salted fresh in all its fat richness, is better than one that has taken the exhausting journey around the Mediterranean and been caught in the middle, that is, Sicily or Italy, and then shipped back. This seems ultimately to be what *Archestratus* was saying in his poem. Travel to where the best fish can be caught and freshly salted, do not wait for it to come to you in a "stinky old jar from Byzantium" (Horace, *Sat* ii.4.63–71).

The mackerel fishery

It is clear that mackerel has the potential to generate all four varieties of sauce: *garum* from the blood and viscera; *liquamen* from the flesh with retained viscera, *allec* from the fish sauce residue, and a salted fish product which in turn generates a *muria*. The process of making *garum per se* is discussed on page 163. Mackerel of Atlantic (*Scomber scombrus)* and Spanish (*Scomber japonicus colias*) varieties are readily caught in vast numbers at specific times and places during the year when they travel west to east to spawn in the Mediterranean, and as they travel, they can swim in vast shoals quite close to the shore. In ancient times, mackerel occurred in a range of 20 to 40 cm; the upper range is rare in modern mackerel fisheries. Galen praises the mackerel from the Spanish coast at Sexi, which was known as *saxitana* after the town in Spain (Galen De alim. Fac. 3.40.1–6; Van Neer *et al.* 2010:171).[4] The bonito (*Lakerda*) is also a *Scombridae* that resembles both mackerel and tuna and was probably fished widely though

we know far less about this. There is much less evidence for the actual capture and processing of mackerel. Seine nets could have been deployed from
boats or from the beach after the random spotting of shoals, or "bobs"
dropped from a boat or a rod and line; the latter method can still result
in large catches, but is far more time-consuming and does not yield the
kind of bulk catch needed to harvest enough mackerel blood to supply the
demand for a mackerel blood/viscera *garum* for the elite market. It may
be that vast catches of mackerel are much more difficult to find, and this
is demonstrated by the fact that there were only a few particular areas of
the Spanish coast where they could be caught in abundance. It is equally
likely that catching mackerel in large numbers was difficult throughout
the Mediterranean and required special techniques and a good deal of
luck. It is of note that the best-preserved and most well-researched fish
sauce factory in Baetica, Bealo Claudia, did not apparently catch mackerel in any volume, as the diverse fish species found on the site does not
include any mackerel bones. The site is dominated by tuna and later by the
sardine and anchovy typical of the small-fish sauce fish (Bernal-Casasola
et al. 2018). Pliny cites Carthago Spartiae or properly Carthago Nova,
modern Cartagena in Murcia, as the best source of *garum*. There is a small
island off the coast which was called *Escombraria* in ancient times after
the abundant mackerel that could be caught there. Strabo tells us "next is
the island of Hercules, near to Carthage, called *Scombraria,* on account
of the mackerel taken there, from which the finest *garon* is made" (Strabo,
Geographia 3.4.7). Strabo, writing in Greek, does not use the Latin terms
garum or *sociorum*, or the descriptors "bloody" or "black." The dating of
Strabo's work, normally given as early in the first century AD, is a debat-
able issue (Dueck 2000:146), and the blood viscera *garum* was a relatively
new product at this time. Thus, we cannot be certain that by "best garum"
Strabo meant a blood viscera sauce; it is possible that he is still referring
to a *garos/liquamen*. Though Pliny mentions many sites where he believes
garum was valued and made, this does not necessarily mean that mackerel
was caught in bulk in these places (Pliny the Elder, *HN.* xxxi.94). Many
of the processing factory sites contained fish sauce residues representing
the products they were making when they ceased to function in the mid-
to late-empire and there is little evidence for mackerel, but mackerel had
already become much less visible in the archaeology at this time according to Arturo Morales (Morales-Muñiz and Rosello-Izquierdo 2016:38).
There is a distinct possibility that overfishing devastated the mackerel
population in order to meet the demand for elite *garum* and *liquamen* in
the 1st century AD, such that later Roman consumption of this fish was
greatly reduced and even elite consumers had to be satisfied with the mixed
small-fish sauces, with random catches of mackerel mixed in with sardine
(Desse-Berset 1993). Substantial amounts of mackerel bones survive in
archaeological sites across the Mediterranean and northern provinces, and
as we have noted, zoo-archaeologists consider them solely to be the residue

of a solid salted fish that is believed to have been traded widely (Van Neer *et al.* 2010:168). It is believed that the various species of mackerel, *colias or Japonica,* also known as chub mackerel, Atlantic mackerel *Scomber scombrus* and *Trachurus trachurus,* horse mackerel were salted whole and shipped widely as the evidence for their bones throughout the empire is very high and particularly in the 1st century AD (Van Neer *et al.* 2010:169). However, as we have noted, it may be possible to understand many of the examples of mackerel cited by Van Neer and currently identified as *salsamenta* as a fish sauce "discard event" rather than evidence of the trade in a salted fish product (discussed in Chapter 11; Van Neer *et al.* 2010:168). The *tituli picti* form *lac/lacc/lacat/lacatu* has now been securely identified with *lacertus* and defined as a desirable small mackerel (Djaoui 2016:118). However, we do not seem to be able to identify the term associated with the apparently widespread trade in the larger mackerel species such as *scomber scombrus.* This term is linked to *garum* production through the regular use of *scombris* on *garum* labels and by the claim from Pliny that *scomber* mackerel was not highly valued as an eating fish and was only used for fish sauce (see page 27). Whether we can differentiate mackerel *salsamenta* more specifically by the individual species is impossible to clarify. It may be that one of the two mackerel species fished in the Mediterranean (*scomber scombrus* or *scomber japonicus*) was taken as *lacertus* and the other taken when larger and utilised as a cheaper *salsamenta* and also for both forms of fish sauce. It does appear as if the small forms of both mackerel and tuna are worthy of praise and elaborate *salsamenta* amphora *tituli picti* while the larger are not (see page 257). It is apparent from shipwrecks that a pure mackerel *liquamen* was traded widely in the early empire (see page 218) though when being traded as a fish mash it did not retain or require a label. There are relatively rare *tituli* for a pure *liquamen scombri* and they are all found in *urcei* (*liquamen scombri flos* (CIL.iii.12010,48; iv.2588, 5716, 941). Most fish sauce is a mixed variety that uses varying quantities of mackerel and is well documented archaeologically (see table 11.2). It is of note that in the open central Adriatic, modern fishermen using lights found large mackerel below sardines, the two groups differing by size, while small mackerel appeared thoroughly mixed with sardines (Mužinić 1977:153). This corresponds well with the evidence of the Grado wreck from the Adriatic, as we find that the sardine had some mackerel mixed in with them and vice versa (Auriemma 2000:43). The *garum* on the other hand is largely always identified as *scombri,* "of mackerel," and we cannot judge to what extent this would have actually been correct, that is, other viscera may have always crept in!

The fishery for the small species

The name applied to the unknown species of small fish that was used to make the simple Greek sauce was *garos* according to Pliny (*HN.* xxxi.93), though this term for a small fish is not found in early Greek sources. The Greek term generally used is ἀφύη/*aphye*, meaning a number of different species, all of very small size, what in English is called whitebait, but the term is more specifically attached to fish that are tiny and transparent, newly born, foam like (ἀφρός/*aphros*, from Archestratus, *Frag.* 11.2; Olson and Sens 2000:53). The term *apua* in Latin has the same meaning (*Apicius* iv.2.12). There is surely an association between *garos* and "gavros," which is the modern term for anchovy in Greek (Demir 2007). In Italian, anchovy is *alici*, and in Turkish they are called *hamsi* (Davidson 1981:48). The smaller fish are dominated by anchovy (*Engraulidae*) and sardine (*Sardina pilchardus*), along with many other small *Clupidiae* species such as *Spratus spratus*, pickerel and *Boops boops*. These species are prolific in the Mediterranean in all regions. The small fish also dominate the archaeological in both processing sites and transport amphorae, where they are identified as the residue of fish sauce production and would appear, from our earlier look at the original recipes, to form the basis of most of the *garos/liquamen* type fish sauce (Tables 1,2,3 and 4).

In the northern Aegean sardine and anchovy were more common, while the *Clupidiae* were more common in the south (Mylona 2008:43). Wherever small fish like these are caught in large quantities, they can be consumed within a few hours; fresh and cooked simply, they were appreciated by ancient elites. Writing from the perspective of Sicily, Archestratus treats all ἀφρός "small fry" with contempt, except those from Athens (Archestratus, *Frag.*11, Athenaeus iv.285b), and we know that they could be served at elite banquets, including the conspicuously modest one served at Thebes in 335 BC according to Cleitarchus (Athenaeus iv.148d). These small "fish sauce fish" were caught in their season, largely the spring and summer months, and for short well-defined periods; they were not regularly available (Slim *et al.* 2007:35). As we have noted, many different species often swim close together in nearby shoals and intermix depending on their size and relative strength: the smaller of each species swim at the same speed. Residues from the processing sites also reflect this mixing of species which must reflect what the fishermen were able to catch on any particular day. Vat number III at Neapolis in Nabeul contained a residue comprising 60% anchovy, 20% sardine, 12% pickerel, 5% mackerel, and 3% bream (Slim *et al.* 2007:35). This kind of fish is fatty and is normally not suited to drying as the fat oxidizes and spoils the product. We have heard that a dried salted anchovy/sardine product called *saperdês* was made in Greece and was very under-valued due to its smell which we may attribute to an oxidized product (Wilkins 2018;231; Archestratus *Frag.* 39.3–4; Lytle 2018:410). Wet salting is considered the best technique to preserve these little fish (Curtis 2016:165). Many different preservation techniques will have spontaneously developed to counter the decay and preserve the haul, and we can see this in the many different ways that are available today. All you would need would be a small net thrown from rocks into a visible shoal but they could also be caught with nets thrown from boats, as described by Oppian in the 2nd century AD (*Halieutica* iv.468). The use of seine nets from the beach and from small boats could lead to massive catches difficult to process, and it is this difficulty that is undoubtedly the source and origin of fish sauce manufacture. Access to salt will have been the essential economic factor that distinguished random subsistence fishing from a step up to an attempt to preserve and retain the catch for future use and consequent small trade. Brining fish without dry salt is a safe albeit less common technique and such practice would not necessarily be recognized archaeologically. The Turkish techniques of salting uneviscerated very small grey mullet known as Fasikh is salted in levels below 5% and can be stored for a number of weeks. Fasikh does have some health risks from botulism, though this is largely due to poor knowledge of the traditional methods (Soliman *et al.* 2017).

If one wants to keep the fish solid then the kench method of salting can be used, whereby often uneviscerated fish are salted in open piles with layers of salt between each layer of fish and the resulting brine, blood and oil are allowed to drain way. The result is a product which has lost its water content and in this semi-dried state, bacteria cannot thrive. This technique was used to preserve uneviscerated pilchards for hundreds of years in a thriving industry at Newlyn in Cornwall.[5] Whole kench salting is quite feasible on a beach, using wooden planks to allow the brine to drain. The alternative method, which is much more labor-intensive, is to eviscerate the small fish, wash and salt them in small vessels for local consumption. Size then becomes the determining factor, as any small fry of whatever species below a certain size, that is, 2 to 5 cm, can be too small to eviscerate in bulk, and fit the category of what we call whitebait in English which are eaten cooked but uneviscerated. There are limited options in order to preserve these tiny fish. They are so small and delicate that if you salt them and attempt to hold them in the small amount of brine that forms around them, the enzymic digestive process begins to dissolve the fish within hours and an embryonic fish sauce begins to be formed quite quickly. It would be this early stage in the process that led to the development of purpose made *allec*. The technique applied to make pissalat does involve the labor intensive process of deboning and eviscerating each fish before salting (page 230) and would appear to be a way of processing small fish caught in bulk relatively easily into a paste that can be held for many weeks and provide protein, taste and novelty (pers. comm., pissalat maker, Antibes market, October 2009; Delaval and Poignant 2007:62). This deboning technique was also applied to the production of *colatura di alici,* which is still made in the bay of Naples (page 231; Carannante *et al.* 2011). Clearly little fish of all varieties were either salted, while retaining their integrity by eviscerating or one allowed them to disintegrate into various consistencies of sauce.

Notes

1 The meat should be pale pink, while any dark meat is generally a sign of blood pooling within the flesh. Commercial fishermen tend to pierce the arteries located just behind the pectoral fins and place the fish back into the water to bleed out over the next 10 to 15 minutes. https://tunafishingcharters.com. au/processing-bluefin-tuna-caught/ Sports fishermen catching single fish are encouraged today to use the Japanese techniques for slaughtering and butchering tuna which is known to be humane, called *Ikejime*. It involves stunning and destroying the spinal cord quickly but allowing the heart to continue to beat for long enough to bleed effectively and this maintains the quality of the meat. The normal sports fisherman's technique for bleeding is to either sever the pectoral fin arteries or punch the gills through and then put the fish in a bucket of salt water, which prevents the blood from coagulating.

2 https://www.yellowfintunaloin.com/2018/01/why-yellowfin-tuna-cheek-is-best-part.html

3 In 2007 to 2010 fisheries in the Mediterranean were forbidden to catch and sell Blue Fin tuna (*Thunnus thynnus*) under 30 kg, but could permit catches by individuals of <10 kg in certain circumstances (data from the *Recommendation by ICCAT to establish a multi-annual recovery plan for Blue Fin tuna in the Eastern Atlantic and Mediterranean*).

4 *Saxitanus* can represents small Sp mackerel in a rare *titulus* from the Gondolfo wreck (Almeida and Liou 2000:14; Van Neer *et al.* 2010:171, *Pliny NH* 32.146; *Ath Diep* 3.121a).

5 The Newlyn pilchard was shipped in wooden boxes to Spain and Italy in the 18th and 19th centuries, though it did not have a market at home: the pungent, strong flavor of the bacterially fermented fish was not to local tastes (Noall 1972:39). The industry only ceased when the process became subject to modern hygiene regulations which did not trust that the preserving techniques were safe.

10 The infrastructure of fish sauce manufacture

Fish sauce production and commerce, small-scale and large-scale?

We have seen that it was under Roman influence that the trade in fish sauces became widespread. During the mid-1st century BC, industrial-scale factories begin to develop where the older facilities existed but also in many new places and spread in areas where there was a guaranteed supply of one or more of the three dominant species associated with sauces, namely tuna, mackerel, and the multiple small fry species. There was a well-organized chain of supply evident in the archaeology from Spain, Portugal, and North African areas under Roman control in the early empire which indicates large-scale fishing, bulk manufacture, transfer to amphorae, shipment, marketing, consumption and disposal of salted fish products. I hope to be able to trace the entire journey from fish processing, transportation, consumption, and disposal through a discussion of the most important sites. To reiterate, this approach cannot be a fully comprehensive one, and only those sites that illustrate the phenomenon and contribute to the model created will be dealt with. Many of the sites display similar characteristics and there is a uniformity to the sites that means we can concentrate on the most well documented. We have seen from the ethnographic studies that there appeared to be a local, largely small-scale production process, and an industrial process largely manufacturing in bulk. The former could be an enclosed process in relatively small vessels and the bulk process open to the elements and the sun's heat. The distinction between these two types of processing is not, however, that clear-cut, as a small-scale process could still have involved an open container subject to evaporation, as the vessels may not have been sealed. Equally, a large-scale process could in theory have been generated in closed containers. It is also clear that small scale containers do not necessarily result in low volume of finished sauce and this is particularly true of evidence from Arles and the Little *Latium* pots (see below).

The large-scale industrial processing sites

The regions of the ancient Mediterranean that exploited fish processing extensively were initially the sea of Marmara in the Black Sea, the coast of Spain between Cadiz and Cartagena (Bernal-Casasola and Sáez Romero 2008), the opposite coast of North Africa in Morocco and Tunisia, the Atlantic coast of Portugal on the Sardo (Trakadas 2005), the entire coast line of Sicily (Botte 2008) and in Gaul at Ploumanac'h in Britany (Sanquer and Galliou 1972). These six regions were unique in providing such rich marine recourses, and though these resources were exploited before Roman intervention, it was only under Roman influence that the infrastructure was set in place to exploit them industrially to produce vast quantities of food, building industrial-scale fish-salting factories which we can identify because of their distinct characteristic, largely mirrored in each region. The masonry salting tanks (*cetariae*) come in a range of shapes and sizes with a capacity ranging from 1–2 m^3 to 10 m^3; with a depth range of ±2 m; they were lined with a watertight render which is more often found to be *opus signinum* (a lime sand and crushed tile mixture that is entirely impermeable: Arévalo 2007:200). *Cetariae* were also lined with less durable surfaces using hydraulic lime render that contains minimal levels of crushed tile or even none at all, and there is some doubt as to the long-term permeability of these renders and therefore their suitability for fish sauces (Bernal-Casasola *et al.* 2018:78).[1] The original Punic fish salting facilities existed with just a pair of tanks, and there is some doubt as to how much fish sauce was manufactured in these early tanks (discussed in Chapter 2). We may speculate that the dominant product in the pairs of *cetariae* were designed for *salsamenta,* while at the height of the fish sauce industry, the tanks existed in extended rows of up to ten. The orientation of the tanks can also vary greatly, as noted by Garcia Vargas and Bernal-Casasola (2009:147). The tanks can run in a parallel-lines or in a U shape and also in an L shape. It seems unlikely that the different orientations would necessarily have a single consistent purpose within the production process across all sites, given their diversity, as the structure could depend on variable production methods and the topography of the site may have dictated the structure. At *Troia* on the Sardo river the processing tanks are arranged in a particularly interesting way. There were groups of 4 small tanks alongside a larger one with double the capacity which is repeated in long rows (see Fig 10.1). Similar configurations are found in many sites such as those at Bealo Claudia (see Fig 10.2). Given our knowledge of the potential to rebrine and dilute the sauces that we saw in the ethnographic study from Chapter 7, these structures would seem ideally suited to the transferring sauces from one tank to another, in order to rebrine the residue but also maximize the volume of sauce of a suitable consistency, taste and salinity.

Figure 10.1   Cetariae at Troia in Portugal. Copyright: Sally Grainger

We have seen that in modern Southeast Asia empty tanks are often used to hold filtered sauces which continue to mature away from the bone residue, which could be re-brined. The experimental work by Driard has also provided strong evidence for the transfer and storage of liquids free of residues from one tank to another in the process of refining the sauces through residues adhering to the sides of the tank (Driard 2012:58; page 150, and see below). In the immediate vicinity of the tanks there were open working areas, preparation areas and rooms for storage. Often a hypocaust or furnace area is to be found, but they are not typically designed with cooking in mind and may have been more to do with cleanliness both of the equipment and the workers. There is a possibility that the storage of filled amphorae in a hypocaust-warmed room would have allowed for a greater

Figure 10.2 Cetariae at Baelo Claudia Copyright: Sally Grainger

nutritional yield. A need for either a water supply if in land or at a location close to the sea, which is the more common, would also seem to have been essential. The processing area was open to the air but with some sort of roof structure supported by pillars. Many are close to the shore with easy access to the catch, as it came in while others, particularly those at Baelo Claudia were constructed in the center of the town and the fish would have been transported from the beach or dock. The shapes of the *cetariae* vary greatly. Botte has noted the early incidence of round and truncated tanks in the Greek south east of Sicily and the later introduction of oblong tanks in the Punic north west of the island, though there does not seem to be any discernible practical reason for the differences (Botte 2009:100). Tanks with rounded sides, described as a truncated cone (i.e., wider at the top) occur at Baelo Claudiae in a single group of four in Factory VI and Botte recognized a single round tank at Lixus (Arévalo and Bernal 2007:79; Botte 2009:100). This truncated cone may have suited a *salsamenta* as the brine could have drained into the space below while the fish remained elevated by wooden structures that would remain undetected. The modern bulk manufacturing process takes care to ensure that the product can be stirred and agitated and there are no dead spots where the fish remain static, such as in the corners, where fish may get stuck and cannot move within the brine; the complete free movement of the fish in the brine is one of the essentials for a safe product. This is why a dry product such as that made from the Gargilius recipes or one where evaporation loss is not been replaced, is so undesirable. Oblong tanks often, but not always, have *ovulo*, rounded angles to their corners, the purpose of which is not understood but may be connected to the same need to stir the product (Curtis 1991:54). One might conjecture that an oblong tank is technically easier to construct and then convert to a rounded tank by filling in the corners. The small depression in the base of the rectangular tanks which are assumed to be connected to cleaning has no other obvious purpose.[2]

The archaeological evidence for the actual processes involved is scarce. Any wooden structures are all gone, and much of the day-to-day infrastructure is missing. It is likely that there was a wooden covering that could be slid into place. One can also imagine rush mats placed over the fish and weights used to hold the fish under the brine. Many large stones which may have been used for this purpose are often present on the sites. Dolia have been found in abundance at Baelo Claudia, and we may conjecture that these were used to make smaller amounts of more expensive sauces, that is, *garos* using specific species or *garum haimation/sociorum*, while the large tanks were for the basic *liquamen*. In the later period (late 2nd–4th centuries AD) a small number of facilities have unusual tanks with drainage channels and lead piping which may also be associated with the production of *garum* made with blood and viscera. These are described below. The scarcity of these types of tanks would suggest that they were developed for a particular type of sauce, and there is certainly the possibility that the blood viscera sauce was originally the intended product. Pottery finds associated with fish salting factories are not sufficiently distinctive to illuminate the process; they are typical of small cookware jars and jugs, and were probably used within the process to sample and market local products in the same way as the *Latium* pots (see below) or by the workers in their own food preparation/storage. One particular form, a straight-sided jug not unlike a giant tea cup with a single handle, was identified at the earliest excavation in Cotta in Morocco by Ponsich and Tarradell (1965: plate 19, fig 6; illustrated in graphic form in figure 12.6 on page 271). The shape is remarkably similar to the traditional jug used to salt domestic quantities of anchovy to make the original *colatura di alici* (when this product was the *muria* from a cleaned deboned salted anchovy) at *Cetara* on the bay of Naples (Carannante *et al.* 2011:74; Fig. 8).[3] We may also equate these small wide jugs with the small-scale production process epitomized by the little *Latium* pots on the Rhone, described below. Funnels called "Cadiz *infundibula*" have been identified through residue analysis as having some connection to fish processing by Dario Bernal-Casasola and Romero (2006:212)

Their precise function is hard to discern, as they do not seem suitable for the process of primary extraction from the tanks as they are so small. It would seem likely that the fish were layered with salt, as suggested in the recipes, though the ratio of 8:1 would not have allowed for a thick layer of salt between the layers of fish, as this would have resulted in too high a salt ratio and too dry a mixture which could not generate enough brine for the mixture to be stirred. One may assume that the workers were experienced in gauging the correct salt concentration through long experience. Evidence of the addition of herbs, spices, and colors are rare indeed in any of the bulk *cetariae*. There are suggestions of meadow flower pollen in residues from an amphora in Augusta Emerita but there is some doubt that this was added intentionally.[4] The Gargilius recipe apart, there is just one reference to oregano in a cooked sauce; otherwise, it seems most likely that with the early forms of fish sauce that were mass produced, herb flavorings were added at the point of sauce preparation in the kitchen. Evidence of the intentional addition of other sea foods is also rare and it is possible that much of the current evidence alleged to show it, can be attributed to the random inclusion of shellfish caught at the same time as the fish by seine nets pulled across sandy banks and landed on the beach.[5] The addition of various dyes has been suggested to change or enhance the color of the sauces, a practice which is mirrored in modern techniques where caramel is used to darken sauces. Evidence of elevated levels of ferrous oxide in samples found at Baelo Claudia have been interpreted as a means of changing the color of sauces (Bernal-Casasola and Domingues-Bella 2012). The implication is that to imitate a blood *garum*, the ancients chose to enhance the color of sauces when they were not necessarily made from pure blood and viscera, or to enhance one that was not the correct color. However, Rodríguez-Alcántara notes that the unusually high presence of ferrous oxide (found also in the dolia at the *Garum* shop in Pompeii) could have been due to the addition of viscera and residual blood to a sauce, rather than as an intentional additional colorant (Rodríguez-Alcántara *et al.* 2018:159).[6]

Essentially in the bulk *cetariae* production process, salt and fish, and extra brine/wine when necessary, were all that was used. After 24 to 48 hours, if sufficient brine had formed, the mash was probably agitated to encourage the production of water from the fish, which in turn allowed the salt to dissolve. Over the next week, a brine formed and gravitated to the surface, but It is of importance to note that this brine was not a separate sauce but the embryonic *liquamen*. Over the next month, or even less depending on the size of fish, the brine would have become saturated in particles of muscle tissue and sunk below these particles, which formed a layer of fish paste on the surface. The top layer would have had some random bones floating free, but from my observations the majority of the bones remain in the middle and at the bottom of the processing tank. The *Geoponica* recipe suggest that a ratio of 2 wine to 1 fish mash was applied to generate a sauce. If this is an accurate figure, we can possibly envisage a process where the fish mash – semi dissolved fish pieces fish paste and fluid – was removed at this early point and placed in adjacent vats so that the required amount of extra fluid could be added and the liquefaction process continue. This provides an explanation for the rows of tanks at many fish sauce facilities and also accounts for the expectation of evaporation. How the finished sauce was removed is difficult to envisage. The timescale of three months in the recipes may refer to the entire time required for a small-scale process, while in the bulk process, once sufficient sauce had formed, it could have been taken from the *cetariae* much earlier. A large flexible rush basket could have been pressed into the sauce to allow the viscous emulsion to flow through while leaving any residual bones behind, though It is not clear to what extent it was actually necessary to filter it at this stage. One can also only imagine a large pan with an extended handle scooping out quantities of sauce from one vat to another. Observations of modern sauce production suggests that the majority of bone is retained towards the bottom of the tank, and would not contaminate the first harvest of sauce. Once removed, it could simply have been placed in an adjacent tank and left to continue to ferment, to allow the fish paste to continue to dissolve. This *liquamen flos* could have been put in small jugs and sold in the local area while amphorae were for wider distribution, either inland or by ship. As for bone contamination, it is clear, when *liquamen flos* is found in a sealed environment like the Grado wreck, that there were generally no bones present (page 219). We may conjecture that the remaining fish mash, which would have constituted perhaps half the tank, was initially re-brined in situ and the second sauce removed in the same way as the first. Over time, it seems, the discovery may have been made that keeping the residue in the tanks was not necessary to continue fermenting the second and subsequent sauces. We have seen from the experiments that even with three months of total fermenting time in the *cetariae,* there would still have been substantial *allec* residue at the bottom of the tank which could be re-brined in situ or potentially removed and placed in amphora, and re-brined for transport. The residue evidence in the form of large quantities of discarded small-fish bones that are associated with fish sauce found in amphora and also in piles discarded in ditches discussed in Chapter 11, clearly represent a fish sauce mash. That this product, effectively *allec,* was put inside amphora and traded is clear. Logic dictates that the residue was traded, what has been difficult to discern until now was why.

There are a number of conclusions we could take from these ideas. One consideration is that the system in place may represent an intensified production process as the initial fermenting time could be a little as one months before a sauce was available to take and decant into amphorae so that it can continue to ferment in transit. This could have led to an even greater volume of sauce being produced than we have to date imagined. We can also see that there could be much more complex business connections between manufacturers, traders and merchants in fish sauce, as this kind of product would not be fit for sale when it left Spain or North Africa and would need constant reprocessing in order to reach a state suitable for sale to the customer. In these circumstances one needs to question how the economic system functioned? Was the *allec* sold on in its unfinished state, or was the entire product owned by traders who engaged with business both in the Western Mediterranean production sites and also where it was further processed for sale either in Italy or the wider empire?

The small-scale production processes

When evidence of fish sauce processing is found in coastal regions and urban settings in the western Mediterranean, outside of the industrial area already noted, production would appear to have been in either dolia or very small vessels, little pots or bottles, perhaps for consumption by a single family or a small group. The literary evidence for a small-scale domestic production of fish sauce is scarce but significant. The 5th century Greek fishermen were clearly making *garos* in small vessels such as small dolia on or near the beach (page 46). Marzano (2013:93) notes many letters from Egypt which discuss preserved fish in small vessels. In the letters of Theophanes there is a reference to the purchase of mullet in order to make a *garos* while a group of travelers are on a journey in Egypt. This is of great interest, as it suggests that *garos* could be macerated and dissolved in small enough quantities to be easily transported and presumably accessed for use on that journey (Pauni 6 lines 84–102; Mathews 2006:110; see page 98). Experiments have allowed us to see that a small amount of fish can generate quite a lot of sauce, which rather suggests that small scale does not necessarily mean low volume over time. A number of interesting archaeological finds, epitomized by two sites, the little *Latium* pots from the Rhône and the pilgrim flask from Jordan, appear to illustrate this very small-scale product. The relatively small-scale dolia production at Pompeii is also of relevance here.

The Little Latium garum pots from the Rhône at Arles

Underwater investigations in the river Rhône at Arles found numerous small pots dated between 60 and 80 AD, buried in the silt. This type of vessel, pear-shaped at first and later more ovoid, with a capacity of 00.30 to 2.25 liters, was made in and around Rome from the Augustan period onwards. They have a strong association with fish products as a number have been found to contain fish bones, one at Pompeii in the house of Caius Iulius Polybius and another with sardine bones at the Narbonne harbor dump, dated to the end of the 1st BC/early 1st AD (Djaoui *et al.* 2014:185). For other vessels including amphora with fish sauce residues see Table 11.2. Many of the vessels at Arles contained a mixture of small fish and scales, and eight vessels have now been analyzed. Many of the sample consisted of whole, that is, uneviscerated fish of typical fish sauce species *Clupeidae* and *Sparidae* along with small numbers of lagoon and estuary fish species. Mackerel was found alone in some of the vessels, and as a smaller percentage of a typical mix of small species used for *liquamen*. Vessel AR3 2001 51. c. 20 cm in height, was subjected to analysis and was estimated to contain 8,000 bones of good preservation; a smaller selection was made from the whole, and an estimation of an MNI calculated as follows: c. 14 individual mackerel of c. 20 cm, 30 shad <18 cm, 2 sardine, 9 sprat <14 cm, 3 anchovy unsized and 1 sole. The size of the species utilized are in the upper range and the resulting sauce would have been a valued *liquamen*. The analysts considered that this number of fish were put into the vessel individually in the specific quantity indicated by the remains; it was concluded that because of what was perceived as a coherence of NMI based on an estimate of the cranium-to-vertebrae count, the product was made inside the vessel from the individual fish listed. However, it is equally possible (and, I think, much more likely) that the product had been blended outside the vessel and then placed inside as a fish mash. In this scenario the fish would have begun to breakup and disarticulate before being transferred. The cranium-to-vertebrae count would not represent entire fish, and given that the sample was not fully analyzed, the estimate reached by the analysts could not be sufficiently accurate to justify their conclusion. There are numerous other vessels with the more familiar combinations of *Clupeidae* and *Sparidae* that made the everyday *liquamen*, details on Table 11.2.

It seems much more satisfactory to interpret all these examples as a fish sauce mash, as described in Chapter 8, that had probably been pre-fermented/hydrolyzed in a *cetaria* or large dolium before being distributed randomly between a number of small pots. There is a possibility that when this mash was in its original fermentation vessel it provided an initial sauce, a *liquamen flos,* which was harvested first and traded as various grades of sauce locally. Djaoui *et al.* (2014:179) defines the product found in the fully analyzed vessel as an *allec,* which it clearly is in a sense, but it would have been far more valuable as a fish sauce concentrate which would have gone on to generate more *liquamen* when re-brined. This was actually a classic example of small-scale *liquamen/garos* production. It is likely that there were numerous occasions when various batches of semiprocessed fish sauce mixture were placed inside these vessels and at various times. They may have been stirred, brined and sealed and transported by a number of small ships or the same ship over many trips, able to navigate the Rhône, up the coast of Italy to Gaul. The mouth of the vessels is wide, while the neck is relatively narrow, and this would have facilitated the placing a cloth or lid over the mouth so the sauce could be poured out while retaining the bones cleanly inside. Over time, a quantity of fish mash in a vessel of a 2-liter capacity could have generated 2 or 3 liters of a reasonable quality *liquamen* which the ship owners could sell at every port and stopover as they moved north. After taking the liquid from the fish mash a quantity of brine could be added and a further sauce generated. It would seem that the journey taken with these vessels was predictable and regular, so that at the end of the journey or shortly afterwards, when the fish mash was exhausted of its potential to make a sauce, sixty of them were discarded over a given period of time in the same place in the river. The act of discarding them would appear to be remarkably regular and may either suggest repeated return journeys to the river site or a large supply of the product all discarded at once. It is also possible that other deposits of a similar nature may have been made along this route. In one Latium pot AR3 2001 51 Djaoui found that cranial elements cut from other mackerel were present. The estimate was six entire mackerel, with the addition of the front part of the cranium from eight further mackerel (Djaoui *et al.* 2014:178). The extra cranial bones resulted from the technique whereby the front of the nose of the fish was cut to facilitate bleeding, primarily for the purposes of preservation, but also, as we have seen, for the production of pure blood *garum*. This cut has been recognized in residues from a number of shipwrecks including Sub Perduto II (Desse-Berset and Desse 2000:80; Desse-Berset 1993:343). In vessel Latium pot AR3. 2031. 9. (c. 12 cm in height) was found only mackerel of 25 cm that had been cut clearly to bleed them and the cut elements of the crania were absent (Piqués *et al.* 2015). The vessel is too small to contemplate the preservation of whole mackerel, yet it is described as such. I would define it as a fine mackerel *liquamen,* rather than a *salsamenta*. The fact that these mackerel were butchered in this way led Djaoui *et al.* (2014:178) to conclude that some of the mackerel were bled to harvest blood for a *garum haemation/sociorum/melan*. I agree with this, but it does not follow, as they claim, that the mackerel were bled into the same pot as the fish mash. It is more likely that the extra mackerel cranial elements were added to the principal dolium or tank where the fish mash was first made, along with the rest of the fish carcasses. The cut crania would likely point to a blood viscera *garum* being made, but this I would argue would be a separate sauce. From this evidence we can conclude that the manufacturer of the Little Latium Pots was clearly making both a *garum* and a *liquamen* somewhere close to Rome. A more complex content was found in Latium pot X- 16373 (c. 14 cm in height). A small quantity of residue of sardine, 10 cm in length and representing just 6 specimens along with the gill rakes of at least 2 mackerel. The analysts (Piqués *et al.* 2015) concluded that this was a *liquamen* sauce that had been made at some point with mackerel viscera, which may be likely. It does not follow though that mackerel blood was necessarily added to small sardine to make a bloody *garum*. The disproportionately small numbers of fish may be due to post depositional conditions, or they could represent an inadequately filtered first *liquamen* sauce which may also be marketed in these pots. It is my belief that the bones inside these vessels at the time of their disposal were in every sense of the word rubbish and worthless. Djaoui concluded that the product in the Latium pot AR3 2001 was an *allec*, actually eaten as a bony fish paste with bread. It seems that others have accepted this interpretation without questioning its validity at all (e.g., Marzano 2018:442), supposing that the bones would have been softened by the fermentation or salt and as a result digestible. Though they consider that *allec* may have been a residue from which a fish sauce had already been taken, they also consider that this *allec* was purpose-made, yet do not question why it was discarded in such large amounts both in terms of the number of pots and the quantity of the product. There does not seem to be any acknowledgment that there is a problem with the idea of a low-value *allec* being made with fish blood which is elsewhere used for an elite fish sauce. The harvested blood is apparently considered waste matter suitable for low value sauces; but that makes no sense as bleeding is immensely complex and difficult and the blood serves no purpose in terms of hydrolyzation and liquefaction so one has to ask why bleed at all and why only some of the mackerel? Other scholars have made the suggestion that products like these have "gone off" and that is why they were discarded (Bernal-Casasola *et al.* 2016:737; Hamilton-Dyer 2008:4). This seems unlikely: there is simply too much of this kind of disposal evidence and it cannot all have "gone off." Further discussion of the problems raised by this interpretation are discussed on page 229. Djaoui *et al.* (2016) concludes that the product represents a food item made for the crew on the various ships that passed this site. This I assume is due to the assumption that the *allec* was a cheap food that sailors would have to eat. I would instead consider this to have been a product made on land and then "finished" by the ship's crew, as local traders, for a large middle-rank local native and Roman market up and down the coast on Italy. As the ship moved north, the fish mash would have yielded a quantity of *liquamen* sauce which could have been sold at each port. This find of multiple pots of what must be "washed out," "used up," and "exhausted" fish sauce residue provides a rationale to interpret other forms of evidence that survives, both in amphorae and discarded in middens. We should note that the *Latium* pots are of a similar capacity to the small wide jugs found at many of the production sites in Spain and North Africa noted above. These vessels may have been used in a similar fashion in the large processing sites, that is to market a semi-processed fish sauce mash for local consumption within the town and surrounding area. The process of harvesting a sauce from this mash may have been undertaken by the individual purchaser, who took the mash home and further re-brined and harvested sauces from it as and when required; or, as with the *Latium* pots, it may have been managed by a trader who sold filtered *liquamen* to his customers on the quayside. This we can gather from the fact that the trader stayed in control of the pots and disposed of the them regularly, potentially over many journeys. There is further evidence for a kind of individual processing of this fish mash from Palestinian Rabbinical sources discussed on page 233.

The pilgrim flask from Petra

The so-called pilgrim flask was found in a Roman house in what was believed to be the kitchen. The site is just 100 km from Masada and is dated to the late 4th century AD (Studer 1994). The flask was 20 cm high, with a long narrow neck with a 2.5 cm opening and a round flat body in the shape of a disc. Inside were numerous tiny bones from the smallest of small fry just a few centimeters long (sardine and anchovy) which Studer considered to be a Red Sea species. The product has been identified as either a purpose-made *allec* and/or a by-product of fish sauce manufacture, but nevertheless a product in its own right containing bones which would have been consumed in that state. The shape of this bottle is particularly relevant here as it is ideally suited to the process that has been described above where the liquid is poured off and the bones remain inside. The fish and salt were probably placed in the bottle whole and the sauce formed within the bottle. When sufficient liquid had formed this could have been poured out while the bones remained in the bottom of the bottle. Given the size of the fish this sauce would have been low value and quickly made and only yield one sauce before the bones become worthless and are discarded along with the bottle. This particular vessel and its *modus operandi* would seem to have been ideally suited to the production of a simple *garos* described by Theophanes when he was traveling in Egypt, as noted above.

The local fish sauce manufacture in Pompeii

The fish sauce industry in Pompeii at the demise of the town in AD 79 appears as a very complex system. There was a local manufacture of a simple *liquamen* type sauce, demonstrated from the *Garum* shop, an apparent local manufacture of expensive *garum* according to Pliny (*HN.* 31.94), and a widespread distribution of *garum and liquamen* from the *tituli* labels on *urcei* in the town and a wider international Mediterranean trade, shown from amphorae distribution in the wider Campanian area, of an imported Spanish *garum, liquamen* and presumably salted fish. It is very difficult to distinguish each of these trades and determine how they were interrelated. The most successful trader in fish sauce was Aulus Umbricius Scaurus, whose name is found on many of the *urcei* that were used to sell fish sauces in Pompeii. Scaurus' name and those of his freedmen associates rarely appear on amphorae *tituli,* and this may be quite significant (Curtis 1991:91; Peña 2007:82).[7] I think it is very probable that Scaurus imported expensive Spanish products and further processed them before selling them in the small purpose-made vessels, though he may have made a local product too. The perceived qualities of these sauces are discussed in relation to fish bones and amphorae in Chapters 11 and 12. Stephen Ellis has proposed that an earlier thriving industry in fish sauce existed in Pompeii which was driven out of the market by the imports from Spain. The archaeological evidence is discussed in Chapter 2 page 58 (Ellis 2011).[8] I am less inclined to consider that the local industry was destroyed by Spanish imports as I consider that Spanish products, which may have been dominated by mackerel, were always relatively expensive (see page 31) while the residues from the *Garum* Shop indicate a locally caught ordinary *liquamen* (Carannante 2019; Vargas *et al.* 2014; Bernal-Casasola and Cottica 2017). Scaurus, from the *tituli,* was clearly selling high end products to the cooks of elite households. This would rather suggest that the local industry using dolia need not have been destroyed by Spanish imports as there was a ready market for both, i.e. a cheap local *liquamen* for everyday use made from small fry caught in the local area, and an expensive and imported *liquamen* and *garum* made from either all mackerel or a higher proportion. It was certainly the Spanish products that were the most valued according to Pliny. Curtis disagrees and thinks that Spanish products were widespread and accessible (Curtis 1991:59). It is also clear that other people in Pompeii engaged in the fish sauce industry as there are a further eight names linked to the trade from the *tituli* not connected to Scaurus (Curtis 1988:33). Lowe cites a number of further places in Pompeii where fish sauce may have been made or sold, though the attribution is based on a concentration of *urcei* finds rather than dolia or residues (Low 2016: 313). At one small house (*Regio* IX.9.6–7) seven vessels were found with all four varieties of sauce inscribed on their labels, but this does not necessarily mean that in this modest house all the fine sauces were consumed. It may in fact be a space where empty vessels were gathered for subsequent reuse elsewhere (see page 199). Evidence for the consumption of small fish sauce species (*Sparidae* and *Clupiedae*, that is, mainly sardine, bogue, and horse mackerel), in a state of good preservation has been found associated with the ordinary working people of Herculaneum in the fish remains from the *Cardo* V sewer which lay below many small shops and apartments. Rowan found many otoliths in the sample and concluded that fish were consumed fresh rather than as a sauce residue (Rowan 2014:62), assuming that the fish sauce process would have destroyed the otoliths completely. However, the bones and otoliths are much less likely to survive the human gut in normal circumstances, which does seem to contradict this theory (Nicholson *et al.* 2018:277).[9] The sewer otoliths had smooth edges which are interpreted as evidence of human digestion rather than degradation from fish sauce, though the distinction in terms of how much damage these processes do is by no means clear. A recently concluded experiment by the author (2020) to determine the survival of otoliths has demonstrated that they do survive fermentation for up to 12 months (Rowan 2014:62, n.16). The bones are believed to have been put down the latrine shafts in the corner of each room, either as fresh waste or as an exhausted *allec*; the latter is more likely in my view (Rowan 2014:62). The residues from the sewer revealed scarce evidence for mackerel of any size, indicating low-value fish sauce consumption mainly horse mackerel, bogue and sardine are known to shoal together in their juvenile stage, which suggests a random catch. A rare example of herring is particularly interesting as it does suggest that such species could have reached Mediterranean waters (Nicholson *et al.* 2018:276).

Many find Pliny the Elder's suggestion that Pompeii was famous for its *garum* a difficult statement to square with the lack of evidence for a thriving bulk fish sauce industry. However, it is clear that a quality product need not be made in bulk. The question is also posed by scholars who think in term of a single sauce. There is a possibility that Pliny was referring to a local blood/viscera *garum* production using multiple species and in small dolia which we cannot see in the archaeology, or that he is referring to the thriving trade in both sauces, derived from Scaurus' many *officinae,* which may not in fact have been a local product at all but customers may not have known this. Scaurus never claims a Pompeian origin for his products; they are from his workshops, but otherwise the regional origin is missing from the *tituli,* which is odd as many other fish amphorae do include specific regional origin (Curtis 1991:195). Given the vast quantities of Spanish amphora in the area one has to ask who was consuming the Spanish fish sauce, which must have been inherently more valued, if Scaurus was also using local fish. Others continue to consider Scaurus the manufacturer of the products that he sells (Flohr and Wilson 2017:125).

The *Garum* Shop in Pompeii is also known as the *Bottega del garum* and the *officina del garum degli Umbricii (Regio 1, insula 12,* doorway 8).[10] The building appears to have been a residence that was later reused for industrial purposes after the earthquake of AD 62 (Bernal-Casasola and Cottica 2017:238). Peña (2007:86) assumes that the shop is associated with Scaurus, but the absence of any *urcei* in the shop and any other link with Scaurus would rather suggest that it was independent from the Umbricii clan. The *Garum* Shop has been described in increasing detail in many recent reports (Bernal-Casasola *et al.* 2014; Vargas *et al.* 2014; Bernal-Casasola and Cottica 2017). The facility comprised a front room that may have functioned as a shop but without the usual counter, a courtyard containing dolia with fish sauce remains, and a number of stacks of reused amphora and dolia, 150 items in total, piled in many corners, a kitchen area with a platform hob, and a garden at the rear with many other forms of amphora, illustrated in Fig. 10.3 (Flohr and Wilson 2017:37).

There were also fish remains in and among the upturned amphorae in the north eastern corner of the back of the garden, identified as *Regio* 13 by Bernal-Casasola *et al.* (2014:220). These were Dressel 21/22, a type recognized as a salted fish amphora made both in Italy and Sicily (Botte 2009). Their shape is significant as they have a wide-open mouth, virtually no neck and a straight body and short spike, and are small and relatively easy to maneuver (Botte 2009: 132). The fish bones within this site have been analyzed extensively (Carannante 2019; Bernal-Casasola *et al.* 2014: Bernal-Casasola and Cottica 2017:241). It is also of note that wine amphorae were present in the shop (Dressel 2–4) and this may indicate the use of wine when fabricating fish sauces, as noted in Chapter 1.[11]

There were numerous different mixes of fish species being used to make or further process five different but separate batches of fish sauce. The most recent work by Carannante (2019) has illuminated how complex the species selection was in the make-up of these sauces. Carannante combining his and Bernal's analysis, showed that apart from the six dolia with a pure anchovy sauce there were 'eight amphorae filled with a pure anchovy *garum* (his term) three others were filled with pure pickerel and another six amphorae contained a mix of both taxa. An additional six amphorae contained

Figure 10.3   The *Garum* Shop courtyard with dolia and amphora stack (*Regio* 1, *insula* 12, doorway 8) Copyright Sally Grainger

different mixes of anchovy picarels, and other *Sparidae* (sea breams and porgies), *Clupeidae* (herrings and shads), *Carangidae* (horse mackerels), and Scombridae (*scomber japonicus*)' (Carannante 2019:3.1–3). In sum, "five different batches of fish (a pure anchovy, a pure pickerel, and three different multispecies sauces." It is possible and likely that the pure anchovy sauce in the dolia came from the amphora that contained a similar anchovy residue and the fish were brought into the shop in the Dressel 21/22 vessels.

The analysis carried out by Carannante (2019) on the pickerel was on remains removed from the site by Mauri in the 1960s and stored, so that a direct link with particular dolia or amphora is now lost. What is most obvious in terms of the perceived quality of these sauces is that there was hardly any mackerel (two amphorae, A85 and A79, contained evidence of a small amount of mackerel caught locally: Bernal-Casasola *et al.* 2013:333), and all the species utilized were very small (15 cm or less) and this confirms the general impression that the shop sold a product that was a common-place, everyday *liquamen* undoubtedly using a locally caught catch. In the process of analyzing the residues from the Dressel 21/22 amphorae stack in the corner of the yard, Bernal-Casasola and Cottica (2017:243) have attempted to associate the specific individual species associated with each vessel with the particular *titulus* found on the vessel. There are brief acro-nyms on the Dressel 21/22 amphora (*CE, COP, AB, MAL, SPA*) which are assumed to be potential sauce names, while I and others disagree (Botte 2009:140; Bernal-Casasola *et al.* 2014:231). Thus, in the case of "amphora n. 18, type Botte 2, with *titulus* SP contained anchovy and pickerel while the amphora has an origin in Calabria." The report asks "How to explain the presence in the same archaeological context of amphorae of different origins with similar or identical content"? (Bernal-Casasola and Cottica 2019:243). The *titulus* demonstrates no indication of relabelling, which Bernal-Casasola and Cottica assume is necessary for the purposes of a comprehensible trade. Relabelling may not be necessary if the people who handled the vessels after refilling were entirely familiar with the content and accompanied the vessels to their place of sale. It was noted that all the vessels protected from the ash bore a *titulus,* which is taken to indicate that the practice of affixing a *titulus* was the norm in the ancient world, at least for certain classes of amphorae with a specific functional destination; however, it seems fairly certain that the *tituli* are not connected to the content, and Bernal-Casasola and Cottica do acknowledge that this is probable (2017:241). The origin of the amphora and the original salted fish product may have been a Calabrian, but the fish residue represents a storage reuse using a *liquamen* fish sauce after the salted fish had been removed and consumed. They conclude that the *Garum* Shop and the pile of amphora represent an "anomalous case in the archaeological record" (Bernal-Casasola and Cottica 2019:244). I would rather see this site as typical, and entirely consistent with the normal processing of fish sauces of the *liquamen* type using what appears to be a random selection of reused amphora to store and process a fish mash. Auriemma concludes that this facility should be seen as a "*scenario virtual*" for the kind of system of reuse of amphorae seen in shipwrecks such as Grado (see Peña 2007:86; Auriemma 2000:45).[12]

Both Curtis (1991:94) and Peña (2007:87) consider the different possible modus operandi for the site. Curtis considers that product was made entirely on the site and sold to extra-local markets in amphorae, due to the lack of *urcei* (Curtis 1979:15–18). Peña also considered that the product could have been made elsewhere and brought in to be finished on site and then decanted into vessels brought to the establishment by the customers, which I concur with entirely as we can be certain that elite sauces were marketed in their own purpose made vessels, while this product appears to be a more commonplace product for the general populace. Peña also suggests that the amphora stack could represent the vessels that the fish sauce was brought to the shop in (Peña 2007:87), and this is surely the most likely interpretation of the modus operandi of the site. I am unsure that the sauce produced in the shop was traded further afield in these amphorae as Curtis does (1991:95) as I suspect the volume generated, though substantial over time, would not have been large enough to justify an amphora trade from the shop. I think rather that *liquamen* made here was sold to local Pompeiians of middle and lower rank, who may have brought their own vessels to take it away. In a detailed analysis of the anchovy bones in the dolia, Rodríguez-Alcántara *et al.* (2018) concludes that the residues were at different stages in the production process and suggests that there was both purpose-made *allec* and *liquamen* production side by side. This was deduced from a decrease in the total nitrogen of the *allec* reference in Dolia 4 and 5 compared to a fresh sample, and this was attributed to loss of nitrogen when the sauce was taken, leaving a weaker *allec* (Rodríguez-Alcántara *et al.* 2018:157). They conclude that "D1, D2, D3, and D6 were containers being used for *garum* production; while D4 and D5 were being used to store the finished product: they had liquid *(liquamen)* and *allec* or both already separated" (Rodríguez-Alcántara *et al.* 2018:161). The researchers clearly consider that *garum, liquamen, and allec* could have been derived from the same batch of fish, and as noted I cannot agree with this.

How was the blood viscera garum made?

It is plausible that the first version of this sauce was made in the region of Rome or possibly on the Bay of Naples. Mullet was undoubtedly bred for the table in the famous fish ponds associated with many gourmets living in and around Baiae (Pliny the Elder, *HN.* 9. 168; Marzano 2013:199–210f. for the fish ponds and their structure). We may imagine that the sauce was conceived of in just such an environment, with gourmets gathered around a fish pond discussing the color changes of a dying mullet and then observing the blood flow. The resulting harvested juices were then mixed with salt and left sealed until the desired color and taste was achieved. One may imagine that when mullet was the only species used the volume that was made was tiny and insignificant. This is all we can say about the early product. Eventually, mackerel were used to make the *garum sociorum* and some kind of systematic procedure was set in place. The industry moved eventually to regions where mackerel was prolific, such as Cartagena in Murcia, but many other sites will have made use of their relatively abundant mackerel to make the elite and expensive sauce as it became more popular. If a blood garum was to be made in bulk, vast quantities of mackerel needed to be cut and bled very rapidly, while they still swam in the traps. How this was managed is difficult to imagine. We have the description from Manilius of what I believe was mackerel being cut into pieces to make both a *garum* and a *liquamen*, but he is silent on the issue of how the blood was obtained separately. I suspect that the initial bleeding process occurred at sea before the catch was brought ashore, where Manilius picks up the story (see page 35). Perhaps a small boat sailed among the shoals of compacted fish and scooped them up on board and cut them before putting them aside to drain. The fish either had their crania cut (as described on page 174) or the gills were punched through to allow blood to flow. After bleeding the viscera would then be removed in a separate process. The amount of viscera in a mackerel is estimated at around 20% of the weight of each, from the author's experience. Most of the blood within mackerel is in the head and does not gather in the visceral cavity.

Once the boat had made land the gutting of the mackerel could have taken place as a separate procedure. One may imagine that the blood and viscera were initially blended in a dolia and other relatively small vessels, while it would only be with the production sites where they could guarantee a bulk mackerel harvest that purpose made *cetaria* were built for this sauce. It does become clear why this kind of sauce was so expensive. It could only be made in bulk where the infrastructure for catching huge quantities of mackerel could be guaranteed. Anyone could make a small batch of *garum sociorum* from their haul of 50 or 100 mackerel, but a guaranteed bulk product required fish in the thousands if not hundreds of thousands, and substantial resources. We have discussed in Chapter 8 how *garum* and *liquamen* appears on completion as a creamy emulsion. I wish to suggest that the term *flos* is ideally suited to convey this emulsion, though we note that others consider *flos* to refer to an aspect of its quality, with its meaning of "flower of" (Berdowski 2003:26; Curtis 1991:163). *Garum* occurs frequently with the term *flos* (52 times in tituli cited by Curtis 1991:195). And I feel it is far more likely to refer to consistency, especially as experiments have demonstrated that consistency was a fundamental part of the manufacturing process in both sauces. From the recipes we learn that it had to be harvested when the vessel is pierced at the bottom, and the sauce flowed out already filtered of the semi-dissolved visceral matter, but not of the finer particles that made the sauce cloudy and slightly viscous. The used-up and useless semi-dissolved viscera may have been left inside the production vessel and/or discarded or formed part of the on-going fermenting process in the amphora. Peña has noted that there are a number of amphorae of the Pompeii VII form with holes drilled in the spike and also African amphorae with the tip of the spike cut off, and this can be interpreted as an indication that sauces of both types might have been drained out from beneath while leaving unwanted matter: bone or visceral matter, inside (Peña 2007:66). It certainly appears from *tituli* that *garum flos* could have been stored and sold in this unfiltered state, and it was at the point of use that it was finely filtered through many layers of cloth to remove the remaining particles and render it clear like wine for use at table. There are indications that *garum* was aged. Ehmig cites two aged sauces from Mainz: a [*garum*} *scombri flos* AA in a Dressel 7–11 jar, though this may have been a *liquamen,* and a *garum scombri flos* AA in a Vindonissa 586 (Ehmig 2003:63), both seemingly held for 2 years.

Direct evidence for *garum* production is scant, and depends entirely on one site, Aila Aqaba, on the Red Sea coast of Jordan, which has a continuous stratigraphic sequence from Roman through to early Islamic periods. The site is a domestic complex of indeterminate status and among the finds was a jar containing bones associated with gill rakes of large tuna. The bones residues also include a species of Nile fish which has clearly been consumed by the tuna and will represent stomach content. The jar was a "baggy jar," which one can define as large and wider in the middle with a slightly smaller ribbed neck and mouth with a capacity of c. 8.5 liters (Van Neer and Parker 2008). The product was identified as a *garum haimation* from the references in the *Geoponica* to such a sauce being made from tuna viscera and blood. The date of this find is crucial (late 1st century AD), as it may reflect evidence of the switch from a sauce made with the difficult-to-extract blood and viscera from mackerel to one that undoubtedly generated a great deal more of the sauce. It is tempting to assume that it was only a mackerel or mullet *garum* that acquired the designation *sociorum* at first, and this is the currently prevalent opinion (Leon 2001), while *haimation* was only associated with tuna, but I am less convinced by this. *Haimation* means "bloody" and must I think refer to the concept of a bloody sauce rather than the blood of a particular species. The term is not rendered in Latin at all and as we shall see, in the late empire this sauce was always rendered just as *garum or garum excellens/primum* without being associated with a particular fish when it appears in didactic recipe texts alongside the still common use of *liquamen*. (see page 92. Most fine *garum* in the early 1st AD was apparently made from only from mackerel, but this may be a manufacturers' fiction, as there would be no way to prove that the viscera utilized was purely mackerel and was very likely always in part a mixture of viscera from multiple species, and as the demand increased, there is even an expectation that when it was possible to catch tuna and mackerel together, the blood and viscera from both would have been used to make a generic *garum*.

It would seem logical to consider capacity of the tanks to be the major factor in trying to distinguish the different types of sauce in the processing factories: a smaller tank for the *garum* and the bulk *liquamen* in the vast, high-capacity vats common in the Spanish and North African sites. Noting that *garum* was harvested from the bottom of the vessel, there are a number of sites with distinct smaller processing tanks which are constructed to allow for drainage from beneath, as they have various forms of channeling at the base. The first is at Rhode (Rosas), north of Barcelona. Curtis (1991: 55 citing Nolla and Nieto 1980, translated from the Catalan) describes them thus:

> Five roughly square vats (between 1.60–65 and 1.80–85 metres) ... arranged in a semi -circle about a central space. The floors of the vats were higher than the room itself and sloped slightly in the direction of the centre of the room. At the bottom of at least three of them ... a conduit extended through the front wall ... this conduit was capable of being plugged and, when opened, emptied into a small round catch-basin cut into the floor of the central space.

This site is dated to the mid-3rd century AD. Curtis notes that a similar arrangement was found at a site at Alcazarsegher, between Tangiers and Ceuta in North Africa which is described as a 1.3 by 2.8 m tank with a conduit running in the front to a semi-circular basin. There were ten other *cetariae* on the site with the normal sealed walls. The date of the site is unclear though, it was estimated to have been built in the 1st century AD and abandoned in the mid-3rd (Ponsich and Tarradel 1965:72). At factory A at Nabeul, also in North Africa, dated to the end of the 2nd/beginning of the 3rd century AD, there were three vats (XXIX–XXXI) placed in a line, and the side vats communicated with the central one, 0.08 metres deeper, through lead pipes (Slim *et al.* 2007:35). A rare example of tuna and bonito bones have also been found in these special *cetariae* and as this can be associated with the last products made in the vats, it may have been a tuna *garum*. The bones also show evidence of cutting (Slim *et al.* 2007:35). It is not clear that these structures are only associated with *garum*, as a *salsamenta* brine may have been drained out through these channels. It is also not impossible that a *liquamen* was also drained out this way. Nevertheless, these structures seem to have been specifically designed to allow sauces to be drained out of the tank from beneath and it would seem logical to associate these vats with a *garum* production, given the instruction in the *Geoponica*. A *liquamen*, as we have seen is easily and conveniently removed with its *allec* from above, while *garum* has no *allec* and can be drained out from beneath already semi-strained. There are other interpretations for these unusual structures. In the analysis of the fish salting industry Garcia Vargas (and Bernal-Casasola 2009:146) discuss the Nabeul evidence and suggest that the drainage is associated with the removal of the oil from a sardine fish sauce, but we note that the oil in fact floats on the surface of the sauce and not sink to the bottom as suggested. These structures are quite rare and are all distinctly late in date, which certainly suggest they were not generally associated with *liquamen*. If there is a link between them and *garum* it may indicate that the amount of pure blood viscera garum being produced in the late empire was relatively limited? Certainly, the literary evidence suggests that *garum* was not readily available and rarely mentioned in texts at this time (Chapter 76). It may be that it was only in the mid to late empire that structures were built specially to accommodate *garum* production, which would seem to be at odds with its apparent popularity in the 1st century. If these structures indicate a true *garum* production, we must ask why they don't occur earlier. The capacity of these rare and uniquely designed vats was relatively small in comparison to the majority of vats across the entire *cetariae* corpus, many of which were constructed from the late-1st BC to mid-1st centuries AD at a time when the fish sauce trade was apparently at its most active (Wilson 2007). One must conjecture that most of the true *garum* was made in small volume in *dolia* (such as the 8.5 liters jar at Aila Aqaba: see above) and therefore largely invisible. This disconnect between the size of *cetariae* and the kind of sauce being made may suggest that we should possibly doubt how popular and prolific the blood viscera *garum* was in terms of its actual consumption even in the early empire. This view is contradicted by the prevalence of *tituli picti* for *garum* which appears to demonstrate that *garum* was widespread and in comparison, to *liquamen,* the more popular product. The relative popularity of *garum* over *liquamen* from the Pompeiian epigraphic sources are discussed in page 254.

There are many possible ways in which fish sauce factories could have been organized so they could make fish sauces. A fish enterprise could have specialized in alternative processes. One might have concentrated on *liquamen* from its predictable catch of *Clupeids* and *Sparids* with occasional Scombrids and with the processing of sporadic and unpredictable catches of table fish and *salsamanta,* which incidentally provided any extra viscera, if required, to make the *liquamen.* This kind of system probably produced little or no blood viscera sauce. Other establishments, with access to modest or large quantities mackerel shoals, would concentrate on making mackerel *liquamen* and *garum* and utilized advanced processing techniques to take and harvest the blood and viscera from large shoals of mackerel in sufficient quantity, and would also salt the empty mackerel as *salsamentum.* It is certainly possible to suggest that it is only when tuna began to be used to make *garum haimation,* documented by the Aila Aqaba find, that the extensive tuna fisheries became associated with the manufacture of a blood viscera *garum,* though we cannot see this process at all well in the archaeology and we note that others largely disagree and consider that tuna blood and viscera were utilized to make fish sauces from the Punic period onwards (Vicente *et al.* 2009:94; Aguilar *et al.* 2011; Bernal-Casasola *et al.* 2003) This whole process was very labor intensive and it is not surprising that few places were geared up to make *garum* in bulk. Pliny the Elder states that Carthago Spartaria (Cartagena) was the most renowned for its *garum sociorum,* made from the particularly abundant mackerel that could be caught there. It would seem that Carthago Spartaria had precisely that level of investment in the techniques required. If sufficient blood/viscera *garum scombri* was to be made to meet the apparent market demand for it then the quantities of mackerel required are immense.

Notes

1 For the render used to line the proposed *cetariae* in Pompeii see page 58. It is not always clear why a vat or tank is given a watertight surface. There is a tendency to leap to the conclusion that fish sauce was the intended product.

2 I note that recipes include the instruction to add or leave a portion of *allec* from the previous fermentation in the tank to seed the next batch of fish sauce, like a yoghurt culture, the theory being that this would retain sufficient enzyme activity to ensure that the new batch of fish begin to dissolve efficiently. The depression may have ensured that enough *allec* was left, but this is conjectural.

3 This small *commune* in the province of Salerno in Campania specialises in anchovy today. The name of the commune would seem to be linked to the Latin word *cetaria,* which is associated both with fishermen and the structures they used to preserve their fish (Berdowski 2013).

4 Pollen in unusual clumps has been found in the residues of a standard *liquamen* fish sauce in a Beltran 2B at Augusta Amerita. The pollen is from the *Asteraceae* family, which is largely associated with wild flower meadow and

is surely a reflection of the surrounding vegetation. Bernal-Casasola *et al.* (2016:746) considers that the pollen clumps reflect intentional addition of the flowers to flavor the sauce, but this is not a secure conclusion. This family does produce green herbs of culinary use but if and when gathered as herbs they are normally specifically not allowed to flower. The clumps reflecting large quantities of pollen could easily have been produced by large swathes of ground left to meadow in the surrounding area. For the rare inclusion of spices in a fish sauce residue see Van Neer *et al.* (2015:574) where two coriander fragments were found along with fragments of wheat and grape pips in fish sauce residues in an amphora from Oxyrhynchus, however these minute fragments are likely to indicates something invasive.

5 See Chapter 7. There is plenty of shellfish evidence on processing sites, but its role is not clear and could simply be incidentally captured as Dario Bernal has acknowledged (Bernal-Casasola 2018:341). Shellfish are common at Bello Claudia but not in sufficient number to suggest that they were added intentionally. Bernal-Casasola (2018: 340; Table 3) cites the MNI for each species, and the majority exist in very small numbers. In other sites at Baelo Claudia, limpets and other marine gastropods have been found discarded in piles, but they are not in abundance in the *cetariae* mixed with the fish sauce residues (Bernal-Casasola *et al.* 2018:347). Such evidence as exists suggests that the shellfish were processed separately to make bone-free sea food pastes, that is, *allec*. For a Dressel 6 A with examples of nine species of shrimp included with a *liquamen* residue, probably caught from a brackish lagoon, see Mazzocchin and Wilkens (2013). For a different view suggesting that oysters were added intentionally, see Bernal-Casasola *et al.* (2016: Garnier 2018:320; see page 126 in the Literature Review).

6 The researchers note that the samples demonstrate a reddish-orange color, compared to their reference samples, which have a grayish color (Rodríguez-Alcántara *et al.* 2018:159).

7 Of the 200 vessels with a *titulus*, 36 are amphorae and 140 are *urcei* (in the Schoene Mau classification, Pompeii VI). Scaurus or his family appear on over 28% of all fish containers (amphorae and *urcei*) found in the Pompeii region (Curtis 1991: 92).

8 A paper dealing with this issue, examining samples of the render used to line the tanks in question for absorbed residues, is in preparation and the results are pending (Botte 2018:382).

9 The otoliths are calcium carbonate structures in the inner ear.

10 I retain the name *Garum* Shop here but prefer Fish Sauce Shop because the former use maintains the idea of the single sauce.

11 Wine residues have also been discovered in the dolia fabric (Pecci *et al.* 2018), which as noted by Bernal-Casasola and Cottica (2017:245) could be an indication of previous use of the dolia for wine fermentation or that wine was added during the manufacturing process. Bernal-Casasola and Cottica 2017 consider that the wine was added in the production of *oenogarum*; it is more likely to have been added as an integral part of the manufacturing process to replace loss to evaporation and to increase acidity and as an antibacterial agent; see page 157.

12 I note the lack of any herbs residues that might link the production process with the Gargilius recipe that these scholars use to fabricate fish sauce to compare to the ancient residues (see page 24. See also Vargas *et al.* 2014:76 and Bernal-Casasola and Cottica 2017).

11 Fish bones as evidence of sauce and salsamenta

Methodology and interpretation

In what follows, the principal evidence for the various forms of residues associated with ancient fish sauces and *salsamenta* are reviewed. Much of the evidence for the ordinary type of *liquamen* is very similar. There is a remarkable taxonomic homogeneity (or the presence of very few taxa) in the residues. The size, state of degradation, and find position is also similar. This similarity in disposal behavior patterns is to be found across the entire Roman Empire during the period of the most active fish sauce trade, up to the late 4th century AD, and it suggest that ancient peoples engaged with, consumed and disposed of, the residues of fish sauce in similar ways where ever they were consumed. The residues of the small-fish *liquamen* sauce have been found in a variety of different combinations. These include *Clupeiforms*, such as sardines (*Sardina pilchardus*), sardinella (*Sardinella sp.*), anchovies (*Engraulis encrasiculos*), and *Sparidae*, including sprats (*Spratus spratus*), sand smelt (*Atherina presbyter*), pickerel (*Spicara smaris*, previously known as a *Centracanthidae*), bogue (*Boops boops*), sea bream (*Sparus aurata*), herring (*Clupea harengus*), and occasional small mackerel (*Scomber japonicus*). It seems clear from our look at the experimental approach that the quality of a mixed sauce would not be determined by species selection. The resulting sauce would be, from observing the process, a homogenous brown creamy viscous liquid with no distinguishing characteristics to separate one sauce from another in terms of appearance, or, I would suggest, taste. The addition of mackerels such as (*Scomber japonicus* or *Scomber scombrus*) or horse mackerel (*Trachurus trachurus*), in small numbers must be linked to a random catch and was not intentional. A higher mix of mackerel may have improved a sauce's quality simply because there was more flesh to dissolve, resulting in higher nutrition and superior taste. Literary references indicate that a pure mackerel sauce was the best and most valued *liquamen* in the mind of the consumer. Other considerations of quality are discussed below but these issues may not have been relevant if the ancient customer perceived mackerel as the best and that is what the literary and epigraphic evidence suggests. The same question of course applies to *garum* itself: how much viscera and blood from other fish could be added before it was no longer considered a pure mackerel *garum*? There are of course other views which consider that individual species selection and minute size differentials made a different to the nature of the finished sauce, its perceived value to the consumer and determined its name (such as Cotton *et al.* 1996, discussing Masada, and Bernal-Casasola *et al.* 2016:747) on the *Augusta Emerita* evidence and the *Garum* shop (Bernal-Casasola *et al.* 2014:230, see below). As we have noted throughout the book, it difficult to distinguish between finds of fish bone which represent *salsamenta* and those which might be evidence of fish sauce residues. Emmanuelle Botte stresses that we should be thinking of *salsamenta* and sauces as radically different, which he attributes to the fact that sauces were not accessible to all, while salted fish was. There is an assumption here I think that sauces were the more valued product with reference to Pliny, exclusive and not widely available. I would suggest it is the other way round, that is, sauces of the small fish variety were available to all and *salasamenta* was prolific in the early period but increasingly became rare as fish stock declined (Botte 2009:177). The survival of fish bones in archaeology is dependent on many environmental conditions. After consumption, the large bones from a fresh or salted mackerel could be discarded in a rubbish area where animals scavenged, trampled underfoot in mineral rich soils which could erode bone, and ultimately deposited in acid soils which completely destroyed bone. The process of salting fish is believed to soften the bone: this may be exacerbated by long term storage and while this is assumed to make it possible for the bones to be consumed,[1] it also may contribute to lack of faunal evidence. Fish bones survive best in damp or waterlogged conditions and enclosed sites such as wells and middens which were quickly covered with other rubbish, or when deposited inside amphorae which were then subsequently discarded and buried to seal the environment (Wheeler and Jones 1989:63).

Morales has noted that the survival of the *Clupeiforms* (sardine and anchovy) is very poor in normal Mediterranean archaeology and points out that the finds of these species in large quantities are usually in the sealed environments and it is these that are associated with fish sauce (Morales-Muñiz 2016:32). The size of the mesh or water float used to sift the sediments for fish bones is fundamental, as the smallest size of mesh can often reveal evidence that otherwise would not be found. There was in the past is a marked inconsistency in the use of such fine mesh and the use of water flotation, and consequently the evidence is significantly distorted by retrieval bias and the subsequent absence of *Clupeiforms* on ancient sites has been termed a "methodological construct" (Morales-Muñiz 2016:33). Van Neer and Ervynck (2002:208) consider that fish sauce can only be identified where "fish bones are present," which is a major problem as the fish sauces of quality are believed to be a clear free-flowing liquid without residue and therefore would not be identifiable in the archaeological record at all. The only product that can be identified is the residue of the sauce. which as we have already seen is a very contested product, but nevertheless prolific in the archaeology. When the retrieval bias in relation to *Clupeiforms* is taken into account, there may have been widespread deposition of fish sauce residues which simply do not appear in the archaeology. When sealed environments are absent, the practice is to sample the fish bones from a spoil heap or ditch and unless the bones are particularly prolific, this will inevitably represent only a tiny fraction of the actual fish consumed on the site as the majority will have been lost to the various forms of degradation. It may be possible to identify species and size of the fish; when the species is not local, one can say it is an imported and therefore a salted product, but we cannot determine how much actual fish in terms of meals the bones represent. Fish sauce residues from sealed sites are very different as they are invariably deposited in different and unusual contexts. They are invariably found in large quantities deposited or left in thick layers of disarticulated bone and they are also found packed into amphorae and apparently discarded.

Fish sauce residues

Ancient fish bones associated with fish sauce can be found in many different environments within the archaeology of a site.

1 In production – in *cetariae* throughout the empire. Many fish processing sites are found to have thick layers of fish bone residues left in place, buried or silted up. The sites are listed in Table 11.1.
2 In production – on a small scale in dolia, such as the residues at the *Garum* Shop in Pompeii (discussed in Chapter 10).
3 In transit – in amphorae preserved from ship wrecks. Many wrecks contain disputed *salsamanta* or fish sauce residues, discussed below and listed in Table 11.3.
4 In prime use in small jugs and *urcei* – a number of vessels in Pompeii contain fish sauce residues which must have been in use at the time of the eruption, discussed (see below).
5 Discarded after use in small vessels in bodies of water – The little Latium pots from the river Rhône represent a discarded *allec*, discussed on page 191.
6 Discarded after use in broken amphora – in Sardinia at Olbia amphorae with the neck and handles removed contain substantial amounts of fish bone residue from fish sauce, discussed on page 228 and in Table 11.2.
7 Discarded after use in whole vessel – such as the amphora at Augusta Emerita referenced page 226 and in Table 11.2.
8 Discarded after use in refuse piles near ports and entrepôts – when the sauce has either been made or been extracted from imported *allec* and the bones discarded as refuse discussed page 227 and in Table 11.4.

Table 11.1 Examples of fish processing factories with *cetariae* containing residues of fish sauce

Site	Date	Site details	Species and size	Preservation	Reference
Quinta do Marin Alhoa Algarve Portugal	Early 3rd century AD	Fish processing factory on the Bank of river Formosa Basin 5 layer 5 25 cm layer sampled	Young sardine and anchovy 6–14 cm (majority 6–8 cm) *Clupeidae*	Poor, small, and fragile	Desse-Berset and Desse 2000:87
La Travessa de Frei Gaspar Setubal Portugal	Late 3rd / early 4th century AD	Fish processing factory 14 *cetariae* 7 have residues. Basin 2 layer 3	*Clupeidae* and *Sparidae* inc. Sardine 6–19 and pandoras (*Pagellus*), 10 cm.	Very small 2–5 mm	Desse-Berset and Desse 2000:89
Troia Setubal Along banks of Sardo river	Mid. 5th century	Fish processing factory Many basins with residues	*Clupeidae* sardines 8–19 cm majority 14 cm plus many different *Sparidae*	Fish bone flour	Desse-Berset and Desse 2000:91
Neapolis in Nabeul Tunisia	Built 60-80 AD	Fish processing factory Vat number iii and xxxiv	60% anchovies 20% sardines, 12% blotched picarel 5% mackerels, and 3% small pandoras (*Pagellus*) *Scombridae*, *Sparidae*, and *Clupeidea*	Small and pulverulent	Slim et al. 2007:35). Sternberg, 2000, 135–153)
Bealo Claudio Baetica Spain	5th century AD	*Cetaria* C.I. XI Vat P3 / Vat P9 / Vat P5 ; *Cetaria* C.I. XII Vat P5 / Vat P3	- sardines Clupeidae / - anchovy / - a mixture of the two / - sardines / - Pandora *Sparidae*	4–8 cm deep layer of residues	Bernal 2018a: 340

Table 11.2 Amphora and ceramic vessel with residues of fish sauce found on land

Site	Date	Vessel form	Species and size	Notes and preservation	Reference
Setúbal 1. Rua Francisco. 2. Rua Antonio Joaquim	Late 1st C AD/early 2nd refuse dump. 1st century AD-fragment in pit	Dressel 14 Variant C Dressel 14	Sardine 16–20 cm in articulation Sardine 17–18 cm Low frequency in sand disarticulated *Clupeidae*	i.d. as *salsamenta* I.d. Ambiguous potential post depositional event	Gabriel and Tavares da Silva 2016.
Saltsburg urban ditch	Undated Amphora late 1st BC	Dressel 6	Sardine, anchovy 4–12 cm plus 22 other low value species, incl. *Atherina, pagellus, boop boops moena, mullus* <10 cm	Considered cheap fish sauce, i.e., *allec*	Lepsiksaar 1986; Von den Driesch 1980, Cotton et al 1996
Masada Urban fortress	Late 1st century AD	Local Syrio-Palestinian baggy jar	Sardine, anchovy, herring, sprat 4–5 cm *Clupeidae*	Bottom of broken jar half filled with sand and bones	Cotton et al 1996:223
Vicenza Reused as Road foundation	1st century AD	Dressel 6A Small mouth, long narrow neck	Picarel, Minnow (Cyprinidae) Lagoon Shrimp (*Palaemon*, Crab (*Brachiura*) Marine and fresh water mixture indicate lagoon fishing.	Considered discarded while part-full, i.e., due to decay. Note i.d. as *allec* purpose made with intentional addition of shellfish but a cheap liquamen. more likely.	Mazzocchin And Wilkens 2013
Aila Aqaba Jordan Humble dwelling.	Late 1st Century AD	Rib necked jar, 9 litres capacity	Frigate Tuna gill apparatus (*Auxis*) SL 40–50 cm Lizard fish	Lizard fish found in stomach content of tuna. *Garum haimation*	Van Neer and Parker 2008
Petra Ez Zantur	Early 5th century	Pilgrim water flask 20 cm height, 2 cm neck opening	Sardine and anchovy Clupeidae c. 2.5 cm	Possibly from the Red sea. I.D. *allec* or fish sauce	Studer 1994:191–196

Site	Date	Context	Fish species	Notes	Reference
Augusta Emerita *Merida* in an underground ice house	175–200 AD	Beltran 2B (smaller form) 23 litres capacity	*Sardinella aurita* 4-5 cm anchovy, 4-5 cm *Clupidiea* horse mackerel 10 cm *Trachurus trachurus*	Fragile, much in powder from. …S(COM)BRI F(LOS) *titulus* Daisy pollen = weeds	Bernal et al 2016
Olbia Sardinia Shop destroyed by fire	4th/3rd century BC	Local undefined and fragmentary, bottom only	Picarel (*Maena Smarnis*) v. small	Handle and neck cut off Potential earliest evidence for *garos*	Delussi and Wilkins 2000:53;
Olbia 1 Sardinia Urban infill from well	2nd century AD	African type but unidentified. Bottom of vessel only	Seven different species incl. bream, wrasse, pickerel, grouper, *Sparidae* 15-20 cm	Handle and neck cut off Each species can vary in size greatly but all of similar size range	Delussi and Wilkins 2000:56; Bruschi and Wilkins
Olbia 2 Sardinia Urban infill from well	2nd century AD	Africa 1/Keay 3 Bottom only	Top layer: *Clupeidae* 5 cm articulated Middle layer *Sparidae* 10 cm broken up Bottom layer: *Clupeidae* 5 cm	Handle and neck cut off Three distinct layers of residue represents separate disposal events	Delussi and Wilkins 2000:59; Bruschi and Wilkins 1996:169
Arles Rhone 3 Little Latium pots	50-80 AD	Latium pot AR3 2001 51. c. 20 cm height	*Clupeidae* 18 cm Mackerel *Scombridae*12-20 cm, Sprat 14 cm Anchovy 18 cm Sole *Soleidae*15 cm Mackerel crania 20-27 cm	Identified as a fish mash = *allec*, which has also the blood harvested from the extra 8 mackerel crania.	Djaoui, Piquès and Botte 2014
Arles Rhone 3 Little Latium pots	50-80 AD	Latium pot AR3. 2031.9. c. 12 cm in height	Mackerel *Scombridae* 25 cm, MNI 3 Each have cut to cranium	Identified as *salsamenta* preserved in small pot when whole?	Piqués et al 2015 (poster presentation)

(Continued)

Table 11.2 Amphora and ceramic vessel with residues of fish sauce found on land (*Continued*)

Site	Date	Vessel form	Species and size	Notes and preservation	Reference
Arles Rhone 3 Little Latium pots	50-80 AD	Latium pot X- 16373 c. 14 cm in height	Sardine *Clupeidae* 10 cm MNI 6 The gill rakes of mackerel MNI 2	Identified as a form of *garum* sauce in which mackerel viscera had been added to the sardine?	Piqués et al 2015 (poster presentation)
Arles Rhone 3 Little Latium pots	50-80 AD	Latium pot AR3. 3020.1 c. 20 cm	The scales of mackerel With coriander dill and fennel seeds	Identified as a form of *salsamenta* flavoured with spices	Piqués et al 2015 (poster presentation)
Arles Rhone 3 Little Latium pots	50-80 AD	Latium pot AR3 1012.8 c. 15-18 cm	Sardine MNI 55, 8 cm 95% Anchovy MNI 2, 10 cm 3.4% *Clupeidae* Pagellus MNI 1, 10 cm, 1.6%	Identified as a small fish mash = *allec*	Piqués et al 2015 (poster presentation)
Arles Rhone 3 Little Latium pots	50-80 AD	Italian pot AR3 3001. 173 c. 20 cm	Multiple species of *Sparidae*, *Clupeidae* sardine (29%) anchovy (27%) plus a random single shrimp	Identified as a small fish and prawn mash = *allec*	Piqués et al 2015 (poster presentation)
Arles Rhone 3 Little Latium pots	50-80 AD	Spanish pot AR3 3001.322 c. 20 cm	*Sardinella Clupeidae*	Identified as *salsamenta* or *allec*	Piqués et al 2015 (poster presentation)
Arles Rhone 3 Little Latium pots	50-80 AD	Spanish pot AR3 2007.126	Sprat MNI 65, 78% Anchovy MNI 18, 22%	Small fish mash = *allec*	Piqués et al 2015 (poster presentation) see page 191 for discussion here

Table 11.3 Shipwrecks with amphora containing residues of *liquamen* type fish sauce by region and date

Wreck, region, origin	Date	Amphora and basic characteristics	Species selection size	Notes	Reference
Sud Perduto II Corsica Spanish	Mid. 1st century AD	30 Dressel 7 1 Dressel 9. Wide neck and hollow spike	Full of whole mackerel 40–48 cm. 0.6–1 kg *Scombrus japonicus*	Cut to the front of the cranium i.d. as salted fish. Potentially a high quality *liquamen*.	Desse-Berset and Desse 2000:76; Desse-Berset 1993:343
Cape Bear III Port Vendres, North Med. coast near border with Spain Spanish	Mid. 1st century AD	16 Dressel 12 Narrow neck elongated body	Mackerel 28–36 cm, 350–550 gm. *Scombrus japonicus*	Vessel not full, Disproportionate No. of cranial to post cranial bones. I.d.as salted fish potentially *liquamen*	Desse-Berset and Desse 2000:80
Port Vendres II North Med. coast near border with Spain Spanish	Mid. 1st century AD	10 Pompeii 7/Dressel 7 Wide neck and hollow spike	Mackerel no details *Scombrus japonicus*	Heavily pitched, with mackerel bones stuck in resin	Desse-Berset and Desse 2000:81; Parker 1992:331
Chiessi Elba I Chiesse of the coast Of Tuscany Tunisia	1st century AD	Vindonissa 583 Wide open mouth and neck, wide cylindrical body, tapered spike	Rare and fragmentary from tuna, 200 cm, Ricciola, 125 cm, mackerel 30 cm Scombridae, Carangidae,	Cranium bones cut up. I.D. difficult. Either muria from salted tuna with the others as interlopers or a doubtful *liquamen*	Delussi and Wilkens 2000:53
St Gervaise III Fos sur Mer France Spanish	Mid. 2nd century AD	Beltran 2B Vessel has narrow hour glass neck and piriform body	4 Horse mackerel 40–50 cm. Carangidae,	Cranium and vertebrae but not a full vessel	Desse-Berset and Desse 2000:81; Parker 1992:373

(Continued)

Table 11.3 Shipwrecks with amphora containing residues of *liquamen* type fish sauce by region and date (*Continued*)

Wreck, region, origin	Date	Amphora and basic characteristics	Species selection size	Notes	Reference
Chiessi Elba 2 Chiesse cf the coast Of Tuscany Tunisia	Mid. 1st – mid 2nd AD	Unknown Stored before analysis	Mackerel 30 cm whole *Scombrus japonicus*	Resin lined Whole fish. Some were cut across the front of the cranium	Desse-Berset 1993 Delussi and Wilkens 2000:55 Bruschi and Wilkens 1996:167
Grado North Adriatic coast Italy 600 reused and pitched amphorae	2nd century AD	Africa 1 - Knossos a53/Dressel 5 - Forlimpopoli (78) - Tripolitana 1 (14) – Dressel 19/6B	½ filled with sardine " Mackerel 30 cm *Clupeidae, Scombridae* Empty	Sardines well preserved with articulation and organic preservation. Mackerel and sardine, sometimes intermingled Titulus on the Dres. 19 *Liq flos*	Delussi and Wilkens 2000:55; Auriemma 2000:31-49
Randello South coast of Sicily Portugal	Early 4th century AD	120-200 Almagro 50 As above	Full of sardine 10-17 cm Occasional mackerel *Clupeidae, Scombridae*	No clear i.d. either a salted fish or *allec* or connected to production of fish sauce	Wheeler and locker 1984 Parker 1992:975
Anse Gerbal Port Vendres I North Med. coast near border with Spain Portugal	c. 400 AD	Almagro 50/51c Narrow hourglass neck long wide cylindrical body	Whole sardine 22-25 cm *Clupeidae*	Cranium attached to vertebrae	Desse-Berset and Desse 2000:92
Cala Reale al Asinara Sardinia	4th/5th century	6 Almagro 51A/B 1 Beltran 72 Sealed with cork bung Not pitch lined	(amphora 675,885, 804, 801, 802) full of Sardine Clupeidae amphora 894 had few bones	whole and complete of unknown size and number -Either unfiltered *garum* or penetration from outside	Delussi and Wilkens 2000:53-65; Spanu 1997: 109-119
Cala Rossano Ventotene	AD 30 - 60	Dressel 8 and 9	Unidentified fish bones representing *allec*	Unreported	Ritondale 2014:28 Parker 1992: 153

Table 11.4 Refuse sites in the Roman empire with evidence of discarded fish sauce residues (Van Neer et al. 2010:176)

Site	Date	Species and size	Associated vessel/ceramics	Notes	Reference
Foss Gaul Urban port warehouse	1st century AD	Sardines 6–8 cm whole and articulated *Clupeidae*	Wooden barrel	Staves covered in resin with bones stuck in it. *Salsamenta* or sauce?	Desse -Berset and Desse 2000
Beddingham Fill in pit at villa	3rd century AD	Sprat and herring *Clupeidae* MNI >600	non	Sauce residue	Van Neer *et al.* 2010:176
Dorchester: Oven in yard of Roman building	Mid- to late-2nd century	Sprat and herring 7–8 cm	none	Both i.d. as *allec* and also potentially 'gone off'.	Hamilton Dyer 2009:4
Dorchester latrine pit	4th century	Sprat and herring 9–10 cm *Clupeidae*	none		
York, urban, under *opus signinum* floor in Roman building	Undated but Late	Sprat and herring *Clupeidae*	none	Thick layer of discarded remains on inland site with estuary species	Jones 1998; Hamilton Dyer 2009
London Southwark Peninsula house water front fish processing	Mid-2nd – early 4th Century	Sprat and herring >8 cm *Clupeidae*	Camuladunum 186 (Pelichet 46/Beltran 2B	8 cm layer of salt and bone with pieces of reused amphora shards	Bateman and Locker 1982:204–207
Cero del Mar Malaga Sea port at river mouth Ditch assoc. with dock for unloading amphora	Early 1st century AD	Sardine, anchovy *Clupeidae*, also *Sparidae boops boops*, *pagellus*, mullet, 10–20 cm, horse mackerel, Tuna (2 m), mackerel 30 cm	*Terra sigilata* amphora Unknown form	A 10 cm layer of fish bones underneath the broken amphorae. Tuna and mackerel potentially separate to fish sauce residues	Von Den Driesch 1980:151–154, Lepsiksaar 1986: 163–185 Cotton *et al.* 1996:223–238

(Continued)

Table 11.4 Refuse sites in the Roman empire with evidence of discarded fish sauce residues (Van Neer et al. 2010:176) (Continued)

Site	Date	Species and size	Associated vessel/ ceramics	Notes	Reference
Tienen Belgium Refuse dump in urban street	Mid-2nd century AD	Herring and sprat *Clupeidae* 3–14 cm, also flat fish, sand eel, hooknose, lesser weaver, whiting, stickleback, smelt, sole, eel	none	Deep layer of discarded fish remains from numerous taxa taken from an estuary. Cheap fish sauce	Van Neer et al. 2005; Van Neer et al. 2010 :175
Tienen Belgium Mithraeum	2nd half of 3rd century AD	Herring and sprat *Clupeidae* >10 cm	none	Relatively small numbers	Van Neer et al. 2010: 176
Braives Belgium Fill of well in urban street	3rd century AD	Herring and sprat *Clupeidae*	none	Relatively small numbers	Van Neer et al. 2010: 176
Arlon Belgium	2nd half of 3rd century AD	Herring and sprat 3–14 cm, *Clupeidae* also flat fish, sand eel, hooknose, lesser weaver, whiting, stickleback, smelt, sole, eel	none	Relatively small numbers Cheap fish sauce	Van Neer et al. 2010: 176
Lincoln Water front Refuse in urban	3rd /4th century AD	Herring and sprat *Clupeidae*	none	Abundant	Van Neer et al. 2010:176
Chichester Urban garden waste	4th century AD	Herring and sprat *Clupeidae*	none	abundant	Van Neer et al. 2010:176

The criteria for defining and distinguishing fish sauce residues from solid products were set out by Desse-Berset and Desse (2000:91), and they essentially say that when sparids and clupeids are fragile, chalky and heavily broken up, it is more likely that they are a sauce residue (Nicholson *et al.* 2018:277).[2] Many of the salting vats in the Spanish and North African processing sites retain the residues of previous salting events. The evidence is often in the form of deep layers of "fish bone flour" made up largely of *Sparidae* and *Clupiedae* (sardine, anchovy, bream, mullet) some c. 5 to 19 cm in size, but the majority are in the 6 to 14 cm range, found at Quinta do Marin Alhoa (Algarve); La Travessa de Frei Gaspar (Setubal) and Troia (Setubal) (Desse- Berset & Desse 2000:86, 90), and various more recent discoveries such as residues at Baelo Claudia (Bernal-Casasola 2018:340) and at a mid-1st century to 3rd century factory at Neapolis (Slim *et al.* 2007:30;35). An important question to ask of the *cetariae* data is, "are they traces of *salsamenta* abandoned in place, or of an autolysis (of sauce) in progress?" (Sternberg 2000:148). Slim concludes that "the fabrication of *garum* can begin with the same initial salting process as the one used for *salsamenta* thus it is impossible to distinguish between the residues" (Slim *et al.* 2007:35). There is also a question as to whether these layers of bone found in *cetariae* represent a true residue that is left over or that the final fish sauce was never taken and what we find is the result of leaving fish sauce to desiccate in situ. This may explain why some fish sauce vats have residues while others do not. The thinking is that an establishment that has had a planned closure would have emptied the vats while one closed overnight because of financial collapse or civil or military disturbance would have left sauces and salted products in place to decay.[3] Rebecca Nicholson has noted that the chalkiness is reminiscent of bones that are left for too long in the protease enzyme solution used to prepare bones for research (Nicholson *et al.* 2018:277). Many samples of undisputed fish sauce residue do not display this degradation, while other examples, which are disputed as a fish sauce residue, also do not display the chalkiness. Fish bones generated from one of my experiments, a batch of sprat and sardine fish sauce, which were examined by Rebecca Nicholson "indicated no macroscopically detectable effect on the bones, although the otoliths were not preserved" (I feel sure that this was due to my own inadequate preservation of the residues; Nicholson *et al.* 2018:277). The differences may be connected to the degree to which the sauces that are prepared are subject to enzyme activity. It is of note that in the recipes it is stated that *allec*, that is, a fish sauce residue, could be added or left in place, and I have suggested that this would have seeded the next batch of sauce, providing a burst of enzyme activity which was necessary to break open the fishes' cavities and expose the viscera to the salt in circumstances where no extra viscera was added or available. There must have been many occasions when it was not. The addition of extra viscera would clearly have done the same thing: if the tanks were simply not emptied fully but left with a percentage of fish bones and *allec* in the bottom, this could easily have resulted in bones going through the process many times resulting in the chalky texture noted by Nicholson *et al.* (2108:277). Alternatively, the residue that display great chalkiness were exposed to the elements for some time before being silted over. In relation to the residues found in amphora and other vessels, which have varying levels of degradation there is a great deal of variability and thus dispute as to what they represent. In Chapters 7 and 8, where I have reviewed modern fish sauce techniques and experimental archaeology of fish sauce, it has been possible to challenge the reasoning used to distinguish these products. We note that in a very recent experiment completed in 2020, a sardine sauce left undisturbed for a year still resulted in the full range of disintegration, that is, some of the bones remained in articulation. This means that the bones would never have been subject to the same degradation as those that are found in the *cetariae,* which, as we have seen, may also have been subject to multiple rounds of fermentation. A well-preserved mackerel or sardine skeleton inside an amphora, or other vessel, or in a *cetaria* or discarded in a ditch, which still remained articulated, may have been a sauce residue. This is also true of those mackerel bones that have been found in shipwrecks where, because of the quality of the preservation, they have been identified as *salsamanta.*

Shipwrecks

Amphorae removed from ship wrecks that contain bones associated with some kind of traded fish product are now numerous. The wreck sites that are most relevant here are those that contain bones from large (c. >30 cm) well-preserved mackerel in amphorae that were clearly designed for liquid sauces as they have long narrow necks: Sud Perduto II; Cape Bear III; Port Vendres II; St Gervaise III; Chiessi Elba 2 (Table 11.3; Desse-Berset and Desse 2000). Mackerel dominate the finds and some of these have been exceptionally large specimens which because of their size are naturally interpreted as salted fish but it would have been impossible to get them in or out of the vessel, which rather militates against this conclusion.

Desse-Berset and Desse analyzed the content of amphorae from Sud Perduto II and identified the product as salted mackerel. One of the vessels (Cat. No. 608) held 50 relatively small mackerel with a cut through the front of the cranium that has been recognized as a bleeding technique. This cut has also been found in specimens in amphorae from the Elba wreck (Desse-Berset and Desse 2000:79; Desse-Berset 1993:345) and in the Little Latium pots on the Rhône (Djaoui *et al.* 2014). The same cut has also been found in relation to six small tuna (*cordyla*) of 40 to 80 cm length in an amphora at the Hellenistic-Roman town of Dion in an elite town house. We may have evidence of a blood viscera sauce or we may simply have cleaned *salsamenta* (Theodoropoulou 2018:396). In the case of the Little Latium pots the cut portion is gathered in with the *liquamen allec*. The number of amphorae (mainly Dressel 7) at Sud Perduto with this kind of cut to the frontal bone is small (one or two amphorae); one must ask, if the technique was so effective why weren't all the mackerel prepared this way? It may be possible to interpret the low incidence of this cut as an indication of a low volume of blood required in a relatively small true *garum* industry at the site where these mackerel were processed.

Desse-Berset considered that she was not able to fully evaluate the content of amphorae from Cape Bear III and Saint Gervaise III as it was assumed that some of the content was missing or that bones from more than one amphora had been incorporated (Desse-Berset and Desse 2000:79–81). In discussing the bones from one of the 16 Dressel 12 amphorae from Cape Bear III, they note that the number of individual bones (NISP:328) does not correspond to the estimated MNI (9) from the hyomandibular count. These amphorae were emptied by the analyst, who would expect that the MIN and NISP would at least be close to a match if the content was both secure and had been a whole salted fish product. These amphorae have long thin necks and small mouths (Table 12.2) and do not represent the ideal vessel for salted fish. The wreck at Saint Gervaise III at Fos-sur-Mer, near Marseilles, also contains an unusual number of horse mackerel bones, which does not correspond to what would be expected from a full amphora (Desse-Berset and Desse 2000:81). These anomalies are readily understood if one considers that these bones may represent a fish sauce mash, semi-worked and intended to produce a fish sauce at a later stage, which had been processed in a *cetariae* and transferred to the amphora already in a sloppy pourable state, in which, as we have noted from the experiments in Chapter 8, there would be a range of disintegration and liquefaction. We note that this process is also suggested for the *Latium* pots (page 191). Many of the skeletons would have been cleaned of their flesh through hydrolyzation but still articulated, while others would have become fragmented and lost some or most of the flesh from the bone. Many would also have already disintegrated fully and some of the bones become disarticulated. In these circumstances the NMI and NISP could never correspond to whole fish as they were not whole when they went into the amphora.[4] The amphora would not necessarily be "full" of fish as the amount of bone contamination would vary depending on how far down the *cetaria* the sauce was taken. It would also contain the fish paste and the liquid sauce taking up capacity within the vessel, which would in a shipwreck disappear over time. A wreck that provides the most comprehensive evidence for the transportation of fish sauce is that at Grado, on the North Adriatic coast in Italy, dating from the mid-2nd century AD (Delussi and Wilkins 2000:59; Auriemma 2000:43). The wreck contained 600 reused and pitched amphorae from five forms (illustrated in figure 11.1).

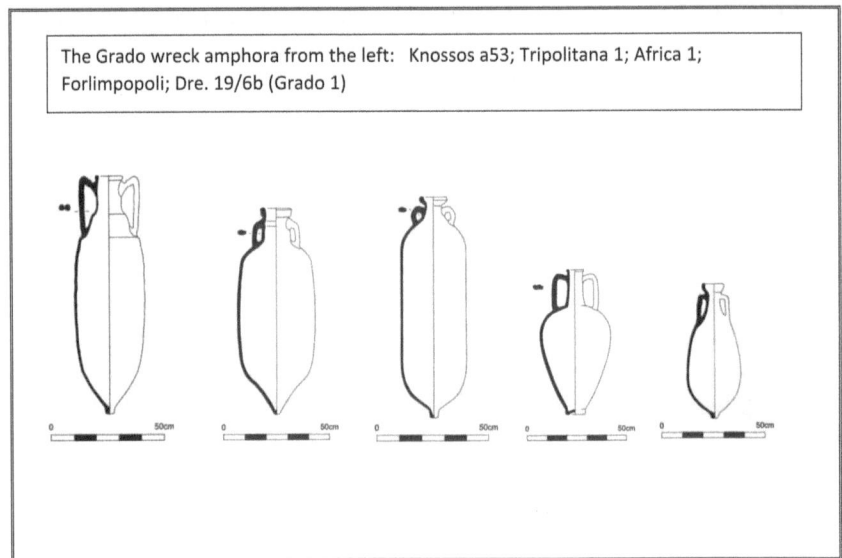

The Grado wreck amphora from the left: Knossos a53; Tripolitana 1; Africa 1; Forlimpopoli; Dre. 19/6b (Grado 1)

Figure 11.1 Grado amphora

Africa 1.	200, not pitched, 44 l. capacity	½ filled with sardine
Knossos a53.	150, pitched	½ filled with sardine
Forlimpopoli a53.	unknown, pitched, 21 l. capacity	½ filled with sardine
Tripolitana 1.	13, pitched 85 l. capacity	full with mackerel and horse mackerel 30 cm
Dressel 19/6B.	large numbers, pitched, 17 l. 55 cm high	Empty but with titulus *liq flos* on 20

The Dressel 19/6b from the Grado wreck were a unique form of small ovoid amphora. They represent the only amphora in the wreck that was purpose-made for the product it carried. The remaining vessels are a disparate collection of wine and oil vessels from Greece, Italy and Africa and in the case of the Forlimpopoli, very unsuitable for any kind of fish product as they appear as wine jugs with a long narrow neck. The sardines and mackerel were well preserved with articulation and organic residues, that is, undissolved flesh. The species were sometimes also intermingled, sardine with the mackerel and mackerel with the sardine (Delussi and Wilkens 2000:55; Auriemma 2000:31–49). The vessels with remains were sealed with stoppers, which had been cut from shards of the same amphora form. Not all the vessels were pitched, but those that were had a thick lining poorly applied, which is rare. The preservation of pitch in amphora from wrecks varies greatly and it is therefore not clear to what extent fish vessels were in fact pitched as a rule. It may be that some pitch dissolves in sea water and fails to be preserved, while in certain conditions where it is particularly thick, it remains (Peña 2007:69). The current view is that the fish probably represent a *salsamenta* but as the vessel would need to be broken to get the product out, it has also been suggested that the fish could be connected to something to do with a fish sauce and represent a semi-worked product that could later be turned into sauce (Auriemma 2000:45). If it was a semi worked fish sauce it is also very probable that the fish mash in these vessels had already given up its first sauce, that is, a *liquamen flos* while it was still in the processing vessel, a *cetaria* or numerous *dolia*, where it was first prepared before it went into the amphora. In my view this first sauce was the product traded inside the Dressel 19/6b *parva,* which has the *tituli picti* naming the product as a *liquamem flos* and which was being transported along with the residue, meaning that the products were intrinsically linked as they were derived from the same batch of fish. It is of note that Delussi and Wilkens (2000:6), consider the sauce within these small ovoid jars as a *garum* despite the label for *liquamen!* This scenario provides a model for ways in which potentially all fish sauce could have been dealt with when industrially manufactured, that is, that the first *flos* sauce and its *allec* could be traded together. The Grado wreck may represent an unusual situation where the fish were acquired in a sudden glut, and the vessels were randomly acquired to accommodate the sauce that was produced. The products had to be transported in whatever vessels the manufacturer was able to acquire at the time. This scenario, where the *liquamen flos* and the second-stage fish mash were removed at the same time and traded together, rather suggests that the length of time that the sauces were left in the *cetariae* could be greatly reduced and represents an intensified production process. Grado is a mid-2nd century AD wreck. The wreck evidence from the 1st century discussed above also displays the same characteristics of mackerel of large size in vessels unsuited to a *salsamenta*. It remains to be seen whether we can also attribute the same intensive production process with shortened time in the *cetariae* to this period. Rita Auriemma has also linked the situation of the *Garum* Shop to that of the Grado wreck as together they provide a scenario whereby reuse of a disparate collection of what seems to be unsuitable vessels appears as a fundamental component of the fish sauce industry (discussed in Chapter 12; Auriemma 2000:46).

Residues on land

The land-based evidence of mackerel bones that may represent a sauce residue rather than a *salsamenta* is very scarce in Europe and the northern empire. Van Neer *et al.* (2010:169) cites over 40 sites, where mackerel bones have survived but the majority are assumed to be evidence of salted mackerel being consumed across the military sites and by also the local populations. Given that there was the potential for a fine quality *liquamen* sauce made with just mackerel to be traded into the north we should expect to find evidence of it in the form of fish sauce residues. Some of this mackerel is in small vessels, understood to be open mouthed globular pots and *doliola* at a fort on the Kops Plateau at Nijmegen (Van Neer *et al.* 2010; table 1 and pages 167; Hüster Plogmann 2006; Lauwerier 1993). I have not been able to access many of the reports and as a result I have no data on MNI or size. It seems to me very likely that a fish sauce residue is possible, given that there does not seem to be a good reason for mackerel bones to have been left inside a vessel and that vessel discarded unless a sauce had been washed out of them in the style of the other evidence and the vessel discarded in order to dispose of the residue cleanly. A recent report from the latrines at the Kops Plateau fort indicates that mackerel of a size range 15-40 cm, which were eroded and had also been cut in the typical way associated with bleeding, were well represented (25). The authors, in light of my own research, agree that a expensive mackerel *liquamen* residue i.e. *allec* is the most likely product (Dütting *et al* 2020: 82). Ehmig points out the rare examples of fish sauce amphorae with mackerel bones (*Scomber japonicus*) from the 1st century AD in Mainz. A Pelichet 46 (Kat-Nr. 4686) contained entire mackerel of 20 to 35 cm and the Dressel 7–11 (Kat.-Nr. 4735) contained only whole cranial bones of mackerel (Ehmig 2003:80). Ehmig has conjectured that the residue from the Pelichet 46 represents *allec* of mackerel and notes it seems incomprehensible and that the "archaeological, philological and cultural-historical sources give no clear indication that *salsamenta* was being traded instead of fish sauce" (Ehmig 2003:87). Mackerel finds in Roman Britain are also scarce: Locker records six sites, at Gorhambury villa, Skeleton Green, Elm farm, Great Holts farm, Winchester Palace, and Chester, all in the more Romanized south (Locker 2007:151;155). An example of mackerel cranial bones were found in a Dressel 16 amphora in Roman London at Winchester Palace. It may attest to the transportation of expensive forms of mackerel fish sauce. The amphora has a detailed *titulus*, *LIQUAM·ANTIPOL EXC I TETTI AFRCANI AFRI*, which is rendered as "*liquamen* from Antibes of excellent quality of/from Lucius Tettius Africanus" (Locker 2007:153; Alcock 2001:80). The bones have subsequently disappeared and it is not possible to review the find, but the initial report suggested that vertebrae were attached to the cranium. which would rather suggest that some post cranial bones were present and the find may actually represent an amphora with a primary fill, that is, a mackerel *liquamen* (with minimal bones contamination, with a corresponding *titulus* (Yule 2005). Amphorae with just cranial bones are in fact quite common. A Beltran 2B amphora from a wreck at Fos-sur-Mer (St Gervaise III) contained the well-preserved crania of 4 horse mackerel of 40 to 50 cm (*Trachuris trachurus*). A similar attachment of vertebrae to cranium was noted by Desse-Berset and Desse (2000:81).[5]

A Pompeii urceus containing a bone residue

A unique example of fish sauce residues in the process of being used to harvest a fish sauce is to be seen in the fish remains found in a small *urceus* from Pompeii. I am extremely grateful to luigia Melillo and Grete Stefani for allowing me permission to visit the Pompeii archaeological store and the Museo Archeologico Nazionale di Napoli to view the small fish sauce vessels known as *urcei* and to take pictures for this publication. One particular vessel, a tiny *urceus* (Inventory no. 81744) had a clear *titulus pictus* which is transcribed on the inventory card as *G(arum) F(los) scombr scauri ex officina agathopodi,* "Flower of *garum* made with mackerel, a product of Scaurus from the workshop of *Agathopodi?*" (illustrated figure 11.3 and 4; Curtis 1991:167). The bottle's height was 23.5 cm; max diameter of body 7.5 cm, internal width of neck 1.5 cm; length of neck 10 cm; base diameter 6 cm. It was not possible to measure its capacity but it has been noted, from a separate group of vessels studied, that the majority were of a size range of 1.9 to 7.4 liters, while as many as 95 *urcei* (66.9%) of those studied were between 1.9 and 3.4 liters. The *tituli picti* entries in *CIL* sadly do not record the capacity and size of the *urcei* associated with the different types of sauce. I estimate that this vessel held no more than 0.5 to 0.75 liters and represents one of the very smallest bottles of this kind. The size range of the vessels was extensive, as can be seen from the those stored at the Naples museum/Pompeii store (fig. 11.2). For a full report on

Figure 11.2   Urcei in various sizes. Copyright: Museo Archeologico Nazionale di Napoli.

Figure 11.3 The tiny urceus (Inventory no. 81744) from Pompeii archaeological store. Su concessione del Ministro per i Beni e le Attivitá Culturali e per il Turismo – Paco Archaeologico di Pompei

these small bottles in Pompeii and the Bay of Naples region see Cappelletto *et al.* (2013). Inside the tiny *urceus* was a small amount of fish bones, kept separately in a plastic bag and delivered to my table in the archaeological store. I am extremely grateful to Dimitra Mylona for assessing these bones for me based on photographs I took at the time:

Dimitra Mylona reports that the whole assemblage of bones is mostly one species: anchovy (*Engraulis encrasicolus*), with possibly a few elements of at least one more, chub mackerel (*Scomber japonicus*) illustrated figure 11.5. We were in no position to assess size by any normal method but they were in the smaller range. There is an obvious conflict between the *titulus* and the content. The bones reflect a standard small fish *liquamen,* while the *titulus* suggest that the bottle once held a fine *garum* of mackerel. It is very clear that *liquamen* and *garum* existed as separate commodities in Pompeii

Figure 11.4 The tiny urceus with titulus. G(arum) F(los) scombr scauri ex officina agathopodi, "Flower of garum made with mackerel, a product of Scaurus from the workshop of Agathopodi." Su concessione del Ministro per i Beni e le Attivitá Culturali e per il Turismo – Paco Archaeologico di Pompei

and its environs. If one were still adhering to the opinion that *garum* and *liquamen* were essentially the same and this "single sauce" could be a fine bloody *garum* and at the same time contain fish bone residues from small anchovy then the label and the content are in accord and there is no conflict. If on the other hand the dual nature of fish sauce is accepted, then the only way to interpret these residues is as evidence of reuse. Once these bottles had been used to dispense the fine mackerel *garum,* which could have been made either in Spain or locally, they could clearly be collected up and reused, and in this case refilled with a local or imported *liquamen.* This would appear to be a second-stage *allec* residue which, as we saw with the *Garum* shop evidence, was readily available within the town using locally caught *Clupiedae.* This *allec* would have generated a second-quality

Figure 11.5　Fish bones associated with urcei (Inventory no. 81744). Su concessione del Ministro per i Beni e le Attivitá Culturali e per il Turismo – Paco Archaeologico di Pompei

liquamen inside the bottle, in exactly the same way as the little pilgrim bottle from Jordan discussed page 194.

Fish sauce residues in amphorae

We can see the same reuse of amphorae to contain other kinds of fish sauce in evidence from Augusta Emerita in Portugal; this is a Beltran 2B amphora containing sardine of 4 to 5 cm (with an occasional horse mackerel and sardinella of 10 cm) buried with its contents in a disused ice house (Bernal-Casasola *et al.* 2016:737). Bernal-Casasola has defined this product as an "exquisite *garum* sauce" based on the reconstructed incomplete *titulus pictus* on the amphora (Bernal-Casasola *et al.* 2013:2529; Bernal-Casasola *et al.* 2016:196). The name of the product on the Augusta Emerita label is missing (it states that the content was an undifferentiated sauce with the label *flos scombri*: the first letter is missing, and could be G(*arum*) or a *liq-(uamen)*. The product is therefore a mackerel sauce but Bernal-Casasola has concluded that the label corresponds to the content and that a sardine sauce could be labelled as one made with mackerel. I would conclude that it is far more likely that the Augusta Emerita product was the discarded remains of an ordinary *liquamen* of little distinctiveness that had been put inside a reused amphora which originally held a fine quality *garum* or *liquamen*. There are other sources of evidence which suggest that fish sauce amphorae and *urcei* were reused multiple times as vessels to dispense fish sauce. A number of rewritten *tituli* described on page 266 reflect what would appear to be an increasingly common phenomenon. The fact that the residues of fish sauce in the form of a semi-processed fish mash are found in small and large vessel of all kinds suggest that the final process of harvesting the sauce itself from the residue could be undertaken by the consumer rather than the trader. The fish bone evidence from the Herculaneum sewer of discarded bones typical of sauce residues that have been thrown away from individual homes above the sewer may also reflect this process (page 196). In these circumstances fish sauce residues that were discarded in amphora represents the end stage, clearly, of a long process of sauce extraction and even when the bones reflect larger species in large quantities, they may well reflect the last extraction of a weak and cheap sauce by people of lower status.

Fish sauce residues after consumption

Bones left over after the sauces had been extracted and the residues rebrined are found all across the empire. Our earliest evidence for a standard, dissolved, multiple-species small-fish *liquamen* can be found in an early 1st century AD site at Cero del Mar site near Malaga (Von Den Dreisch 1980:15155; Lepsiksaar 1986163–185). This fish dump was in a ditch associated with a dock for unloading amphorae, and including three broken vessels with sediment which is identified as a fish sauce residue, and there were also discarded tuna bones. The broken amphorae were unidentified, while the sediment (sample CM-76-28/29) contained elements of clupeids such as sardine, anchovy and sparids such as *boops boops* and *pagellus* as well as *mullus, trachurus trachurus* and *scomber japonicus*. The *Sparidae* and *Clupeidae* were largely in the form of a fish flour which corresponded to fish 10 to 20 cm in length, while the two mackerel species reached 30 cm. It cannot be determined whether the fish were all used together as one product or separately, it nevertheless seems to be in part a deposit of the exhausted residues of an everyday *liquamen* using local fish from Malaga. Von den Driesch defines one aspect of the product as a high-status mackerel *garum*, though with the intended meaning of the "single sauce." (Von Den Driesch 1980:152, Desse-Berset and Desse 2000:76). Lepiksaar and

Cotton both conclude that Von Den Driesch is mistaken and suggest that the remains must be *allec* made for its own sake, which of course fails to comprehend that *allec,* of this nature, was principally intended to produce a *liquamen flos,* before it was an *allec* (Cotton *et al.* 1996:231). Of a similar date and nature are the residues found at Masada, associated with Herod the Great and close to the amphora with the bilingual *titulus pictus* designating a "*garum* for the king" (Cotton *et al.* 1996; discussed in more detail on page 66). In this early 1st century context, a buried broken baggy jar was found to contain the remains of fragmented fish bone and sand. The species present were largely different forms of *Clupeidae* including sardine, anchovy, sprat and shad of only 4 to 5 cm (Cotton *et al.* 1996:223). The micromorphological study of the non-faunal remains concluded that the sand in the base was consistent with a foreign origin and the temper of the amphora was also believed to come from Spain. The analyst concluded that the sauce was an expensive and desirable *allec* made as a paste shipped from Spain and emptied into the local vessel (Cotton *et al.* 1996:229). I would define this as the exhausted bones from a relatively commonplace *liquamen.* These examples of residues on land (whether inside amphorae or deposited in thick layers) do not fit the criteria created by Desse-Berset and Desse, as they tend to be in varying states of degradation but often do not resemble the fish bone flour that are found in many *cetariae.* A classic example would be the site at Olbia 2 in Sardinia, where the urban infill from a well from the 2nd century AD containing an Africa 1 amphora which had lost its neck and handles and contained three separate layers of fish remains representing three separate discard events. The bottom contained relatively well-preserved sardine and pilchard 5 cm in length, the middle layer was bream (*Pagellus*) of 10 cm that were typically broken up, while the top contained pilchard and sardine of 5 cm with evidence of articulation (Delussi and Wilkens 2000:59; Bruschi and Wilkens 1996:169). From our experiments we can be fairly certain that the deposit represents three separate exhausted fish sauce residues, which, after the potential sauce had been taken from each, were discarded together. They represent nothing more than a stinky pile of bones that attracted flies and vermin and needed to be sealed in a buried environment. Numerous sites mirror this scenario and often appear to represent a full vessel when the flesh was present (see Table 11.2). There is considerable doubt as to why these products were discarded as it must be assumed, if they were fit for consumption, that the sauces and residues would be distributed and consumed in an ad hoc fashion with the bones being discarded in a similar way, if not consumed. Either these bones represent a discarded decayed product or the bones were discarded when all desirable sauces had been removed. The sites include the Augusta Emerita Dressel 14 buried full of very fragile sardine bones (Bernal-Casasola *et al.* 2016:737) and two 1st century AD sites at Setúbal with Dressel 14 amphorae containing entire sardine 16 to 20 cm in articulation (Gabriel and Tavares da Silva 2016).

A Dressel 6A amphora from the Adriatic was found in road foundations at Vicenza with a fill of small pickerel (including shellfish: see page 205) and the analysts conclude that it must have been thrown away while still partly full (Mazzocchin and Wilkens 2013:110)[6] and a Romano-British find of small sprat and herring inside an oven at Dorchester (Hamilton-Dyer 2008:4). At Olbia (1) in Sardinia, a 2nd century AD site revealed similar multiple *Sparidae* of 15 to 20 cm in the bottom of an Africa 1 amphora which were identified as a salted fish product because of their good preservation (Bruschi and Wilkins 1996:167; Delussi and Wilkens 2000:56). Even where the particular species selection, the state of preservation and the amphora shape all suggest a *salsamenta* rather than sauce it is still likely that used up sauce residues were being discarded, than that a salted product had become decayed.

The evidence for many small-fish bones like this linked to fish sauce, whether in amphorae or other deposits, is widespread in Britain and Northern Europe. Finds occur at London (Bateman and Locker 1982), York (Jones 1988), Lincoln (Dobney *et al.* 1996), and at a villa in Beddingham, East Sussex (Hamilton-Dyer 2008); there are three deposits from Tienen in Belgium (Vanderhoeven *et al.* 2001, Lentacker *et al.* 2004, Van Neer *et al.* 2005), Brieves (Van Neer and Lentacker 1994) and Tongeren (Van Neer and Ervynck unpub.). This evidence reflects a local production of an everyday *liquamen*-fish sauce which attest to its popularity and desirability in the northern empire. Evidence for the transportation of a purpose made bony *allec* traded widely in the empire seems economically illogical to modern archaeology, as it appears so cheap and worthless. This estimation of its value is largely due to the fact that the bones are still present in what is deemed a fish paste consumed by ordinary people (Van Neer and Ervynck (2002:208). Having now understood that *allec* was potentially a product with many purposes, we can now turn to a fully integrated review of its nature.

The nature of allec. How do we classify allec by characteristics and quality?

When *allec* was a smooth creamy fish paste it clearly had the potential to be highly nutritious and when spread thinly on bread as a relish, remarkably appealing. It is very similar to the concept of *patum piperium*, that modern cooked fish paste made with 60% anchovy, which (despite a similarity to ancient fish pastes it is in fact cooked) was always a product consumed in the West by modern elites.[7] The modern negative view of *allec* are largely due to the belief that it was primarily a fish sauce "residue," a *faex*, and as it is perceived as being full of bones it was something only fit for the poor and slave. The essential dichotomy is that the paste was a potentially desirable food item but the bones were not and there is a real dilemma in trying to conceive how the bones could have been removed given that it was a dense paste. Pliny has told us that when *allec* was purpose-made it was essentially a bone free product (see page 32; Pliny *HN.* 31.95.4). His description does

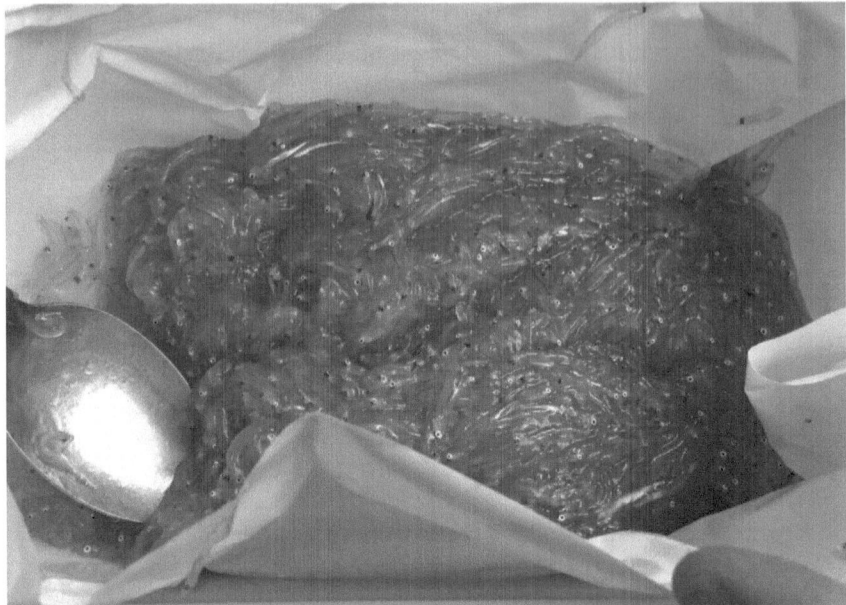

Figure 11.6 Poutine on offer at La Cantina de Lulu in Nice in 2018. Copyright Sally Grainger

have a strong resemblance to the product known as "poutine," the tiny transparent jelly hatchlings of anchovy, which do resemble rain drops and naturally have no bones. These are fished under strict regulation in Nice and used in Niçoise cuisine.

Poutine are often used to make the pissalat of Southern France, which is also a soft fish paste. This kind of paste is also made with juvenile anchovy and sardine of a relatively small size (c. 3–10 cm). In the case of the larger fish within this size range the fish are invariably beheaded, eviscerated and largely deboned in one action as the spine and viscera comes away when the head is removed. They are then salted for a brief period in a pre-prepared brine to wash the blood from them. There is minimal fermentation as there are no digestive enzymes in the mixture. The fish soften, at which time they are agitated and sieved so that the remaining bones and excess brine is discarded and a relatively smooth fish paste is the result. The best pissalat is still made from the *poutine* in Nice (Delaval and Poignant 2007:70). I think it is certainly likely that much of the purpose-made *allec* was made in the same way as pissalat and therefore naturally and/or intentionally bone free.

We have strong evidence that there was a desire in some instances to remove bones from small-fish salted products in residues found in a small vessel at the House of Marcus Fabius Rufus in Pompeii. This is a high-status house with considerable luxurious decoration, and a number of fish sauce amphora with *tituli picti* were found in it, indicating that *garum flos* was stored and served (Grimaldi *et al.* 2011:11). A collection of fish bones, exclusively anchovy (*Engraulis encrasicolus*), have been identified, and it is noted that the majority of them, but not all, have been beheaded (Carannante *et al.* 2006:140). It would seem probable that the viscera was also removed at this time, as noted above. We cannot say precisely what the Marcus Fabius Rufus fish remains were. It still may have been intended to produce a sauce as well as a paste, or a simple relatively bone-free purpose made *allec* or a means of salting cleaned anchovy as a solid product which would generate a *muria*. I favor the latter interpretation, as does Carannante, who believed that the process was "designed to preserve the integrity of the fish" (2011:142). We should note that the technique of beheading is still used when the modern fish sauce from the Bay of Naples: *colatura di alici* is made. Originally this was made with anchovy which were meant to retain integrity, while the brine was harvested as a sauce. It is only in the last decades that manufacturers in *Cetara* have ceased to consume the anchovy as a solid product and allowed the fish to fully dissolve and generate a darker richer sauce (Caranannte *et al.* 2011). The artisan tradition still includes beheading and eviscerating the anchovy, salting and discarding the first bloody brine to clean them, before layering and compressing the fish with more salt. There are no digestive enzymes, and as the salt levels are high the sauce is naturally quite weak in protein, color and taste. This product is regularly associated with *garum* and *liquamen* and assumed to be the natural survival of ancient sauces. We have seen (page 107) that *muria* becomes associated with *liquamen* in Palestine and in the wider Empire, and that the distinction between a clean fish brine and an enzyme-fermented sauce becomes blurred in the literary evidence. In these circumstances *colatura di alici* has strong associations with this later sense of *muria/liquamen*.[8]

The alternative form of *allec*, which, from the description in Pliny (*HN*. 31.92), was the only product he identifies as valued and desirable was made with various forms of seafood oysters, sea urchins and sea anemones which clearly have no bones. This product was valued precisely because it was not a residue, that is, it had not generated sauce first but has been made especially from valued seafood. The terminology used in Pliny is particularly interesting. Though somewhat misinformed he has knowledge from his sources of the residue that is useful. Pliny or his source defines *allec* as *imperfecta*, "something not fully digested," (*HN*. 31.95). Good *allec* may therefore be a rich fish paste and therefore suitable as a food resource or as a source of more *liquamen*. The idea of it not being strained which Pliny also says might imply that some forms of *allec* retained their bones when marketed but straining may also mean that the *allec* is still runny and straining means removing the residual liquid that is remaining in the *allec* in the same way as pissalat is made (see below for further discussion on filtering and Chapter 3).

There is certainly no way to know what kind of fish was used once the paste has been formed and so trust and honesty on the part of the manufacturer was the only way to know what it was made from. The majority of *tituli* indicate that *allec* was *optima*: the best [it can be] (*CIL*. IV.5717-9; 9407–9411. Two of these were in *olla* – pots or *urcei* – and also in the case of *CIL*. IV 9411, in a kitchen. There is one potential *HA(llec) S(combri* (*CIL*. XV.4731), but this certainly need not be expanded this way as *HAS* may correspond to any number of products or names of traders.[9] The references in Cato (*De Agr.* 58) to slaves being given *allec* is unclear but we must assume I think that this was a genuine residue of fish sauce rather than one purpose made but we cannot know whether the bones were present. The latter is assumed, which is an unnecessary assumption and further it is assumed that the slaves were of such low status that a bony fish paste was all they were entitled to! (Van Neer and Ervynck 2002; Van Neer *et al.* 2004). There is an unspoken assumption in all this thinking that the poor peasants and farm slaves must have had no choice but eat this produce, and their low status is almost seen as a justification for the idea that they were offered a paste with the bones left in. There is a circular argument at play here. Zooarchaeologists on the other hand are hampered by the orthodoxy of the essential oppressiveness of ancient societies, such that the

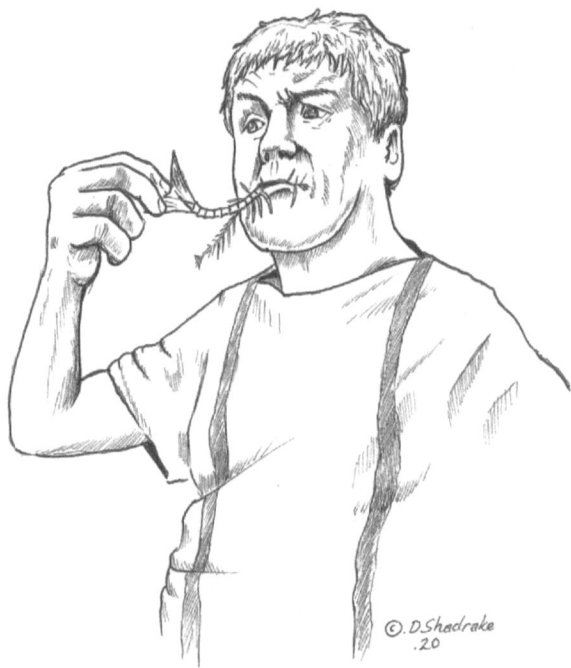

Figure 11.7   Roman peasant with his portion of *allec*! Line drawing Dan Shadrake

image of the downtrodden masses forced to eat a bony fish paste fully fits the Marxist criteria they use to imagine what the ancient society must have been like!

There are other points of view of course. Mylona (pers. com.) has pointed out that eating fish bones is quite a common phenomenon in the Mediterranean and that I lack an empirical perspective on this issue. I would counter that a whole fresh sardine sizzling hot straight of the barbecue would be perfectly acceptable and delicious, while an accumulation of not necessarily disarticulated bones within a fish paste would not. Using bread to get them down, as suggested by Djaoui *et al.* (2014), almost analogous to the idea of bread and butter served with kippers in Britain, just doesn't make the idea of consumption of this kind of thing easier for me to swallow, metaphorically or otherwise. The culinary combination of smoked fish and bread and butter was developed to ensure that the occasional bone did not get stuck and therefore I do not see this an analogous at all. There is no need to envisage a battle to get the stuff down especially as we now know that *allec* can be harvested bone free!

The phenomenon of *allec* is illuminated in a particularly effective way from the Palestinian Talmudic sources. Susan Weingarten demonstrates that two distinct products were available, one where the fish bones were visible, and therefore identifiably of a species that had scales (or would have had scales when mature), and one that did not contain fish bones and therefore could not be guaranteed to be kosher and was therefore prohibited.[10] The Babylonian Talmud says "What is the meaning of *hiliq*? .... It is the *sultanit*. Why is it prohibited? Because its mixed multitude comes up with it" (*Avodah Zarah* 39a; Weingarten 2018:240). The mixed multitude will clearly have included prohibited shellfish. These products were collectively called *hilliq* (after *allec*), but also *sultanit* and *afitz* after *aphye,* the collective term for the little fish used to make not only a purpose-made *allec* but also a cheap fish sauce. In the Palestinian Talmudic sources, it is named *tarit tarufa*, which is translated as "chopped" or "mashed." The sources say that when it is "mashed" it is impossible to identify the fish that has been used and it is prohibited, and Weingarten assumes this is because the bones were too tiny to recognize as coming from a fish with scales. An alternative possibility is that this is a fish paste derived from the manufacture of fish sauce that has been removed from the bone and is therefore a smooth fish paste, which is prohibited precisely because it is absolutely impossible to tell what it was made from. The chosen translated term of "mashed" seems problematic to me. Weingarten says "It is counted as "not mashed" as long as there are recognizable bits of fish in it" (2018:240) and quotes the *Tos Avodah Zarah* ii 6, ii 7: "What is *tarit* which is not mashed? Everything which has the spine and head recognizable in it." Small fish utilized to make fish sauces were frequently of multiple sizes which means that frequently the heads and spines would be visible. It would seem unlikely that this general prohibition was concerned only with tiny fish that had not yet developed scales. When it was not "mashed" it was allowed and therefore kosher, and it seems very likely that the not- mashed state meant that it was a semi-processed fish sauce mash, traded as such and full of identifiable bone, which could be further processed to harvest sauces from it. Weingarten quotes Mishnah Avodah Zarah ii 6 on *hiliq*: "These things of the gentiles are forbidden, but it is not forbidden to have any benefit at all from them ...," which may conceivably mean that the consumer was allowed to take a sauce from a whole residue, that is, one "not mashed', themselves, so long as they could be assured that the fish in the residue was kosher to begin with, which implies "not mashed" was equivalent to the bones being present. It is difficult to define the terms used here, as the author uses "mashed," which has different connotations in English. Clearly, if this is associated with a fish sauce residue, it cannot be chopped or mechanically mashed or minced, as these processes do not happen to the little fish – they are instead enzymically dissolved.[11] The Palestinian sources seem to demonstrate that *hiliq/allec* came in two forms: a smooth paste without bones, and a residue still containing the bones. Given the experience from experimental archaeology It seems logical to extrapolate that in Italy and the wider empire *allec* derived from a *liquamen* fish sauce existed in a similar fashion, in two forms.

In an article by Robert Curtis we learnt of the "*Negotiatores Allecarii*" of *Germania inferior* and *Galia Belgica* who seem to be traders in *allec* from the late 2nd/3rd centuries AD in Colinjnsplat, in a Dutch province of Zeeland. They were discovered when fragments of numerous alter stones with dedications to local deities were dredged from the sea. Curtis concluded, with reference to the later Medieval usage of *allec* to refer to the larger group of clupeids and sparids but particularly the anchovy and the herring, that these merchants probably traded all the four sauces and also these specific species as *salsamenta*. He suggested that these traders had to choose one of these products to be named by (Curtis 1984:152).[12] Curtis also concluded that when *alix* is listed as an ingredient in the fish sauce recipe in the Geoponica it also refers to this generic idea of a group of species and not as I prefer the residue itself (see page 20). The later evolved usage of *allec* to mean the general idea of multiple small fish species suitable for salting and making sauces can be dated to the 6th century and beyond while the late 2nd century traders in Gaul and Holland, I would suggest, were still trading specifically in a fish sauce mash, the layer of semi dissolved fish at the bottom of the tank which would go on to make more *liquamen*. This probably came from either Britany or the south coast of Gaul and potentially Spain, and was sold on locally and also potentially into Britain. They may have acquired small amounts of *garum* and also traded *muria* but their main product I would suggest was always going to be *liquamen* at this time.

Turning to the other form of *allec* as a smooth fish paste. There is no agreement among commentators as to the meaning of the designation *garum flos* and *liquamen flos* on amphorae *tituli picti*. The terms *flos* and *flos flos/floris* are currently defined both by Curtis (1984, 1991) and Berdowski (2003:28) as superlative sales terms directed at the consumer to encourage the purchase of a superior product, that is, "the flower of the flower" or the "best of the best." Given that there does seem to be a superfluity of superlative terms associated with fish sauce already (*primus, excellens, praecellens, optimum, summus*), a different perspective is possible here. It may be that the use of *flos* was not a qualitative term at all, but may refer to some technical aspect of the product itself. A term to designate the *consistency* of the sauce would seem particularly necessary. It has been demonstrated in Chapter 8 that this definition of *flos* as an unfiltered viscous sauce works equally for a blood viscera *garum*. Berdowski (2003:18) has noted that the terminology *flos floris* may suggest that something was removed, "the flower of the flower," and my experiments have clearly demonstrated this to be possible and that sauce with the designation *flos floris* may represent a fully filtered and clear sauce. The incidence of the duplicate *flos floris/flos*, that is, "flower of the flower" and therefore fully filtered is rare, and found only on the *urcei,* and none are found on amphorae (Berdowski 2003:27). It is of course important to recognize that a significant role was played by local traditions in relation to terminology and wording and therefore use of terms like *saccatum* (filtered) found on a few fish sauce amphora would suggest that some fish sauce was marketed already filtered. A fine, translucent sauce was desirable at table, whether it be a *garum* or a *liquamen*, and I suspect this was normally achieved at the last minute by the cook, using fine linen cloth to catch all the fine particles. In *Apicius*, a recipe that calls for *allec* also requires that it is strained (*Apicius* 7.6.14). One can only assume that this means to remove any remaining liquid sauce, leaving a much thicker paste for the recipe. This might suggest that *allec* was marketed in various consistencies, from a solid paste to what might have been its natural semi-runny state when skimmed off the surface of a batch of fish sauce. A possible visual representation of this phenomenon has emerged. There is a remarkable mosaics pavement at the Rio Verde Roman villa at San Pedro de Alcantara near Marbella in Spain. The villa is modest in size and may have been built by someone with a great interest in cooking rather than dining and there is a possibility that it was related in some way to a master cook who had made his fortune. One section of the villa has a pavement decorated with images of kitchen equipment that a cook would use such as knives, *mortaria,* pots, dishes, trivets, ladles, racks of meat, and fish. One image (fig. 11.8) depicts what is clearly a cloth filter bag hanging over a wooden frame with a small dish beneath to catch something dripping through. It is very reminiscent of a jelly bag

Figure 11.8   Mosaic of a cloth filtering bag that may have been used for fish sauce. Copyright Sally Grainger

today.[13] It is certainly possible that this contraption may have been utilized in the kitchen to strain a *liquamen flos* sauce to render it sparklingly clear for service and also to strain an *allec* to achieve a thicker paste when required. It could clearly strain any liquid and so we cannot be certain of this but it remains a possibility.

The quality of a sauce

As there is such doubt about how to identify residues and to which sauce variety to attribute them, it is not surprising that there is no consensus as to how to classify them into elite and commonplace sauces. The richness and superiority of a fish sauce is a complex issue as there are many factors to consider. Few of these factors are recorded on the labels, so we are restricted in our ability to assess quality from this evidence, and have to fall back on the residues when we have them, their species and size and context. The distinctiveness and perceived quality of a given sauce was going to be apparent from its taste and richness, which is reflected in texture and consistency, density and color. However, these factors are not necessarily determined by species or size selection. We have seen that the major factors which determine the quality of the product are the duration of fermentation, degree of dilution, salt levels and length of time in storage and in what consistency that storage takes place (see page 162). The archaeological approach is therefore methodologically hampered by the lack of evidence in their attempt to assign quality attributes to residues they find and study. From the *tituli picti* designating fish sauce, manufacturers do not in fact indicate any species selection other than that of the use of mackerel either to make a *garum* or a *liquamen*. Others disagree, Djaoui (2016:123) considers that a *titulus* that designate *COR(dula)* SARD LIX from a Beltran 2A cited by Ehmig (2007:230) may have represented a *liquamen* from these two species, that is, small tuna and sardine mixed together and dissolved into *liquamen* and this is associated with various *tituli picti* designating "mixed" products *MIXST* and *MISCEL (CIL* XV 4806). As a mixture of different kinds of salt fish, this is plausible, but I do not think it is plausible for a sauce as most fish sauce appears to be mixed in this sense. I would consider the labels for *COD* and *LAC* to always refer to a form of *salsamenta,* as one would need the designation *liquamen* to be present for a dissolved sauce to be signified. When a mixture of small-to-medium fish-sauce-fish, sparids, scombrids and clupeids are all mixed together and dissolved into a sauce information about the size would not be valuable or significant enough to put on a label in my view. A named species and its potential size signified by *LAC* meaning small mackerel and *COD* meaning small tuna on *tituli* are clearly relevant, but only if those fish retain their integrity as *salsamenta* inside the vessel. Clearly, the recipe and method applied to the sauces is important, and that is often only stipulated by named manufacturers or creators of the recipes, as well as specific geographical locations which over time had acquired a reputation for good sauces, such as *garum Hispanum* (Spanish), *Ostiense* (Ostia), *Pompeianum* (Pompeii?), *Antipolitanum* (Antibes), *Luense* (?), *Romulianum* (?), *Lucretianum* (?) (Curtis 1991:195). These names will presumably have corresponded to recipes and even known qualities of sauces but we can say very little else.

A more desirable *liquamen* is almost certainly made from pure mackerel, and *tituli picti* labels found, albeit rarely, on amphorae attest to it (*CIL* IV.2588, *CIL* III. 12010, 48). Labels signifying *garum flos scombri*, that is, a blood viscera sauce from mackerel intestines, are far more common from the *CIL* catalogue but a large majority of them are written on *urcei,* the capacity of which are considerably smaller than amphorae, something which has distorted our perception of the important of this sauce compared to *liquamen*. Shipwreck evidence such as Grado and also the Little Latium pots strongly suggests that sauces were made from a mixture of varieties, including mackerel, and sardine (Delussi and Wilkins 2000:53–65; Auriemma 2000:31–49, Peña 2007:72; Djaoui *et al.* 2014, and table 8 below). A key question to ask of this evidence is, how much mackerel needs to be included in a mix of fish to justify the manufacturer being able to claim that the product is made with mackerel? The Grado shipwreck tells us that a mixture of sardine and mackerel could not be claimed as a mackerel sauce, as the vessels are inscribed with the label *liquamen flos* (Auriemma 2000:31–49; see page 219). The lesser or more ordinary and readily available form of *liquamen* would appear to be just "small-fish" of multiple species and size all mixed together, or combinations of sardine and anchovy or sardine and mackerel alongside the lesser known sparids and clupeids. It is not, however, entirely that simple, as a sauce made with the smaller species that was subsequently left to age for longer under an ideal temperature, 25°C to 35°C, and not diluted, would potentially be superior, in terms of its protein content, reflected in taste, to one made from mackerel if that had been over-diluted or taken too early, as ageing sauces unfiltered would seem to be a crucial factor. The finest fish sauce in the world, "Red Boat," is made from anchovy <10 cm in size, and the quality is determined by low salt and the ageing process (Cuong Pham, owner of the Red Boat fish sauce factory, pers. comm.). These issues are clearly complex, and unfamiliar to us. The discerning ancient consumer would probably have been able to tell how rich his sauce was by taste and the density and the stickiness to touch would be the best test for richness. When lower levels of salt are used, the taste is much more intense and distinctively meaty and cheesy and exceptionally fine, while a cheap, low-grade diluted and sweetened modern sauce just tastes of salty fish.

The remains found inside a dolium at Masada of tiny 3 to 4 cm anchovy and sardine, traded from the Mediterranean, have been considered a valued, purpose-made *allec* product, apparently because of the specific selection (Cotton *et al.* 1996:229, 232). The distance involved and the consistency of species by size and selection have led to assumptions of a high quality. I consider that the Masada product was almost certainly originally a commonplace *liquamen* and the bones an example of the discard practices whereby the residue was simply conveniently thrown away. It is also understood by some that sauces made with tiny fish of all the same species and/or size must have been an expensive and desirable product, because the selection is seen as specific and/or difficult to achieve, and therefore exclusive. Bernal-Casasola *et al.* (2016) have reached the same conclusion for the Augusta Emerita evidence of a sauce residue of sardine 3 to 6 cm (with an occasional larger horse mackerel and sardinella of 10 cm) which they define as an "exquisite *garum* sauce" based on the reconstructed incomplete *tituli picti* on the amphora, but they also define it as a purpose-made *allec* similar to a *pissalat* because of the size. It is difficult to see how the same product could have been both. The scholarly analysis conjecture that the small size of the fish utilized determined that it was of a high quality as the bones would be so small as to be un-noticeable. It is also claimed that when this kind of *allec* was supposedly made with a larger size of fish such as mackerel, the resulting *allec* was necessarily of lesser quality, because the bones were so much bigger and more difficult to consume (Bernal-Casasola *et al.* 2016:196; Bernal-Casasola *et al.* 2013:2529). The idea that large mackerel were used to make a purpose made *allec* is doubtful I feel and it is bizarre to conclude that the bones would have been left in such a paste for consumption. In determining its quality, Bernal-Casasola *et al.* (2016:748) further claim that the difficulty of fishing such a species in such a small size range would also be a factor defining quality. I also find this difficult to understand as it suggests fishing for little fish was difficult and specialized skill which added a premium to the product. There are other way of looking at this. It seems to me that it would not have been at all difficult to catch very small fish of equal size, as they often swim by size-range in shoals close to the shore, and a simple net from the beach can catch thousands of these equal size fish with ease and little expense. Fish of this size were always the poor man's fish, with little meat on the bone, and of little commercial value unless caught and cooked within a few hours. The description of "The trade in salt-pickled *hamsi* (anchovy) from the Black Sea" by Mustapher Demir demonstrates the ubiquity and commonplace nature of this kind of species across the Eastern Mediterranean (Demir 2007).

Notes

1 There has been a recent report of a fish sauce factory in Newfoundland Canada closed overnight in 2001 after the owner went bankrupt. The tanks were full of sauce in mid-fermentation and 20 years later the decaying fish sauce and the gross smell has almost closed the town. The local community have no idea how to deal with the 1000 gallons of putrid sauce. The only practical way to prevent ongoing smells would be to bury the residue and seal the vats which seems to be what has happened in many ancient sites. https://www.eater.com/2019/5/23/18634704/st-marys-irish-loop-newfound-land-smell-fermenting-atlantic-seafood-sauce Monday, 05 October 2020.
2 NISP: The Number of Identified Specimens or Number of Individual Specimens. MIN: minimum number of individuals.
3 Horse mackerel was not generally salted as *salsamenta* according to Van Neer *et al.* (2010:170).
4 If the fish paste was truly eaten with its bone, as many think, then the bones consumed would also have disappeared. The work done by "Bone" Jones has demonstrated that most consumed fish bones do not survive the digestive process (Jones 1986).
5 Piper piperium is made with "The finest Spanish anchovy fillets, which have been packed in barrels of salt and left to mature for 18 months. Once suitably matured, they are rinsed in brine and gently cooked before being cooled and blended with butter and rusk. A secret blend of spices and herbs is then added' https://www.theguardian.com/theobserver/2001/feb/18/life1.lifemagazine11
6 We note that a anchovy brine cannot be distinguished from a fresh *liquamen* from color alone.
7 Amphorae in the Praetorian camp: *CIL* XV 4730, 4731; *urceii* from Pompeii: *CIL* IV 5717-20, 9407, 9408, 9411; *olla* in Pompeii: *CIL* IV 9409, 9410.
8 Kosher fish sauce is well documented from *tituli picti* such as *GARUM CAST(UM)* and *CIL IV* 5660-62 and with reference to Pliny (HN 31.44), discussed page 33 and also see particularly Robert Curtis' article at https://www.biblicalarchaeology.org/daily/archaeology-today/biblical-archaeology-topics/the-garum-debate/
9 A further debate concerns whether these product or related products were macerated and or minced and were associated with what was known in Babylonian sources by the terms *tarita* or *tzahana*, a so-called "foul smelling"

dish of fish that may have been chopped, and had been macerated in salt, oil and spices for a week, which would generate a lightly fermented dish similar to the Egyptian *fasheek*. This is found in medieval Arab cookbooks as *sahna* (Weingarten 2018: 241/2).

10 Isidore of Seville (*Origines* 12.6.39–40) states that in the 6th century *allec* were *pisciculus ad liquorum salsamentorum* a small fish (suitable for) the liquor of salted fish (Curtis 1984:152).

11 It is unlikely that this contraption was used for wine as the wine sieve is quite well documented as perforated metal sieve. It's hard to imagine what else it was used for and given the need for sieving of all the products of fish sauce it seems entirely plausible to associated this with fish sauce.

12 Fish sauce amphorae as functional vessels

Introduction

It is self-evident that the food commodities (wine, oil, and salted-fish products) traded in amphorae "are precisely the types of products whose transport is crucial to an understanding of maritime trade" and the Roman economy (Bernal-Casasola 2015:62; Rice 2016:168). The main aim of this book has been to try and develop a theoretical model of amphora use and reuse and to see whether it can illuminate the trade in fish sauces and thus the economic systems behind it. There were a multitude of different amphora forms from all over the Mediterranean and in all periods that appear to have been used regularly or randomly to carry various different fish products over the period in question; for various reasons, these will not form part of the model. The patterns of amphora shape and function that we are looking for will only be present in the large scale and consistent use of a small number of shapes. Thus, I will concentrate on the main amphorae made in *Baetica* and during the period when fish sauce was at the height of popularity from the mid-1st century BC through to the 4th century AD. It is only in this place and in this period that the fish amphorae display a consistency of form and an abundance in terms of consumption across all areas of the empire. Here, therefore, we may be able to identify a rational mind behind the design and potential function of the various shapes. Discussion will therefore focus on forms Dressel 7 to 16, Beltran 2A and 2B and Almagro 50 to 51. These vessels were apparently designed specifically to accommodate fish products and therefore demonstrate the design characteristics that we should be able to associate with the needs of specific fish products. We have seen that fish sauce could be traded in various consistencies and states of disintegration and it is logical to assume that these different states will have initially determined the design features of the various amphorae under discussion. Now that we can visualize the appearance of the sauces it has become possible to begin to attribute their appearance to the shape of the amphora. This discussion is primarily concerned with how amphora functioned as fish vessels. Readers should, by now, have no expectation of a traditional archaeological approach and I make no apology for this. Readers will not find extensive data on presence and absence of particular types of amphora or any degree of complex statistical analysis, beyond what are simple calculations of widely known significance or lack thereof in terms of *tituli picti* and amphora distribution. These are the tools of traditional ceramic archaeology, while here I propose a radical rethink of that traditional approach and bring to the issue all the information thus far gleaned from the holistic approach in the previous eleven chapters to suggest new and different ways of looking at the topic. There are numerous general studies and specialist reports for amphora production and consumption across the Roman world, and extensive bibliographies are readily available in the on-line resources which have been my main source of information.[1]

Amphorae: a definition

An amphora can generally be defined as a large ceramic jar and its ability to carry liquids or solid product in a liquid base, over a mid to long distance trade, is the most important defining factor. This trade often includes sea travel but not exclusively. Amphorae tend to be characterized by certain distinctive criteria such as width of the mouth, the length of neck, the shape of body and the size of spike. The majority of amphorae have a pointed spike of various lengths and widths which makes them easily maneuverable and ideally suited to sea travel as they are able to move with the rolling of a ship when bound together. When on dry land their pointed spikes are less useful and when stored they were either leaned against a wall in numbers or stacked on their sides in specially constructed wooden shelves. The term amphora itself can often be limiting in any study of amphora capacity and the wider issue of the products being traded and consumed, as some vessels are referred to as amphorae but actually resemble table jugs (*urcei*) with a wide enough base to stand and a low capacity such as the *urcei* illustrated Fig. 12.1, while others are vast and cumbersome containers such as those for oil (Molina Vidal and Mateo Corredor 2018). Despite the disparity in size and capacity, the *urceus* is often categorized as an amphora for the purposes of statistical analysis of the extensive collection of fish sauce *tituli* which can be very problematic. The most detailed fish sauce *tituli* that provide not just the name of the product, but its qualities, species utilized, as well as owner, manufacturer and consignee are often displayed on these small vessels that, in terms of capacity, represent a fraction of the fish sauce in amphorae that was actually traded and consumed around the ancient world (see Figs. 11.2, 11.3, and 11.4).

The incidence of amphorae *tituli* identifying what appear to be dry goods is remarkable small, and includes nuts, chickpeas, rice and flour, which Peña describes as nonstandard content. In many of these cases, the interpretation given to these *tituli* has proved to be uncertain or erroneous. A number of such examples have also been found to be on what might be termed storage vessels and necessarily the result of reuse, rather than packaging

Figure 12.1 An *urcei* of more ovoid shape on display at the Musee Archeologique de Nice-Cimiez. Copyright Sally Grainger

amphorae, and are not therefore relevant for a study of long-distance trade (Peña 2007:103). On land, many amphora forms are unwieldy and cumbersome to handle, and we are unsure as to what extent the contents could have been decanted into smaller or less unwieldy vessels when long-distance travel on land was envisaged. Often the transportation of empty amphorae from the pottery workshops to the fish salting facilities involved a considerable distance (Bernal-Casasola 2016:307). There has in the past been an expectation that moving empty vessels was economically unviable, but as Bernal-Casasola has pointed out this may be the wrong way to look at this issue (Bernal-Casasola *et al.* 2007:61–70). The reuse and redistribution of empty vessels may have been a far more regular procedure. Peña sites instances of a trader engaged in reclaiming wine amphora for reuse for wine and he identifies himself as an amphora broker "*negotians.*" It will be my contention that the systems involved in the reuse of fish sauce amphorae

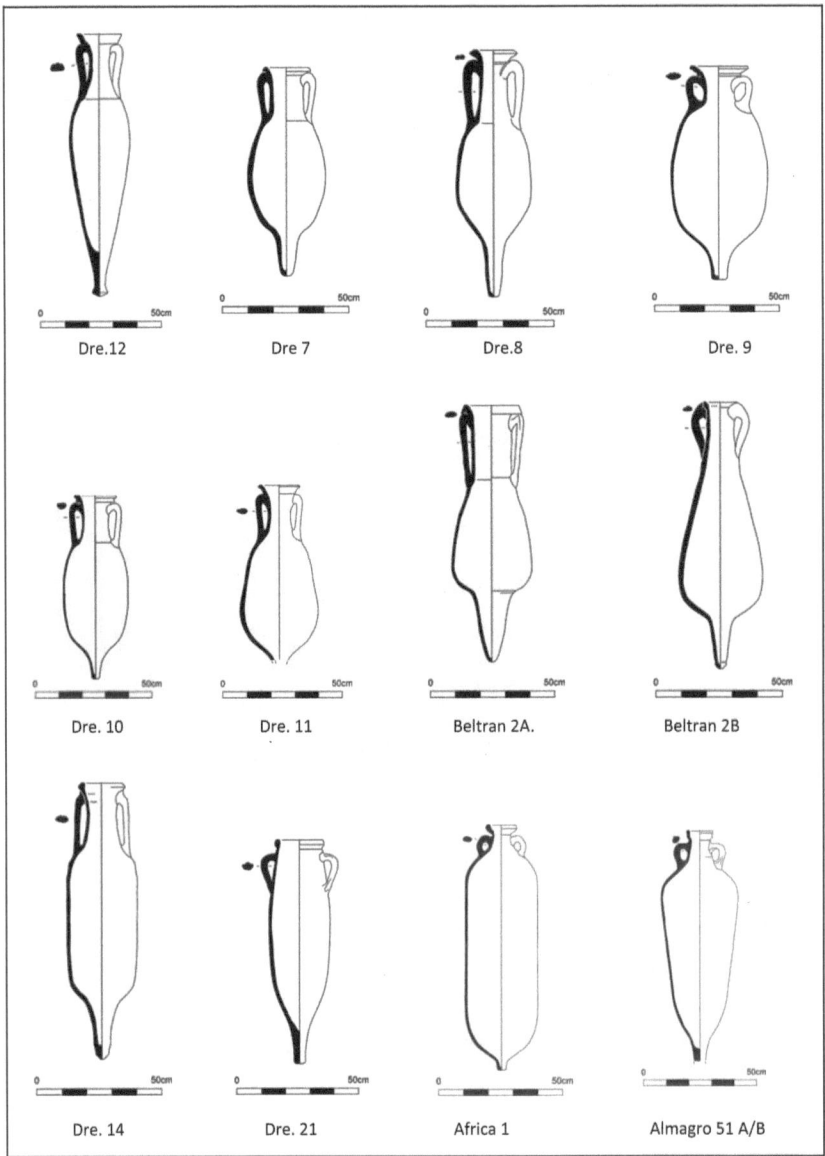

Figure 12.2 Fish amphorae discussed in this volume

will have been managed in the same way (Peña 2007:115). There is also the possibility that barrels were increasingly used to transport salted fish products, with an expectation that the barrel took over from the amphora when they cease to occur on many sites. Products delivered in barrels could also in theory be decanted into amphorae far from the production site. This

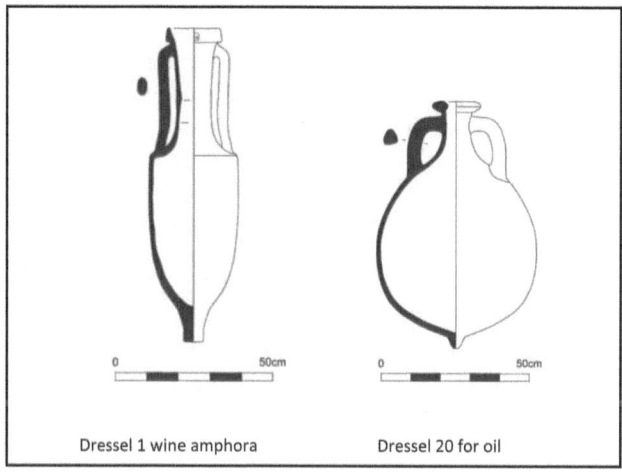

Dressel 1 wine amphora   Dressel 20 for oil

Figure 12.3 Typical amphorae for wine and oil

is posited for the contents of the Dressel 9 and 10 *similis,* the locally made fish vessels from the middle Rhône Valley, from the potteries of La Muette and Manutention in Lyon (Ehmig 2001:69). From their *tituli,* it is clear that fish products were contained in these amphorae, and though barrels may have been used to bring the products to the filling site, there are other possibilities for these Lyon amphorae discussed below. It is impossible to discover how common barrel use for fish products were as they rarely survive (Wilson 2007:5). An example of well-preserved anchovy fish bones caught in the resin of a barrel from the 2nd century AD have been found at the dock at Fos-sur-Mer, and this evidence may suggest that a significant volume of sauce or *salsamenta* could have been processed and traded in barrels (Desse-Berset and Desse 2000).

The group of amphorae associated with the trade in ancient fish products come in an array of baffling and unusual shapes which do not readily fit any rational pattern: the neck can be wide (Dressel 7,14, Beltran 2A) or a narrow and an hourglass (Dressel. 9,10,12,16 Beltran 2B), long (Dressel 7,8, Beltran 2A) or short (Early Ovoid, Dressel 9,10, Dressel 19/6b); the mouth can be flared open (Dressel 8,9,10,11,12, Beltran 2A) or relatively straight; the body a wide ovoid (Dressel 7,8,10) or narrow and elongated (Dressel 12,14,16); pear shaped (Dressel 9,11, Beltran 2A, Beltran 2B) or cylindrical (Dressel 12,14,16); the pivot-spike insignificant (Dressel 9,10) or long and either solid (Dressel 7,8,12, Beltran 2A/2B) or hollow (Dressel 7,8,12, Beltran 2A/2B; illustrated figure 12.2. Many amphora types share characteristics which seem contradictory such as the vast capacious hollow spike which has been found to be subsequently filled in with clay to render it useless and cumbersome or the idea of an elongated hourglass neck, the reason for which is hard to determine.

Amphorae for wine and oil

The amphora form Dressel 20 for oil was a transport vessel for an "uncomplicated" liquid. The spike to these amphorae acts as a pivot but is otherwise insignificant. They have relatively narrow, short, hourglass necks, with a wide enough mouth to allow pouring of a smooth liquid without spillage and a massive ovoid body to maximize volume, estimated at 70 to 75 liters. The reuse of oil amphora was understood to be completely impossible and we believe this was due to the oil penetrating the fabric and oxidizing which could lead to unpleasant odors. The mountain of amphorae at Monte Testaccio in Rome attest to the disposal of thousands of the Dressel20 amphorae broken in situ and they must have been considered worthless to accumulate in such numbers (Funari 2001). Wine was a rather more complex product as it may have continued to ferment during transit and we can see this requirement in the elongated neck on Dressel 1 and 2 to 4 amphorae, and in the common practice of drilling small holes in the neck or bung so that the gas could escape (Estreicher 2006:20). We have ascertained that fish sauce could quite easily continue to ferment during transit and therefore the long neck in fish sauce amphorae might correspond to a similar need to accommodate gases. There is a distinct "shoulder" separating the neck from body in many wine amphorae which may indicate that the product would stay below this line and not necessarily fill the neck, especially if it was narrow. In the current analysis of capacity of amphorae, the tendency is to estimate from just below the rim, but this may lead to an overestimation of capacity in use (Molina Vida and Mateo Corredor 2018). The body of these vessels was largely cylindrical, though Dressel 2 to 4 amphorae tended to be wider, and their average capacity was c. 26 liters. In comparison with oil this corresponds to criteria more concerned with quality rather than quantity. They taper to a solid pivot spike, and as the hollow spike may be associated with some form of a residue, we may conjecture that wine did not contain a residue that needed accommodating, illustrated in Fig. 12.3. Using these design criteria as a bench mark, we can see how simple uncomplicated liquids were accommodated.

Amphora design and content

There has been considerable scholarship published in recent years on the issue of amphora content. An international interactive conference on amphorae content took place at Cadiz in 2015 at the instigation of Dario Bernal-Casasola (2015) which brought together numerous specialists in ceramics, epigraphy, and residue analysis. Bernal-Casasola has defined the parameters for future study into amphorae and their contents which include protocols to ensure that vessels are examined with minute care to determine the nature of the content (Bernal-Casasola 2015:63–64). The important factors include the direct evidence of macro-residues such as fish bones or olive stones, the *tituli picti* which can indicate at least one of the fills in the vessel, and micro-scale residue analysis which can also provide information about a single fill or multiple residues from multiple fills of the same vessel (discussed in Chapter 6); the general link between typology and a known type of content in specific circumstances, and the general knowledge of the economy of the region of production (Bernal-Casasola 2015:63). These criteria can together provide patterns of typology and associated content that are generally true, but not always. These new protocols were formulated to deal with the whole panoply of amphora forms and all manner of amphora content, while we are concerned entirely with traded salted fish products. Residue analysis is providing new and exciting knowledge of particular content, however many issues remain unanswered, as outlined by Bernal-Casasola (2015:66): how do we distinguish the trade in sauces from *salsamenta* in vessels with no labels that seem to be used for both indiscriminately; how can we comprehend the traded in salted tuna, when there is a near complete absence of faunal evidence for it and the numbers of amphorae for this product in the early empire do not correspond to the expected volume of this trade; how do we distinguish fish sauce residues, when they can be both micronized and also completely undamaged, or when there are no residues or *tituli*; how do we estimate the volume of the varying qualities of sauce traded in a given place? Calculations of amphora quantity and volume are made based on various criteria: sherd count and weight, rim count and weight, minimum number of individuals (MNI) and number of identified specimens (NISP). All these forms of calculation have disadvantages and advantages as outlined by Mateo Corredor and Molina Vidal (2016), as ultimately the actual number and capacity of the vessels in question can only be an estimates. Peña has noted that reclamation of amphora and the disproportionate use of body fragments over necks and spike in construction will have distorted the archaeological record (Peña 2007:345). If we cannot know whether an amphora held a solid salted fish product or a sauce as its primary content; how do you determine what constitutes the primary content, if, as I have suggested in this volume, *salsamenta* and *muria* undoubtedly travelled together for a period of time before being separated, and each could be viewed as of equal culinary value; how often were amphora regularly reused to transport similar and different products to the prime use fill and, more importantly, how many times were amphorae refilled before they were discarded? The ground breaking work by Theodore Peña, *Roman Pottery in the Archaeological Record* has opened up archaeology to the uncomfortable possibility that presence/absence data need not correspond to consumption. He has created definitive definitions of types of amphora use which has proved essential for amphora analysis. "Prime use" involves the content for which the amphora was designed; "type A reuse" involves the same or similar content; "type B reuse" involves an entirely different content and "type C reuse" involves modification for structural purposes. Using these criteria he has recognized that many factors influenced the deposition of amphorae in the archaeological record and that large quantities of amphorae were reclaimed and reconditioned for building work and hydrological purposes, and their final deposition may have been a long way from where they were emptied, and to carry this forward the numerous consumption *loci* of smaller quantities of product may be a considerable distance from the emptying *locus*. The anomalies that are seen in amphora use where nonstandard content can be found in vessels and in the reuse of vessels to carry untypical products means that it can never be possible to make large-scale assumptions about widespread and extensive trade based on the small numbers or single instances of secure knowledge about a particular typology and its content. The concept of reuse will always place barriers in the way of ascertaining the larger patterns of amphora use that we want to see. I think that little progress can be made in the area of research into amphora content unless a new approach is taken. In what follows, my approach will be to temporarily set aside the intricate concept of typology and instead to view the vessels simply in terms of their functional characteristics. I want to "deconstruct" the fish amphorae and look at individual functional characteristics, regardless as to type. I think it is possible to pick out key design features that correspond to function from the huge variety of subtle differences between types and sub-types. Many of the design features that are relevant were duplicated by most of the forms and made in numerous pottery kilns all along the Baetican coast from Cadiz to Cartagena, and it is therefore useful to consider the common features to be the most important in terms of function.

It is possible to see that some of the morphological features do not seem to correspond to the needs of the specific products at all and we might legitimately disregard them from a discussion about general function. These are: the shape and position of the handles, the ridges on the handles, the placement of the handles on the shoulder, and the shape of the rim of the mouth, whether triangular or flat, wide or narrow. These characteristics seem to be about the pottery workshop identifying their work through simple regular features and decorative touches and they are also concerned with ease of maneuverability, issues related to management of the vessel in the ship, at the port, and whether they were strung together. All these issues would seem to be true of all amphorae and therefore not necessarily relevant for fish products particularly, illustrated in table 12.1 and figure 12.4.

These design elements are often precisely those that archaeologists use to distinguish the subtle differences in sub-typologies, which in turn signify production areas, fabric and ultimately the dating criteria. If these elements do not correspond to function, we need not necessarily take them into consideration when thinking about function. The design criteria that correspond to function concern far more fundamental issues around the distinct **segmented structure** of all these amphorae. The three crucial segments are mouth and neck as one segment and then the shape of the body

Table 12.1 Design features of fish amphorae not necessarily connected to the accommodation of fish products

Morphological feature	Possible function
1 Shape of rim at the mouth	Can be decorative. Identification of workshop May have a connection to being roped together or hung from the rim.
2 Placement of handles on rim or shoulder	Ease of maneuverability Can be decorative. Identification of workshop
3 Ridges/grooves on handles	Can be decorative. Identification of workshop
4 Length of handles	Ease of maneuverability Can be decorative. Identification of workshop
5 Presence/absence of shoulder	Can be decorative. Identification of workshop When pronounced may correspond to fill level
6 Fabric	Can be decorative. Identification of workshop
7 Color, slip.	Can be decorative. Identification of workshop

and the shape of the spike. These criteria will logically correspond to three issues related to the sauces or *salsamenta*: ease of filling and decanting through the width of neck and mouth, consistency of the product and the related capacity of the body, and the nature and consistency of any residues in the spike, illustrated in Fig. 12.2

Amphora design

it is entirely reasonable to assume that an amphora designer collaborated with a salted fish manufacturer to manufacture particular vessels to accommodate particular products, with some communication concerning the specific shape and its purpose. It is not known to what extent one vessel might be an original design conceived to solve particular content requirements, which could equally be met by different design features at another pottery where the local fish sauce industry either made different sauces or had a different perspective on the products and how they might be accommodated. However, it is not necessary to assume that they were all multifunctional from the outset. They do however appear to have become multifunctional over time, which is problematic. Amphorae with narrow necks and relatively small mouths such as Dressel 7 and 8 and the Beltran 2A, are still apparently used to accommodate what is still largely believed to be *salsamenta,* rather than sauce, though I note that these views are not universal. The uniqueness of the different shapes suggests individual and early

Table 12.2 Functional features of fish amphorae associated with accommodating fish products

Morphological feature	function	Amphora type
1 Wide open mouth	Enables a lumpy or viscous product to be poured in safely	Dr 8,9,10,11,12 Beltran 2A
2 Long neck	To accommodate gases in continuing fermentation. To accommodate froth generated by the force of pouring. To accommodate a residue that floats above the liquid	Dr 7,8 Beltran 2A Forlimpopoli
3 Short neck and wide ovoid body	No residues, no froth, no fermentation, a simple singular product.	Early Ovoid, Dr 9,10, Dre. 19/6b
4 Wide neck	To enable and accommodate the pouring of complex lumpy products	Dr 7,14, Beltran 2A
5 Tron-conical – cone shaped neck, narrower in the middle	Enables pouring while containing product and prevents spillage. Access difficult unless liquid	Dr 9,10,12,16 Beltran 2B
6 Ovoid/elliptical body	To accommodate and maximize the volume of a bulk liquid in transit	Dr 7,8,10
7 ovoid/pear shaped body	To accommodate a bulk liquid at the bottom and another of different consistency or characteristic at the top	Dr 9,11, Beltran 2A Beltran 2B
8 straight cylindrical body	To accommodate an uncomplicated product of regular consistency	Dr 12,14,16
9 Short hollow spike	Acts as pivot in terms of maneuverability Accommodates small amount of residue	Dr 9,10
10 Long, narrow or wide solid spike	Acts as pivot Identifies a product that may have a residue, but in this case does not	Dr 7,8,12 Beltran 2A/2B
11 Long, narrow or wide hollow spike	Acts as pivot Accommodates various quantities of bulky residue	Dr 7,8,12 Beltran 2A/2B

attempts at accommodating the particular characteristics of particular products. Some of these design features were no doubt uniquely relevant for the first use of the amphora at the place where the vessel was made, but thereafter, if shipped empty or reused elsewhere, particularly if it was used to transport a slightly different or entirely different product, the particular feature noted may have had no relevant application to the subsequent utilization of the amphora. As Peña notes,

"There is no way of ascertaining either the extent to which Roman potters had assumptions regarding the ways in which the vessels that they had manufactured, would be used, or the extent to which those who acquired newly manufactured vessels actually employed them for these purposes."

(Peña 2007:10)

There are any number of variables could come into play once the vessel had left the potter's workshop and been sold on or transferred to another site. This is particularly true of fish amphorae made in the Cadiz region of Baetica and in the factories at Troia, which were made in an inland rural agricultural setting alongside oil and wine vessels but were then transported to the coast for filling at the processing sites. There seems to have been less opportunity for exchange of ideas or ongoing dialogue between fish sauce manufacturer and potter in these circumstances (Maganto-Martinez 2005:199). One can even imagine a situation where, once a group of functionally specific vessel had been passed on to new owners, they could have been used in a completely random fashion from the outset. It has been pointed out by Opait (2007) that amphorae that we can categorically identify as *salsamenta* vessels from the Black sea region, mainly for tuna, are quite distinct. These have large open mouths and either an absence of neck entirely or an exceptionally wide one merging with a huge tube-like capacious body. This is optimized by the Zeest 75 (Opait 2007:108). In the Punic period in Cadiz, where it is probable that forms of salted fish, mainly tuna, was the main product traded rather than a bulk fish sauce, (though we note that others disagree: see page 50). The vessels are similar, that is, the many forms of Ramon T, which are characterized without the segmented structure we see with the later fish amphorae. These are largely a straight cylinder with minimal neck and a very wide mouth. The Dressel 21/22 forms now recognized as vessels for salted fish made in Italy and Sicily also have this simple neck-less tube structure (Malfitana 2008; Botte 2009). The *tituli picti* on the Dressel 21/22 confirm the association with large *salsamenta* as a number from Sicily have the term *cet(us)* which can be securely associated with a large species of tuna (Botte 2009:150–151ff) Given that the products being traded were likely to be solid salted piece of scombrid species, the need to access the fish would seem to be fundamental, and the design of the amphora is entirely suited to accommodating that product. There appears to be a rational mind at work behind the design of the solid-salted-fish vessel and the same rationale can be seen in the vessels utilized for oil and wine.

Amphora labels

We cannot judge whether amphora labels were common but fail to survive, or alternatively if many amphorae did not need to be labelled because they held a "standard" content, and only when products were nonstandard was a label was necessary. A similar idea has been suggested by Lyding Will who considered that vessels may only have been labelled when reused (Berdowski 2008:264). Most amphorae do not retain a label and there are a number of reasons for this discussed by Peña (2007:100) and Bernal-Casasola 2015:67–68). Conditions at the time of deposition or shortly after destroys the label, as noted with the amphora stack at the Garum Shop in Pompeii, where labels exposed to the ash were destroyed but those inside the stack survived. A tag called a *pitaccium* was also commonly attached to the vessel, but these rarely survive (Petronius *Sat.* 1.34; Callender 1965:4). It is not clear to what extent this separate label was a more up-to-date indication of later content than the *titulus* which may have only indicated the first or one of many fills of the amphora and as we will see may be an old label not indicative of subsequent fills. Relabeling is apparent on *urcei* (see below) but less apparent on amphorae, but as so many do not retain a label or may never have been labelled at all, this is less significant. Reuse of amphorae is an unknown quantity; it has been described as insignificant by Parker, who suggests that "the chance of amphora reuse contaminating the archaeological record is slim" (cited by Rice 2016:167), but this view would seem to be valid only if form could be guaranteed to correspond to content most of the time. On the other hand, reuse as discussed by Peña (2007) was actually quite common in relation to all kinds of amphora-born commodities. There is an understandable desire to want to find *titulus* labels that correspond to the content identified from residues in a closed site so that amphora form and content can be irretrievably linked but this is very rare and there is often doubt as to the validity of the identification (Bernal-Casasola *et al.* 2014, 2016; Bernal-Casasola and Cottica 2017:243). The lack of a *titulus* might be because everyone in the chain of supply from the filling to the retail outlet knew what was inside, and may even have accompanied the vessels to the actual sale of the commodity in question, while the product itself was sold on, using small jugs or vessels brought by the consumer. We also do not know to what extent amphorae with labels were subsequently reused without there being any need or inclination to alter the *titulus* to comply with the second or subsequent fill. As we shall see this situation has been noted with a number of vessels.

The use of resin to line and waterproof vessels was once accepted as a guarantee that oil could not have been transported in them as the oil was believed to dissolve the lining. It was concluded that Africa 1 and 2 forms were typically assumed to be used for oil but when waterproofed must have been used either for wine or fish products. We now know that this assumption is less certain, as recent residue results have suggested that Dressel 20 oil vessels were waterproofed, but with far less resin which often fails to survive, possibly because it does in fact dissolve in the oil (Dorrego *et al.* 2004; Bernal-Casasola 2015:69). We might assume that a resin was necessary in a *salsamenta* vessel as the brine would in theory evaporate through the fabric over time and this may also be an issue for a *liquamen*. The Grado wreck amphora of the Dressel 19/6b form were heavily coated with resin suggesting that a *liquamen flos* was believed to be subject to volume loss through evaporation.

Amphora tituli picti and their meaning

The majority of detailed amphora *tituli picti* from Italy are derived from Pompeii, destroyed in AD 79, and from the 19th century excavations at the Praetorian camp in Rome (Curtis 1991:38). Many more also survive from Gaul and Germany, collated by Ehmig (1996, 2001, 2003). Amphorae from the Praetorian camp cannot be dated more precisely than to sometime in the early 1st century AD which limits their value in interpreting the relative value and distribution of elite forms of fish sauce. There are many problems associated with the amphorae and inscriptions collected in the 19th century under Heinrich Dressel and recorded in the *Corpus Inscriptionum Latinarum*. Many of the vessels are now lost and the *tituli* cannot be corroborated; while, as the inscriptions were often not drawn, but transcribed, there is no chance of a reinterpretation of the inscription. Despite the many substantial and subtle differences in typology of fish sauce amphora, we often have no choice but identify fish sauce amphorae under the generalized Dressel 7 to 11 typology (which also includes Dressel 12, 14, 16, and the Beltran 2A and B). This is particularly so within the Schoene-Mau typology from Pompeii, which classifies all of these under the one class of Schoene VII, which is far too narrow a classification for any useful interpretation of amphora function. The rich resources of evidence from Pompeii of amphorae and detailed *tituli picti* are difficult to use with confidence as we cannot be sure that the content and form are securely identified. The issue is discussed in detail by Peña (2007:100). As many of the commodities named in the *tituli* and the identity of residues found in the amphora are also disputed. It seems more realistic to think of fish amphorae as essentially multi-functional and their content simply fish-based, which greatly reduces their usefulness in any study of the Roman economy.

Van Neer notes the frequency of *garum* on Baetican amphorae generally as 60% (n=79), while *liquamen* appears to represent only 25% of the total. When looking at amphorae generally across the whole corpus (n=167) the distribution is less pronounced, with 29% *garum*, but an even smaller percentage of *liquamen* at 12% (Van Neer *et al.* 2010:72;164; Ehmig 2003). These figures have naturally been used to suggest that *garum* was consumed widely and was the sauce of most significance while *liquamen* was much less significant and consequently consumed less, but these statistics and this perception may in fact be completely distorted by the issue of vessel capacity, the inconsistent presence or need to label, and reuse itself. This entire book has been concerned with raising the profile of *liquamen* and recognizing that it was in fact the dominant form of fish sauce. The combined literary, zooarchaeological and experimental approach has allowed *liquamen* to take centre stage and this has inadvertently reduced the apparent significance of the blood viscera *garum* to what seems to be a rather obscure and bizarre commodity with a relatively narrow and short-lived market. If this is a reflection of the actual situation re *garum* and *liquamen* then the relatively small incidences of *liquamen* compared to *garum* on *tituli* are hard to understand. The predominance of *garum* from *tituli picti* would seem to be irrefutable and as we cannot determine the volume of traded but unlabelled *liquamen* there appears to be a major barrier to further understanding. I have suggested elsewhere in this volume that this impasse can be solved by recognizing that a great many fish sauce vessels initially contained *liquamen* or *liquamen* in a semi worked state and they may not have needed to be labelled as such at all. We find various forms of *tituli* associated with *liquamen* such as *flos* = unfiltered (though note others consider this a quality signifier, see page 201). A similar range of terms is associated with *garum*, but the most common term associated with it is *flos*, (Curtis cites 168 *tituli* for *garum*, of which 134 has the *flos* signifier, against 69 in total for *liquamen* of which only 9 include *flos* and 40 for *muria* including 26 that include *flos* (Curtis 1991:195). Curtis notes that *flos* is too common an epithet across all sauces to be an indication of very expensive sauce, and as noted I consider this to be an indication of an unfiltered consistency and that it is relevant for both *garum and* liquamen (Curtis 1991:163).

Nevertheless, one must particularly ask why it is that *garum* appears to be so significant and especially in Pompeii. From a simple calculation of the presence of the *garum tituli* It certainly appears as if *garum flos* was the dominant sauce, sold more often and in larger quantities. However, as noted Curtis does not distinguish between amphorae and *urcei* when citing the labels themselves, which is crucial as an *urcei* held anything from 0.5 to 4 liters, while an amphora had various capacities from 30 to 40 liters (Cappelletto *et al.*, 2013). Curtis states that of the 200± so-called amphora inscribed with labels from the Campanian region, 36 were actually amphora while the rest were *urcei* (illustrated figure 11.2; Curtis 1991:91; Cappelletto *et al.* 2013). No matter how these data are presented, they are not detailed enough to demonstrate anything significant as we cannot judge how much fish sauce was sold without labeling so that a simple calculation by presence of the two names does not tell us how much was actually made, sold or consumed (Curtis 1991:197). The data can only tell us, I think, that the labels are more likely to be identified with the more expensive sauces and also tells us how many vessels were labelled at any one time prior to the eruption.

It is certainly not apparent from the format of the inscriptions that *liquamen* was in any way derived from *garum*, as Curtis suggests (1991:8; Martin-Kilcher 2003:78). If the former was weaker or a sauce taken second, it seems hard to see how terminologies such as *flos, optimum, primum, excellens* could be attached to both products. The first indication of a *liquamen* as a commodity is very likely to have been found on amphorae, discussed in Chapter 3, and these probably occurred when the two terms seem to be already fully incorporated into an identifiable language of the trading industry in the early 1st century. We can only speculate that the term *liquamen* may have been used for the first time by the manufacturer and trader in order to identify the ordinary and ubiquitous fish sauce. We find rare references to *liquamen argutus* meaning sharp (and often aged as well), the one possibly as a consequence of the other, but as a rule sauces were not aged as much as *salsamenta*.[2] There are two aged *garum tituli* (cited on page 201). Others include a *muria flos AIIIIA* in a Dresse 12 (CIL XV 4724) and a *liquamen penuarium excellens sum AAAA* in a Pompeii VII (*CIL* IV 2596-5621). That *liquamen* could have been aged for four years is quite surprising, but not impossible. I have kept bottles of Thai fish sauce for that length of time, and they have not fared well but this may be due to poor storage. There is an expectation of oxidation and a reduction in quality but this need not happen if the seal is air tight and no light is allowed to penetrate, which we may expect would be the case with amphora storage

I have been in the privileged position to view many of these little *urcei* jugs stored in the Museo Archeologico Nazionale di Napoli (Figs. 11.2–11.4). What was most striking is how rough and casually made they are. Even *urcei* with an elaborate *titulus* are coarse and badly made. One of the small-est vessels, discussed in relation to the fish bones inside it on page 224 (*CIL*. IV.9405) had been thrown roughly at the foot, with lumps of clay protrud-ing from the base. An *urceus* of *liquamen* or *garum* may have been seen in a kitchen or the working area of a *popinae* but not in the dining room. We have seen that the literary evidence suggests that the diner would have more regularly made use of *garum sociorum* than ordinary *liquamen* and it is I think certain that when *garum* was served neat for use by the guests or blended it was decanted into a special silver or glass jug or small Samian ware cups for service. These considerations lead to the inevitable conclusion that, at the point of purchase of these sauces, the *tituli picti* were not meant to impress the elite hosts and guest but the cook, the bar owner, steward or *opsonator*: the private caterer. It certainly appears from the experimental and archaeological perspective that the "essential substance" was *liqua-men*, and that, apart from *salsamenta,* most fish sauce vessels held it, or a semi processed version of it. In these circumstances the majority of vessels would not need to be labelled. The general motivation for the labeling of fish sauce amphora may therefor correspond to the need to identify the first harvest of finished product, that is, a *garum* or *liquamen flos* which was perceived to be of high quality or price, particularly as so many of the labels seem to convey these value characteristics.

Curtis considered that some of the Spanish sauce products may have been expensive, but most fish sauce from Spain could not have been oth-erwise, Spanish exports would not have apparently controlled the salted fish markets in Italy (Curtis 1991:59). The distinction between *garum* and *liquamen* in this argument is critical. Wreck sites with residues originating from Spain are dominated by mackerel in the 1st and 2nd century and while some could have been *salsamenta* a good deal is likely to have been *liquamen* suggesting a predominance of this kind of sauce from Spain. The small species only begin to dominate in wrecks in the 3rd/4th centuries as noted by Morales-Muñiz and Roselló-Izquierdo 2016) and it is these species that are largely found in the *cetariae* when the factories cease to function in the late empire. If we factor in the knowledge – from the Little Latium pot and the Local fish sauce industry in Pompeii - that a mixed *liquamen* using local Italian fish was always being manufactured alongside the Spanish imports in the 1st century AD, this rather suggest, contrary to Curtis, that Spanish products were always relatively superior in the early period whether they be *garum* or *liquamen*. To take this thinking even further, the emergence of the widespread use of multiple species small-fish sauce in the 3rd/4th century corresponds to a lack of the larger scombrids everywhere in the Mediterranean. Curtis posits that Italian waters were in fact depleted of stocks by the end of the 1st century AD with reference to Juvenal (*Sat.* 5.92–98), and we may posit this means specifically the larger scombrids. We may attribute this depletion in stocks to the early increase in *garum* and *liquamen* production in Italy and it is this shortage that resulted in the dominance of Spanish fish sauce and *salsamenta* products. Ultimately it is likely that fish stocks in Spain were also depleted and they had to rely on the smaller species to make the standard *liquamen* fish sauce, which ironically was the first kind of fish sauce and had always been made in a local and small way everywhere from the Hellenistic period onwards.

Tituli picti labels for salsamenta

It is not until amphora *tituli picti* become commonplace that we find labels designating products that many understand to be a salted tuna. This appears as *CORD/COD(YLA)* and is understood as a very small tuna (Van Neer *et al.* 2010; Djaoui 2016; and see page 175). Research by Djaoui has identified a size distinction for mackerel similar to that for tuna. In modern fishery, *lacertus* is understood to be another term for mackerel, as is *lacerto* in western Italy today (Davidson 1972:121; Djaoui 2016:118).[3] In ancient sources, *lacertus* is used of a small fish (Pliny the Elder, *HN.* xxxii.146), while Martial tell us that it is smaller than a *cordyla*, that is, the youngest tuna (Martial, xi.52).[4] It is now understood to be significantly smaller than the mackerel known as *scombri*. There are many amphorae *tituli picti* in many different formats for a product variously called *LAC*, *LACCAT* and *LACCATUS*. There is a combination of *LACC(tum) ARG(utus) VET(us)*, meaning "salted mackerel sharp and aged" found on a Beltran 2B in Maguncia (Martin-Kilcher 2002: no. 9) which we should identify as a *salsamenta* made from this smaller mackerel (see below).

Some form of salted tuna seems to have been held in their brine for as long as four years according to amphora *tituli picti*. The *titulus* formula involves using A as an abbreviation of *ANNUM* followed by a number of IIII and a final A, which indicates clearly that the product was aged for four years (Ehmig 2003:61). An essential but rarely raised question is whether this was an instruction to store or information about the prior storage of the product. When the customer saw this label, what did he or she understand by it? It is curious that researches in this field do not find the idea of four-year-old tuna more disturbing! A wet salted product kept for that length of time would not be particularly palatable, at least by modern standard of food preferences.[5] Athenaeus says that "the older salt fish is, the stronger and more pungent it becomes" (Athenaeus iii.120). The term for the pungency is δρῖμυς meaning "piercing, sharp, keen, pungent and acrid." The Latin phrase associated with the aged tuna and mackerel on *tituli picti* is *argutus*, which has the same range of meaning. However, we cannot claim that the aged fish in this situation is necessarily the same as that recommended in *Athenaeus*. I suspect but cannot be sure that the "old" salt fish was aged for months not years, given the value placed on the three-day old mackerel by *Archestratus,* (page 177). There is certainly a lot of confusion as to what constituted good, fine and elite salt fish and what was just an everyday commodity. Many of the labels for *cordyla* and *lacertus* are also associated with the terminology of laying down products in storage, such as the one on a Pompeii VII amphora, COD -ARG -PENUARIUM SUM AIIIIA (*CIL* iv.9370). The ARG is expanded as *argutus*, with the sense we have seen of "sharp." The meaning of the term *penuarium* is obscure but related to the food supply, from the Latin *penus*, and some kind of instruction to store up the amphora for the stipulated years may be what is going on here. The term *summarum*, which always occurs in the genitive plural, is assumed to be related to quality and "of the best" (Berdowski 2003:32; Ehmig 2003:65). With few exceptions, the *tituli* cited by Ehmig for *cod lix(us), cod port(uensis)* and *cod ting(itana),* which would seem to designate the origin of the tuna, as well as *lympha*, were all aged for between two and four years (Ehmig 2003:64–66).[6] Many believe that these terms represent other forms of sauce rather than a solid salted fish product (Ehmig 2003:61; Djaoui 2006:117). Others consider they represent a solid product, an interpretation which I myself prefer (Van Neer *et al.* 2010:171; Botte 2009:54). Sadly, with few exceptions, the amphorae associated with these labels are classified as either Dressel 7 to 11 or Pompeii VII, which is a generalized typology with little subtlety in terms of its specific shape. Djaoui has noted that there is a marked link between the terminology for the salted *COD(yla)*, that is, tuna, and salted *LAC(ertus)*, that is, mackerel, in the use of terms like *penuarium vetus* and *summur* related to ageing storage and quality, and also the use of the ageing format *AIIIIA* on *tituli picti* for this salted fish product (Djaoui 2016:117; Ehmig 2003:65). However, Djaoui may have made too big a leap in suggesting that these small mackerel were specifically used to make a particular type of *liquamen*, as *liquamen* does not otherwise seem to have been identified specifically by the size of the fish utilized though others disagree (Djaoui 2016:117). The absence of *flos,* meaning "flower," which is strongly associated with liquids and is consistently applied to *garum, liquamen* and *muria,* on the labels which most commentators consider *salsamenta,* such as *lacertus,* is a strong indication that these products were solid fish products, not liquid sauces.

On the other hand, the *tituli* associated with *muria,* that is the brine that salted fish is traded with, are fairly simple and uncomplicated. *Muria* occurs with *flos, flos excellens, excellens, Antipolitana,* (from Antibes on the coast of Gaul) and *Hispana* (Ehmig 2003:64). This can only mean that the salted fish itself came from those regions and had been consumed along the journey that the vessel took, from the fishing and processing grounds, and the brine itself only became a product in its own right after the fish was consumed. The *muria* was seemingly transferred to other vessels at that point. If *cordyla* was aged for 4 years, as would seem to be probable from the *tituli,* does this mean the fish was aged for that time or just the brine, and in fact, does this hint at the possibility that *cordyla* actually designated a desirable *muria*? The titulus *muria* is relatively uncommon in the Mediterranean and occur much more often in the north. *Muria* occurs particularly on Dressel 9 *similis* amphorae made in Lyon, which accords well with this scenario as it will only have been after the fish had been consumed that the *muria* was available to be decanted into the locally made Dressel 9 *similis* and marketed (Ehmig 2003:64). It is of note that a *muria* derived from cleaned mackerel can appear to be the same shade of brown as a refined *liquamen* and could have been mistaken for and possibly sold as a form of fish sauce similar to *liquamen* (discussed page 108). We have suggested that the taste of the *muria* was less pungent and it may simply have been more to the taste of the northern tribes. The effect of these sauces on food is similar, that is, there is umami but it is just less intense.[7] The consumption of *muria,* but not *garum* or *liquamen* is documented as being purchased by auxiliary soldiers on Hadrian's wall in the Vindolanda tablets (Tab. Vindol. 2. 190; 10.25 Bowman 1998:110). The *titulus* which combines the term *muria* with *flos* must give us pause. If *flos* can be defined as a smooth creamy sauce unfiltered of its fish paste (*allec*), that is, unrefined, how can *muria* be described in these terms too as one would assume that the fish pieces did not dissolve? If *flos* is a quality signifier then there is no problem, but if it indicates a unfiltered consistency then we have to assume that over the three or four years of aging often stipulated for *cordyla* and *lacertus.* The fish being preserved in the brine would soften and shed particles which would mean that the brine was cloudy, and therefore require some form of filtering and thus be unrefined.

Fish bone residues of *salsamenta* with labels that identify what the product inside the vessel was called do not appear to exists. We have fish bones that some consider salted fish but, in all cases, there is no label. The rare incidence of fish bone evidence in an amphora with a label can be securely attributed to *liquamen*. An example would be the mackerel bones in a Dressel 16 from London Peninsula house with a label proclaiming it to be a *liquamen* from Antibes (Locker 2007:153; Alcock 2001:80). Problems arise when we find labels for products that are perceived as *salsamenta* that were apparently placed into amphorae with narrow and or long necks which render the product difficult (if not impossible) to get in or out without breaking them – which was clearly not the case in the complete items which survive. The presence of *cordyla arguta vetus penuaria* (young tuna aged and sharp for storage, *CIL*. XV.4741), in a Dressel 7 suggests that the smaller whole salted tunas are apparently accommodated in these nar-row-necked vessels. There is also evidence from the Titan A wreck (Benoit 1958:5–8), suggesting that the Dressel 12 amphora with a particularly nar-row hourglass neck contained bones from these juvenile tunas. Peña has noted the presence of small neat apertures in amphorae in Pompeii allowing easy access and removal of whatever was inside and this may be the logical answer to this dilemma see below. It is always plausible that some vessels were cut down or holes drilled in the neck to facilitate access to solid pieces of fish in some situations. Bonifay recognized holes in an African oil vessel (Africa 1/2) in Hammamet in Tunisia which Peña also concludes was prob-ably to gain access to fish pieces (Peña 2007:68). There are seven vessels with carved windows in the neck or body in Pompeii which are identified as the less-than-precise Schöene VII form which could correspond to any of the Dressel 7 to 11 and Beltran forms from Spain, (Peña 2007:69). If the removal of solid pieces of fish by this means was common, it would seem reasonable to expect to find more evidence of it (Peña 2007:68). It does not seem to me likely that cut apertures can have been a regular or habitual means of accessing salted fish in otherwise inaccessible vessels. We have seen that the widely traded fish sauce mash was potentially a concentrated form of sauce which could generate a further two volumes of sauce after rebrining. In these circumstances the reuse of fish amphorae may well have been a commonplace practice and indeed so necessary to the "system" of fish sauce trade that to damage many this way and render them unusable for future fills of a fish sauce would have been considerably counterpro-ductive (see below; Bonifay 2004:467–468; Peña 2007:67). We have seen that earlier vessels specifically designed for a solid salted fish are distinct in their ability to allow access to remove the product. Given this, it seems odd that as the industry become more sophisticated in terms of organization for sauces there was a reduction in the production of vessels specifically designed for the easy access to *salsamenta* (see below). If one identifies *cordyla* and *lacertus* as well as dissolved special *liquamen* sauces then there is no problem with the apparently continual use of narrow necked vessels, though this is still a conjecture that is unproven (Djaoui 2016:117). Botte exposes the anomaly rather effectively when discussing the different mouth and neck shape of the Dressel 1 wine amphora purported also to be used to house solid pieces of fish: "It would be like wanting today to return pieces of anchovy or sardine in bottles of wine" (Botte 2018:141). The image I took of the amphora from the Musée d'Archéologie Méditerranéenne in Marseille with fish compacted inside (Fig. 6.1) which was also cited by Ehmig in 2003 bears witness to the illogical concept of putting whole mackerel in a Dressel 9 amphora when the product is better suited to an open-mouthed vessel.

The question at the heart of this issue is the size and shape of the fish pieces being traded as discussed in Chapter 11 We assume that there was a large-scale systematic trade in solid salted fish, either large mackerel or tuna steaks of substantial size from around the Mediterranean, but it is difficult to see in the later period as the vessels are not visible. The Punic Ramon T forms ceased to be made and the later amphorae strongly associated with *salsamenta* from Spain, through their wide-open neckless characteristic (Dressel 21/22) are not found far beyond the middle of the 1st century AD. A salted fish tuna product identified by the *tituli* CET has been identified by Emmanuel Botte who recognized the Dressel 21/22 forms as salted fish vessels from Italy and Sicily. This we can particularly associated with the largest tuna, (*ketos* (Athenaeus. vii.301; Botte 2009:54, 142). This form in Sicily also ceased to be made by the end of the 1st century AD (Botte 20119;35;167). The salting industry therefore appears to rely on the Dressel 7 to 14 and later Beltran forms from the 2nd half of the 1st century AD onwards, all of which have the anomalous necks, segmented bodies and multifunctional characteristics that are clearly less suitable for large solid objects. The switch from using purpose made vessels to using ones which seem to be less ideal and certainly not designed to accommodate large pieces of fish is significant. The processing sites both in Sicily and Spain continue into the 4th century and the question remains how much *salsamanta* of substantial size was being traded at this time? The barrel is assumed to be the best answer to this problem in the long run (Botte 2009:167), but the alternative scenario is that the stress placed on fish stocks in the early empire was such that the industry was for a time at least greatly curtailed even in Spain. This is particularly relevant to the obvious stress on tuna stocks from taking too much *cordyla*, but also the seeming reduction in mackerel noted by Desse-Berset (1993). Did the gourmet obsession with the viscera and blood of these species for *garum* and the evident liking for juvenile tuna result in the devastation of the *salsamenta* industry during the later 1st century AD such that there was no great need for specific and appropriate vessel for *salsamenta* when so many fish sauce vessels were available to be used instead for the smaller volume of trade?

It is possible that the remaining *salsamenta* industry may have involved much smaller cuts, the cube (κύβιον: Athenaeus iii.120), which may have been meant to generate two products, a brine and a secondary low-value salted fish. The latter would be of sufficiently small size that ingress and access from the narrow-necked amphora may have been possible. The *Amphorae ex Hispania* site notes that the diversity of content (of Dressel 7 and 8) could be due to difficulties in distinguishing this form within the broader amphora family of Dressel 7 to 11. The comprehensive citing of *tituli* for *laccatum* and *cod(yla)* on amphora forms in Ehmig (2003:65), which includes German, Pompeian and Pretorian examples, does not make any more precise distinction than either Pelichet 46 (Beltan 2B) and Pompeii VII (Ehmig 2003:69). The former vessel is characterized by the wide long neck and so would seem to be a more suitable container, but the more nebulous Pompeii VII may have been a Dressel 7 to 11 with either a long and wide or short and narrow neck, and it is these characteristics that determine how suitable the vessel would have been to carry a 20± cm juvenile tuna (*cordyla*). The lack of faunal evidence for this product, whether it was a form of *liquamen* or a *salsamanta*, will always remain a problem, discussed in Chapter 11. It remains to be seen whether my view that the incidence of other forms of fish products, *cordyla, lacertus,* and the obscure and undefinable *lymphatum*, were on the periphery of what was a widespread trade in sauce or an equally popular commodity and to what extent this was dependant of availability of fish stocks at different times.

Fish sauce vessels in detail

The first vessels recognized in archaeology as specifically designed for a widely distributed fish sauce were the Dressel 9 and Dressel 10 forms, made extensively in Baetica from the early 1st century BC but becoming dominant through until the mid-1st century AD. I have speculated that when this form began to be made the Latin term *garum* still referred to *garos*, that is, the essential substance, as the blood viscera *garum* had not appeared yet and in fact as *tituli* are unknown at this time we cannot confirm or refute this (Martin-Kilcher 2003:77). One may speculate that until fish sauces became more sophisticated *tituli picti* naming specific products would seem to be unnecessary; they certainly cannot be dated to this early period. These vessels have the ovoid belly, short neck and open mouth and short insignificant spike that is typical of oil vessels and reflect the accommodation of an uncomplicated liquid. Our experiments have allowed us to see that this uncomplicated liquid was very likely to be a form of *liquamen flos* derived from the top of the processing vessel or *cetariae* with virtually no bone residue but which was rich in its *allec* floating on the surface. Undatable *tituli* associated with the Dressel 9 type cover virtually all the fish products that were traded and over a substantial time period, but sauces of the *garum, liquamen* and *muria* type dominate.[8] Dressel 10 types also carried mainly liquid sauces,[9] though other vessels seem to carry more complicated solid fish products such as *cordyla*.[10] Identification of amphora is always going to be hampered by problems in determining typology where the distinctions are difficult to determine from fragments as noted with the Masada amphora associated with Herod discussed on page 67. The vessel that is associated with expensive *garum* and *liquamen* in the early 1st century AD is the Dressel 12, which first appears from 60 BC and it is possible that, at this time, they would have been used to accommodate a fine mackerel *liquamen* and only later elite mackerel *garum* (Martin-Kilcher 2003). It has a narrow, elongated and cylindrical body, open mouth, short narrow neck and resembles a wine amphora. Cesteros cites the incidences of Dressel 12 with *tituli* for both types of sauce (Cesteros 2012:20). This vessel is associated with *garum* in five *tituli* from the elite site of the Praetorian camp and its use with mackerel *liquamen* is also attested with two.[11] Such numbers are not sufficient to suggest that *garum* entirely dominated the sauces carried in these vessels. A further Dressel 12 from Ephesus with a *titulus* of *scom* but without the primary alpha identifier (the formula for interpreting amphora *tituli picti* devised by Dressel 1879)[12] cannot automatically be assumed to be *garum* as it could be either *garum* or *liquamen* (Cesteros 2012:11). In other finds, an absence of the primary alpha identifier has led to an overemphasis on the product being identified as *garum*, as noted regarding the Beltran 2B vessel from Augusta Emerita with a *scom(bri) f(os)* *titulus* which was restored by Bernal-Casasola Casasola to *garum scombri flos* (Bernal-Casasola *et al.* 2016:738). It is noteworthy that when Ehmig cites the *tituli* with no alpha identifier (4), she lists them under *garum* as a matter of course. However, these four *tituli* are in Pelichet 46 (Beltan 2A) *amphorae* with the characteristic segmented shape which are associated far more with *liquamen* discussed below.[13]

Fish sauce vessels evolved into more complex shapes during the later Republican period. The new forms have longer spikes, which are narrow with a corresponding relatively narrow neck in the Dressel 8 type, while in the Dressel 7 there is a wider neck and correspondingly wide spike. The belly of these amphorae is either ovoid or slightly elongated and becoming pear shaped. The hollow spike evolved slowly from its simple beginnings as a relatively narrow space in the Dressel 7 and 8 types through to the Beltran 2A, where the spike is a substantial container in its own right, (a Dressel 8 I have been able to examine had a spike capacity of just 250 ml).[14] The commodities that appear to have been carried in Dressel 7 and 8 amphorae over the entire period are not sufficiently distinctive to allow a separation of products according to vessel shape. All types of fish sauce were accommodated in Dressel 7: *gari flos* (CIL. XV.4699), *gari scombri* (CIL. XV, 4710), *liquamen* (CIL. XV.4713), *muriae flos* (CIL. XV.4722), *muria arguta excellens flos* (CIL. XV.4723), which all appear to have been uncomplicated liquids with minimal residues that might be accommodated in the hollow spike.[15] The Dressel 8 form is very similar to the Dressel 7, but with a more piriform belly larger at the bottom. It carries a much smaller variety of products according to the *tituli*, dominated by liquids as always but particularly *garum: g(ari) f(los)* (CIL. XV.4689, 4691, 4693, 4694, 4701, 4703, 4704, 4706, 4707, 4711, 4718); *g(ari) s(combri) f(los)* (CIL. XV. 4692); *g(ari) f(los) scomb(ri)* on a Dressel 8 found at Saint-Roman-en-Gal, *g(ari) scomb(ri) flos* on a Dressel 8 found at Mainz.[16]

Did the increase in the size of the spike of the fish sauce amphora correspond to more of the residue being shipped with the sauce? Experiments have demonstrated that when a fish sauce was transported with a bone residue this generally sat at or near the bottom, while the liquid sauce formed and gathered in the middle of the processing vessel in the belly and the fish paste, which would naturally have floated above the liquid sat at the top of the belly and in the neck. This separation of the three components of fish sauce depends on how much fluid is included in the vessel and the density of that fluid. A concentrated fish mash poured in but otherwise not diluted would simply fill the space, while if the fish mash filled half the vessel, it could be topped up with brine which would turn into *liquamen* while in transit. It seems more than coincidental that the three compartments of the amphorae: neck/mouth, body and spike accommodate the three components of the sauce: liquid, fish paste and bones in such a functional way. An increase in the transportation of the fish mash would in theory correspond to an increase in the size of the spike. This possibility, which in the Dressel 7 and 8 types reflect small amounts of residue is brought to its full and logical conclusion with the design features of the Beltran 2A which has the wide and capacious spike. The Beltran 2A is also known by the designations Augst 27, Augst 29, Callender 6, Camulodunum 186C, Dressel 38, Peacock & Williams 18 and Pélichet 46. Most importantly morphologically the form has strong links with the Dressel 14, due to the wide neck, tube body and large hollow spike. The Dressel 14 form, predominantly made in the Sardo region of Portugal, corresponds to the amphora most often associated with *liquamen tituli*.[17] A number of large-size Beltran 2A vessels that I was able to examine at the Museum of London store had substantial spikes with a diameter of c. 37 to 45 cm at the join with the body and with a length of spike up to 30 cm and an estimated spike capacity of c. 2 to 3 liters. We can estimate that the capacity of the spike would in some circumstances represent c. 10% of the total volume. A number of these substantial vessels measuring upwards of 110 cm also had solid spikes. From close observation it appeared that the filling in the spike may have been added later as a narrow space had formed between the clay of the fill and the wall of the spike. Sadly, there was no opportunity to photograph it. The purpose of this spike is obscure, especially as the added weight made the vessels utterly unwieldy and dysfunctional. One may conjecture that a vessel containing a *flos* sauce had no bone residue and therefore the hollow spike was superfluous and this was the logic of filling it in. Alternatively, it was a means of selling customers short as the weight of the vessel when full corresponded to volume.

We can speculate that as it became clear to the manufacturers that the sauces could be fermented in transit, then it is likely that new designs of amphora would be the result and this appears to be the development of the increasingly segmented structure to fish amphora. This would mean that the small, low capacity spikes accommodated a small amount of bone residue, while as the spike grew in size it represented a change in how the sauces were traded. The evolution of the spike and its increase in size would seem to be the crucial factor in determining when fish sauce became fully industrialized. The manufacturers may have realized that the sauce could be removed from the tank after the fish have broken up into a pourable state and after a relatively brief period of initial fermentation. The mash could then be removed to amphorae, allowing the *cetariae* to be refilled, thus allowing multiple batches of fish sauce to be made where before one batch was made at a time. Add to this the vast increase in facilities and the increase in the number of individual tanks in each facility, and we are compelled to view this as an industrial revolution in terms of scale of the production of fish sauce. The Beltran 2A appears to be a "fully evolved" fish sauce vessel. It has three sections where the neck and spike are equal in width and length, while the belly due to its pear shape is capacious (small form 18 to 20 liters, large form 31 to 33 liters.[18] This tripartite structure allowed the Beltran 2A to be a multi-functional vessel because it had a wide enough, straight neck for easy access, a wide capacious spike for residues, and a capacious belly for a bulk liquid, and it could accommodate all types of product: sloppy fish mash, solid fish pieces of moderate size such as *cordyla*, and when necessary simple liquids such as *garum*, *liquamen* and *muria*. This is what we find to a large extent: the Beltran 2A held the most diverse products, dominated by *cordyla* according to the *tituli* (Ehmig 2003:63;65 note 9 above). Whether it was designed to house a *liquamen* because of the dominant hollow spike and only subsequently became multifunctional is not clear. A fish mash derived from mackerel of large size would necessarily require just such an amphora to accommodate it and we may conjecture that the abundance of this form correspond to the transportation of the quality *liquamen* fish sauce derived principally from mackerel, which from ship wrecks evidence seems to have been traded widely in the early years of the 1st century AD. The Beltran 2A has a relatively short span of production and distribution, though it represents the most prolific amphora type in the north (Ehmig 2003; Martin-Kilcher 2003). Its earliest dates are c. AD 10 to 15 and the latest are mid-2nd century AD. The demise of this form and in fact the reduction in the size of fish amphora generally at this time may also correspond to a reduction in the mackerel stocks and a return to using the smaller fish species such as sardine and anchovy to make most of the traded *liquamen*.

The other form commonly associated with fish products at this time is the Beltran 2B, which has the distinct narrow hourglass and elongated neck merging, with and without a shoulder, with the body. It has a pronounced pear-shaped belly, wider at the bottom and a modest hollow and often solid spike. The date range is similar to the 2A, though emerging a little later and seemingly in smaller numbers in the north. This form emerges when the original uncomplicated liquid vessels, Dressel 9 and 10, have become less visible and may have been made to replace them. The *tituli* are dominated exclusively by uncomplicated liquid sauces with no ambiguity *muriae flos arguta*, on Beltrán 2B found in the vicinity of Fuentes de Andalucía, Sevilla.; *muriae flos*, at Écija, Sevilla; *muriae flos excellens* on one found at the Saint Gervais shipwreck; *liquamen argutum summaur excellens* found at Fréjus. This vessel has such a narrow neck that it could not be suitable for any type of *salsamenta* of whatever size which is strongly indicative of some kind of logical system of use.[19] The neck and structure of this vessel corresponds logically to the accommodation of liquid sauces and it would seem reasonable to assume that these liquids were dominated by *liquamen flos:* that is the first *liquamen* sauce to be extracted from the *cetariae,* even though *tituli* would suggest otherwise. The numbers of labels are simply too small to make any other conclusions.

The final fish vessel of note emerges in the mid-3rd century after the "crisis," in Roman deposits at Ostia and it represents quite a change in amphora design. The Almagro 50/51 A-C amphorae are either Spanish or Portuguese, but regardless they are clearly designed for a fish product and almost certainly a sauce of the *liquamen* type. They are characterized as moderate sized vessels (70–77 cm) with very short hourglass neck, which is relatively narrow in comparison to the body which is a straight wide tube and snub spike. The capacity of these vessels in comparison to earlier forms is low as they can carry as little as 10 liters up to 25 to 30 liters. There are no *tituli,* but from wrecks such as Cala Reale at Asinara, we know that these brought a fish sauces from Portugal over the next century (Delussi and Wilkins 2000:53). The Randello wreck off the south coast of Sicily, derives from Portugal in the early 4th century AD and carried 120 to 200 Almagro 50, they were full of sardine of 10 to 17 cm with the occasional mackerel and represents a fish sauce mash being traded to be reprocessed/refined when it reached a Roman port.

Reuse evidence from tituli picti

There are a limited number of occasions where we can securely identify a reused event either from a change of *titulus* or an obvious accumulation of nonstandard amphorae being utilized in particular ways. The classic examples are the Grado wreck, which utilizes African oil and Greek wine vessels for fish products, and the Garum shop at Pompeii, which gathers various nonlocal wine and fish vessels for reuse (page 195; Peña 2007:72; Auriemma 2000:45). A Dressel 8 amphora with a double label is particularly interesting as it demonstrates the act of relabeling in the fish sauce trade: G F (signifying *garum*) was the first primary content, and its second fill was *liquamen* in another hand and which was probably written in Italy (*CIL*. XV.4718). Curtis notes the incidence of refilling and relabeling of *urcei*: in *CIL* IV.2574, a g(*arum*) f(*los*) sc(*ombri*) from Scaurus' workshop is followed by another similar fill from the same workshop, *g(arum f(los) sc(ombri)*; in *CIL* IV.5705 the first fill is *liq/uamen* but the second fill is ambiguous: *⅃ V.C*; the *urceus* IV.5711 sited by Curtis appears not to be relabelled from my reading in *CIL*; the refilling of the *urceus* IV. 10280 mirrors our first one in that *g(arum) f(os)* is replaced with *liquamen*. The *urceus* with a *garum* label and a *liquamen* filling, I was able to examine (discussed in Chapter 11, page 223) also demonstrates that *garum* vessels were refilled with *liquamen* without relabeling. Curtis has asked a very important question about the differential access to fish sauce in Pompeii. He asks why *urcei* with *garum flos* labels can be found in relatively modest dwellings and bars, while at the same time this product can be extravagantly expensive (Curtis 1991:174; Peña 2007:99). We have mentioned the small house in Chapter 10 (*Regio* IX.9.6–7) with seven *urcei* vessels with *tituli picti* for *garum* as well as all the other sauces. It does not follow that all these sauces were consumed in the house, rather it may in fact be a space where empty vessels were collected so they could be reused. Curtis was thinking of *garum* as the primary product of course nevertheless the question is an important one. Much of the *garum* in Pompeii was probably imported from Spain, though it may have been locally made as well. There is a *titulus pictus* on a Dressel 12 for *G F Poteolani* (*CIL* XV.4688) which we might assume was a fine *garum* that was made in Pozzuoli and Pliny clearly thinks that *garum* was locally made. Labels for a *garum* with the designation *sociorum* are very rare: Curtis cites three, of which only one is a clear citation (*CIL IV 5659*). Their rarity is difficult to equate with the vast quantities of *garum flos* that appears to have been circulating, with or without the additional *scombri*. It may be that *sociorum* in these circumstances still refers to the original idea of a mullet blood sauce, which we must assume was always scarce or that the label is an old one repeatedly reused. It is certainly likely that the numerous vessels found in the small house may have been refilled numerous times before ending up there and it therefore does not necessarily follow that a high quality *garum* was consumed very widely in the greater community of Pompeii.

A proposal for a model for fish sauce amphora use and reuse

I propose the following model for the process of manufacturing fish sauce, exporting the products and utilizing the empty amphora again for more products centered on the mid-1st century AD. The Dressel 8, discussed above (*CIL. XV*.4718) carried a *garum flos*, it was probably one of many with a similar fill and it probably came from Spain and had arrived in a bulk lot at a port facility such as Portus. The final resting place of this Dressel 8 was Rome. The same ship will probably had other vessels containing a *liquamen flos* and also others with *allec* the fish sauce mash. Some of the *garum* amphora will have travelled into the city and been sold to elite households, some would have had their prime use content, the *garum*, decanted at the port and sold on into smaller vessels, and others were taken into Rome and similarly sold on in smaller amounts. It seems plausible given out knowledge of *garum* now that a true blood viscera *garum* would always have been sold in relatively small amounts unless purchased by the very wealthy. Some of the empty Dressel 8s could have been left in a facility in the city to be reused later, and some gathered and sent back to Portus, while others which were emptied at the port remained there and became available to take other sauces arriving in a different form. Our vessel became available to be refilled when the next shipment to arrive turned out to be a fish sauce mash (what is normally called *allec*), essentially a concentrated form of *liquamen,* which probably arrived in numerous Beltran 2B's. The potential yield is hard to calculate. We can estimate that a vessel packed full of fish mash could be sufficiently brined to loosen it to allow it to be poured out and further processed in another vessel while one half full of mash (as was the case with the Grado vessels) that had been brined before shipment, would have initially generate a 2nd *liquamen flos* after the sea journey (the first was taken in Spain from the *cetariae*). This would mean that for each vessel that arrived with a content of *allec*, there would need to be another empty amphora ready to receive that sauce. This *liquamen flos* would ideally have been put into a vessel regularly used to carry an uncomplicated liquid such as a Dressel 9 or 10 if they were available and especially if these vessels were going to be sold into elite households where it may have been necessary to guarantee the association between shape and content, bearing in mind too that this second sauce is not necessarily 2nd quality. Other amounts of *liquamen* may have been put into less ideal vessels gathered at the port of various shapes one of which was likely to have been the Dressel 8. This was taken into Rome and the content sold on. It may have even been reused many times more to transport *liquamen* or *garum* without the need to relabel before reaching its final resting place. To return to the initial delivery and the remaining *allec* inside the Beltran 2B's, this may have been a mixture of sardine and mackerel of small and medium size. It would have been possible to rebrine this mash inside the vessel and leave it to form in a warehouse for about one month, before the final sauce would have been poured off and sold as a second quality *liquamen*. It certainly would have been perceived as second quality, though it need not necessarily have been as this depends on the size of the fish utilized. These Beltran 2Bs may have also been sent into Rome or into other towns and cities to generate sauces for local markets. This final rebrining would have resulted in a weaker sauce with minimal *flos* but would also have resulted in exhausted fish bones inside the vessel which could have been conveniently discarded, as described in Chapter 11. It is interesting to note that if vessels that had held oil were not considered suitable for reuse than one full of fish bones after the sauce had been taken may sometimes also have been viewed as expendable. In some circumstances the disposal of the rank but exhausted bones sometimes became more important than the potential value of the vessel for reuse. It would depend a great deal on the organizational system of collecting vessels for reuse and how efficient it was. All or most of this processing could have happened at the port or at some large retail outlet with in Rome. It is of note that while fish amphorae are often found in large elite rural villas, the majority of finds of these vessels, and the residues of fish sauce, are found in sub-elite urban and semi-urban and reflect the fact that the processing and harvesting took place where the sauces were retailed, not where they were consumed (Clavel and Lepetz 2014; Locker 2007). It might be useful to consider to what extent the segmented vessels (Dressel 7 and 8 and Beltran 2B) which I conjecture held a fish mash were urban vessels, while the ovoid forms and Dressel 12 that should represent the sale of *flos* sauces and therefore more expensive sauces would be more likely to be found where the sauce were consumed in large amounts, that is, elite rural and urban villas.

Using this model, it is logical to suppose that there would have been a great need for empty fish sauce vessels to accommodate the sauces as they became available, and at numerous points along the trade route. This scenario is valid for amphorae from wrecks from the 1st century AD with large mackerel and sardine such as Grado and Sud Perduto II and Port Vendres (table 4). As we have noted, the vessel that was chosen to transport the fish mash need not have been significant in terms of attributing form to content, as the fish mash needed further processing and the resulting sauce from that mash was only put into a vessel for sale once the product had reached its destination. In these circumstances many amphorae must have been labelled not where the sauces were made in Spain and Africa, but where the sauce were further processed. It is very significant that the only fish sauce vessels to be manufactured away from the sauce production sites, in Gaul, were the Dressel 9 and 10 *similis* type from Lyon. It has always been understood that these vessels were made in order to accommodate sauces that had been transported into Gaul some other way but it is now possible to see these ideal locally made fish sauce vessels were potentially manufactured to accommodate *liquamen flos* sauces transported as a fish mash. The *tituli* associated with Dressel 9 and 10 *similis* in Mainz are not dominated by *liquamen* (Ehmig 2001:69): *muria* dominates, followed by *garum* (including Spanish *garum*), and there are just a couple of *liquamen*, which might cause our model to fail. Such thinking presupposes that the statistical data derived from the *tituli* are significant but they clearly only represent a fraction of the products being traded across the Mediterranean. There is simply not enough of them and the presence of abundant fish amphora in a site must be understood to represent the transportation of the primary product, that is, *liquamen*. It may even be possible to go so far as to suggest that some of those amphorae with *garum* labels found in the north could have been refilled with *liquamen* before they left the Mediterranean without having that label changed. High quality *garum sociorum* was undoubtedly traded into the northern empire for military elites but I doubt it was widely consumed. Broekaert (2016:76) has posited the idea that the Lyonese vessel may have been underrepresented in the archaeology and that they probably accommodate fish sauces from Plomarc'h in Britany as well as sauces from Spain. It is however noteworthy that the Dressel 9 and 10 *similis* and any form that might be suitable for *flos* sauces, cease to be found long before the Britany processing sites cease to function, which leaves the barrel as an option for sauces made in Britany or the continuing repeated reuse of fish sauce vessels over many years, back and forward between the processing sites and the consumption sites (Broekaert 2016:84).

Discussion

This model presupposes that fish sauces were continually processed and reprocessed away from the original production site and a prerequisite of the model is that empty reused amphorae were systematically collected so that they are available to take the liquid sauces as this becomes available along the trade route. Vessel may often have been suitable and designed for a given product and we may be able to make this assumption when products were placed in them at the production sites, but when products were placed in vessels at different stages along the trade route, the link between form and specific content would have been less secure as it would always largely depend on which amphora had been collected at a given locus. This view also suggests an alternative to the traditional classical archaeological assumptions about consumption. This largely relies on the view that vessels were made, traded with their products, consumed, as in emptied, and after potentially being reused for some other product or used as a storage vessel at the same place where the content was consumed, it was either discarded close to where the content was consumed or recycled in some other way entirely unconnected to its prime use, that is, as building material but also in a *locus* relatively close to the consumption of its content. The chain of events is uninterrupted and is understood as a linear progression from one use to the next. Peña had placed doubt on this linear progression back in 2007 when he highlighted how significant the reclamation and removal of amphorae from establishments for the wholesale/storage/retail, of goods including fish sauce, for reuse elsewhere as building material, was in interfering with the consumption data (Peña 2007:344). If the idea of multiple reuse of amphora for prime use content in a continuous repeating cycle reflects even a small percentage of the potential trade of fish products within the Mediterranean then a strong and consistent link between specific fish products and specific fish sauce amphorae seems to be beyond our grasp. Amphora consumption data as Peña notes has always been "predicated on the assumption that amphorae recovered at consumption sites represent the distribution to and the consumption at that site of the substance understood to be the principle content of each of the classes represented (and) originating in that region where that class was manufactured" but now such data cannot be relied on to necessarily indicate any of those things (Peña 2007:345).

The potential journey of a batch of fish sauce from a processing facility in Spain to a market in Rome is illustrated as a flow chart in figure 12.4. I have posited a scenario where there were potentially multiple episodes of reuse over a substantial period of time which greatly extends the use-life cycle imagined for many amphorae forms. I have employed and adapted from Theodore Peña's original methodology. He acknowledges that the possibility of determining the actual lifecycle, in terms of years, of any amphora is limited. He uses modelling utilizing the only forms used for fish products with reliable data of reuse, namely the Tunisian and Tripolitanian forms, used for the fish sauce in the Grado wreck. He posits a prim-use use-life-span of a just one year after they were filled with oil or possibly fish products in Africa. As the chances of these vessels being returned to the source of their original content is virtually impossible it is generally assumed that any reuse would have largely been something other than the prime use (Peña 2007:335). This is also clearly a standard view in archaeology about fish sauce vessels in the prime-use stage of their use-life cycle too, as the accumulation of empty vessels might well have occurred where they had been emptied a long way from any source of further fish sauce. In the scenario we have posited the means of generating fish sauce is also transported so that sauces are generated where they are required not at their source.

The case of Grado implies a certain systematic accumulation of vessels, some of which may have brought a fish sauce mash (*allec*) from Africa already, while other were clearly used primarily for wine or oil. Some appear more suitable for a fish mash than others. The entirely unsuited vessels at Grado: Forlimpopoli, held 21 l. and would appear to have been a last-minute acquisition to accommodate left over *allec* that could not be accommodated in the larger vessels, which were chosen simply as they could temporarily accommodate a fish mash. The continued reuse of *salsamenta* vessels in the same way is also possible. With the Pompeii *Garum* shop, the Dressel 21/22 *salsamenta* vessels were likely emptied of their prime use salted fish from either Sicily or the *Latium* coast at an undetermined location but possibly Rome or any of the towns south of Rome before being gathered to be reused for more or different kinds of salted fish. This product would have arrived in Pompeii and been sold resulting in their collection at the *Garum* Shop, where they were probably sent empty to the coast to be used to carry a semi worked small fish sauce mash back into Pompeii. Rita Auriemma has suggested that the Grado wreck and the Pompeii *Garum* Shop were typical of the system of stockpiling vessels for reuse and redistribution and my research has strengthened this conclusion (Auriemma 2000:46). I have posited a situation where the flow of vessels and the flow of products is constantly moving and looping back upon each other and there is only a limited chance that we will ever be able to see the patterns in any detail but

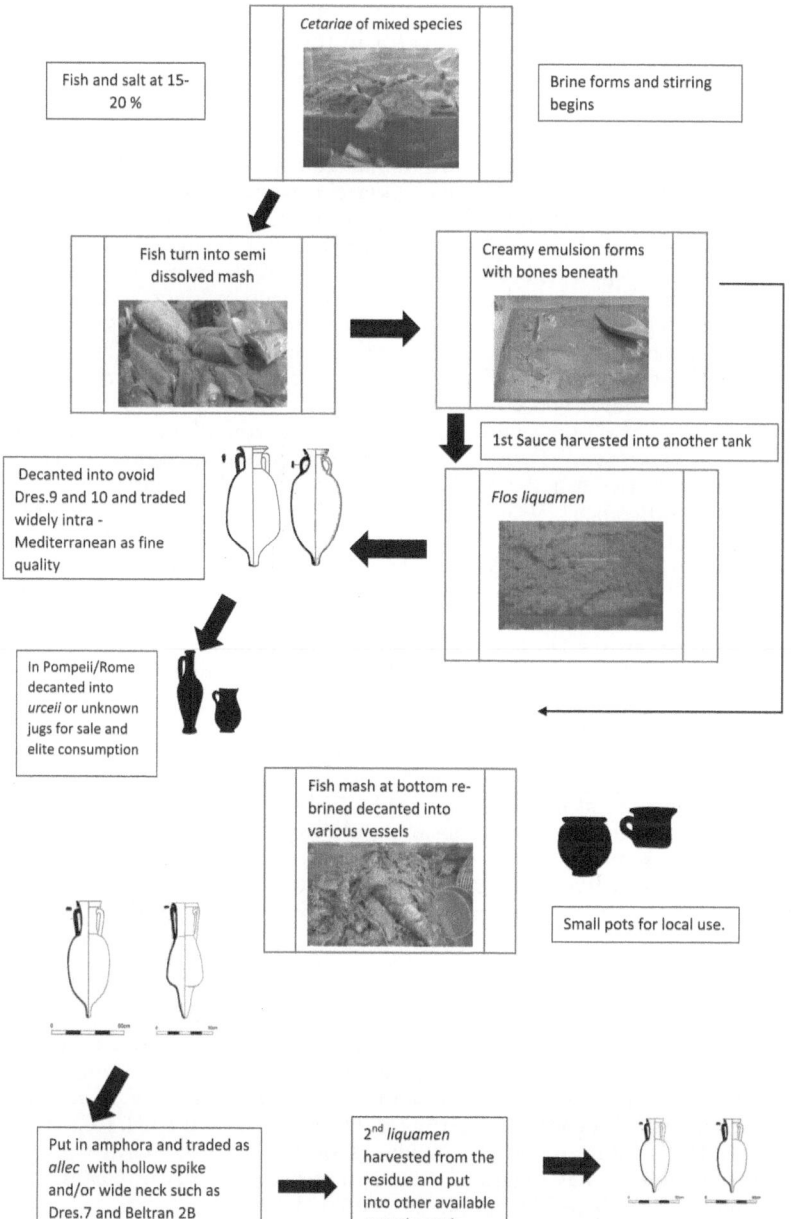

Figure 12.4 Flow chart of amphora function and reuse

the very idea of this random circular motion is itself a pattern that could be identified in current data and looked for in future research.

Notes

1 The University of Southampton's *Roman Amphorae: a Digital Resource*, at https://archaeologydataservice.ac.uk/archives/view/amphora_ahrb_2005/ and *Amphorae ex Hispania*, University of Cadiz, at http://amphorae.icac.cat/.

2 There is a possibility that this is a *liquamen* actually made with something sharp like vinegar, replacing the brine or wine that serves to dilute the fish mash at the beginning of the process. (Athenaeus ix.366c). Alternatively, a bottle of Thai *nuc nam* and even Red Boat, when left for years at the back of the cupboard, becomes rather oxidized and is definitely more pungent. Cf. *CIL* xv.4741 *G V(etus) penuar*, implying an aged *garum* that is to be stored or has already been stored.

3 That *lacerta* was salted whole and traded is clear from Ulpian' statement "*lacerta cum muria sua*" Digest 35.9.3,3 "Fishermen have no difficulty in distinguishing the two fish, (Sp mackerel and Atlantic mackerel) but ordinary men see little difference" (Wentworth Thomson 1932:247) Bonito are referred to as *sarda* in *Apicius* 9.10.1 and mackerel is named only once as *lacertos* in *Apicius* 10.2.7. Today in Turkey the salted slices of sarda sarda are called *lacerda*. *Lacertus* in modern times is a name for Sp mackerel (Wentworth Thomas 1932:246).

4 Initially, following Dressel, the term *LAC* and its many forms were believed to refer to a fish sauce flavored with a herb or a similar flavoring in wine (Djaoui 2016:118). Until recently it was understood to be some sort of mixed fish sauce (Martin-Kilcher 2011:420).

5 Modern research gives brine pickled salted fish no more than 6 months of shelf-life when it is kept at ambient temperatures (Latifa *et al.* 2014). Yet all the evidence seems to point to an aged salted fish being desirable.

6 The issue of the nature and meaning of *lympha* is beyond this study, but clearly has something to do with the addition of water to a fish product.

7 However, tuna *muria* can appear to be quite pink in color, depending on the efficiency of the bleeding, so could not have been mistaken for refined *liquamen*

8 *gari flos, gari scombri, liquaminis flos, muriae flos, lymphatum vetus* (*CIL*. XV.4690, 4699, 4698, 4715, 4721, 4736).

9 Such as *liquamen flos* (*CIL* XV.4720), *muria* (*CIL* XV 4727) and possibly *ha(llec (s)com)* (*CIL* XV.4731), though this may be doubtful as the *titulus* is basically *HAS* and the extrapolation may be incorrect (University of Southampton (2014) *Roman Amphorae: a digital resource [data-set]*; York: Archaeology Data Service [distributor] https://doi.org/10.5284/1028192.

10 On a Dressel 9 from Port-la-Nautique (Liou 1993) It is not clear where *Port* was. Berdowski has suggested one of the principal ports of Gades, Lixus, Tinitana or perhaps Malaga (Berdowski 2003:40).

11 Dressel 12 with *tituli*: Rome, the Praetorian camp, *garum scombri,CIL* XV.4687, 4705, 4709; *garum* 4708; *Garum scombr flos*, 4687. At Pompeii, *liq flos excel scom*, *CIL* IV.5176=2588; Augsburg, *liq scombr excel.*, *CIL* III.12010,48 (Cesteros 2012:21)

12 I am only concerned with the *alpha* and *beta* lines of a *titulus* as they are the only ones relevant to this study. The manufacturer, owner and trader and to a lesser extent weight and /or perceived volume will not be dealt with here (Martin-Kilcher 2003).

13 (*garum*) *scombri excellens* Blánquez u.a. (1998) 290 Nr. 52484, (*garum*) *scombri excellens* Blánquez u.a. (1998) 293 Nr. 80620, (*garum*) scombri *excellens* Blánquez u.a. (1998) 293 Nr. 52483 (*garum*) *scombri excellens* Blánquez u.a. (1998) 295 Nr. 52491.

14 I was able to examine in detail an intact Dressel 8 held at Harlow Museum. The spike was 24 cm and it was equal to the neck length. In total it stood at 91 cm. The spike had a capacity of just 250 ml, while the entire vessel held 15.75 l. to the bottom of the neck and therefore represent a tiny fraction of the total capacity (with thanks to Chris Lydamore for allowing me to put water in the vessel).

15 Enrique García Vargas, Dario Bernal Casasola, «Dressel 7 (Baetica coast)», *Amphorae ex Hispania. Landscapes of production and consumption* (http://amphorae.icac.cat/amphora/dressel-7-baetica-coast), 04 July, 2016.

16 Enrique García Vargas, Daniel Martín-Arroyo, Lázaro Gabriel Lagóstena Barrios, «Dressel 8 (Baetica coast)», *Amphorae ex Hispania. Landscapes of production and consumption* (http://amphorae.icac.cat/amphora/dressel-8-baetica-coast), 04 July, 2016.

17 Ehmig notes that *liquamen* is most often found on Dressel 14, though we note that it is not always easy to tell whether the vessel was of Spanish or Portuguese origin (Ehmig 2003:71).

18 Enrique García Vargas, Daniel Martín-Arroyo, Lázaro Gabriel Lagóstena Barrios, «Beltrán IIA (Baetica coast)», *Amphorae ex Hispania. Landscapes of production and consumption.* (http://amphorae.icac.cat/amphora/beltran-iia-baetica-coast), 08 July, 2016.

19 Wine is also attested at the Saint Gervais shipwreck and at Arles. https://archaeologydataservice.ac.uk/archives/view/amphora_ahrb_2005/details.cfm?id=44

Afterword

This has been a strange and complex journey of exploration in all aspect of an obscure and little understood commodity of the Roman world. Fish sauce was an essential element of the Roman culinary package that all cultures who encountered Rome appear to have embraced enthusiastically. Romanization is a dirty word but the acceptance of fish sauce in all areas of the empire was very much evidence of that process in action. There must have been something exceptionally good about this commodity, despite its reputation, to cause it to be spread so widely and for so long. The image of fish sauce as something rotten and putrid has I hope been shattered. Ancient fish sauces existed in many forms. We have a simple fish brine (*muria*), an equally simple dissolved fish sauce for mass consumption (*garos/liquamen*), a more exclusive species-specific dissolved fish sauce (*liquamen*), and the strange and bizarre *garum sociorum/haimation* itself made with fish rubbish but nevertheless valued above all the others. Despite the huge confusion over the names applied to these sauces, we have achieved some clarity. A detailed and intricate unpicking of the literature has revealed a possible chronology which is complex but very plausible as an explanation for the nomenclature issue.

Initially in Greece we had a simple *garos,* and it seems to have spread slowly around the Mediterranean and may not have become a commercial product widely traded until the Hellenistic period. Regardless as to how controversial this might be in archaeological circles, it is by no means clear to me that *garos* was commercially traded from region to region within the Mediterranean during the Punic period. Our first indication of its manufacture archaeologically would seem to be residues from Olbia in Sardinia where a late4th/early 3rd century BC local amphora contained what may have been a small-fish sauce residue. Residues from Spain of a *garos* are much later in the mid-2nd century BC. *Garos* appears to be a local product probably made in all regions of the Mediterranean. When it was embraced into the Roman culinary culture of the late Republic, quite naturally *garos* and *garum* were equivalent as there was only one type of sauce. At some point a new type of *garum* was invented which was made with just fish blood and viscera and the equivalence of *garos* and *garum* could not continue and the term *liquamen* was coined to signify the original sauce. In the later period when this new *garum* had ceased to be so widely marketed, some ancient commentators began to revert to the earlier usage as they consider that *garum* was always simply a transliteration of the Greek and it had effectively returned to its original meaning. It was once again used to signify the essential substance, hence our and their great confusion.

Recent discoveries suggest that in the late empire *muria* began to be used more widely and named a *liquamen* as the distinction between these two sauces becomes blurred Nevertheless, the blood/viscera *garum* appears to dominate both in our minds today and in the ancient literature and epigraphy, but this dominance is as we have seen somewhat distorted by how much significance we place on the various forms of evidence. There are three areas where *garum* appears to be in sharp focus: in Roman satire: where it is largely held up as evidence of Roman perversion and dissipation; in the numerous labels attached to the various containers it was retailed in and it is particularly visible from the commercialization of *garum* in Pompeii. We see these aspects in sharp focus in ancient archaeological and historical studies because they are writ large in our minds. On the other hand, the popularity and vast commercial trade in the essential substance is largely out of focus and only visible through the fish sauce residues reflecting what is perceived to be a low value bony fish paste as poverty food. This disconnect, between evidences has resulted in a situation where all the various disciplines have been "methodologically constrained" from an understanding of the nature of fish sauce in the Ancient world. The small whole fish sauce has for too long been subsumed under the reputation of that other sauce. It might appear from the incidence of *garum tituli*, that vast quantities of it was made, traded and consumed but this may be a somewhat distorted picture of the popularity of *garum* as we cannot determine how much sauce the labels actually represent. We have seen that there may have been a constant and repeated reuse of bottles and amphora, that were initially labelled *garum,* but were subsequently used for *liquamen,* and in fact this may have been so common that we cannot effectively estimate the quantities of *garum* produced and marketed. Our review of the body of evidence for the black and bloody *garum* has revealed that, though very visible in certain forms of evidence, it was scarcely to be found in sources concerning its practical utility. Radical though it may be, it is just possible that, rather than the labels for *garum* representing a fraction of a much wider commercial trade, as would be expected, they may actually represent the majority of the blood viscera *garum* that was commercially traded.

This has been a truly multidisciplinary approach to the study of fish sauce in the ancient world and it has allowed these products to become a palpable reality. Ancient fish sauce was a complex magical ingredient, that should no longer be a subject of jokes and inuendo but recognized for what it was; fundamental to an ancient Mediterranean cuisine, and used by virtually everyone in their daily lives.

Sally Grainger
June 2020
Grayshott

Bibliography

Adams, J.N., 1995. *Pelagonius and Latin Veterinary Terminology in the Roman Empire.* Brill, Leiden.

Adams, J.N., 2003. *Bilingualism and the Latin Language.* Cambridge University Press, Cambridge.

Aguilar Corrales, P., Prieto, J.M.C., Aguilar, M.C. and Padilla, J.S., 2011. "Salsamenta malacitano": avances de un proyecto de investigación. *Itálica: revista de arqueología clásica de Andalucía,* (1), pp. 29–50.

Alcock, J.P., 2001. *Food in Roman Britain.* Tempus.

Anagnostakis, I., 2013. 'Byzantine diet and cuisine in between ancient and modern gastronomy,' in Anagnostakis A., 2013 (ed.), *Flavours and Delights. Tastes and Pleasures of Ancient and Byzantine Cuisine,* pp. 43–65. Armos Publications, Athens.

Anagnostakis, I., 2013. 'Byzantine delicacies,' in Anagnostakis A., 2013 (ed.), *Flavours and Delights. Tastes and Pleasures of Ancient and Byzantine Cuisine,* pp. 81–105. Armos Publications, Athens.

Ang, C. Y.W., Keshun L, Huang Y.W., 1999 *Asian Foods: Science and Technology.* CRC Press, Abingdon.

Anonymous, 1982. *Small-Scale Processing of Fish* (ILO - WEP, Fish Handling, Preservation and Processing in the Tropics: Part 2 (NRI). Tropical Development and Research Institute. New Zeeland digital library, University of Waikato http://www.nzdl.org/gsdlmod.

Aldrin, J.F., Briand, Y., Verger, B., 1969. 'Études sur les nuoc-mam de poissons de mer en côte-d'Ivoire', *Revue d'élevage et de médicine vétérinaire des pays tropicaux (Journal of International Livestock Science)* 22, 249–70. CIRAD, Paris.

Amores, F., Vargas, E.G., Gonzalez, D. and Lozano, M.C., 2007. Una factoría altoimperial de salazones en Hispalis (Sevilla, España). In *Actas del Congreso Internacional CETARIAE. Salsas y salazones de pescado en Occidente durante la Antigüedad, Universidad de Cádiz, Noviembre de 2005,* p. 33. BAR International Series 1686, Oxford.

André, J., 1986. *Isidore of Seville, Etymologies (Livre XII, Des animaux).* Auteurs du Moyen Age. Éditions Belles-Lettres, Paris.

Arévalo, A., Bernal, D., 2007. 'Historia de las investigaciones de la industria pesquero-conservera en Baelo Claudia.' In *las Cetariae de Baelo Claudia: Avance de las investigaciones arqueológicas en el barrio meridional (2000-2004),* Cadiz.

Asakura, H., 2003. *World History of the Customs and Tariffs.* World Customs Organization, Brussels.

Auriemma, R. 2000. 'Le anfore del relitto di Grado e il loro contento,' *Mélanges de l'École Française de Rome - Antiquité* 112. 27–51. Rome.

Bagnall, R.S. and Cribiore, R., 2006. *Women's letters from ancient Egypt, 300 BC-AD 800*. University of Michigan Press.

Bagnall, R.S., 2011. *Everyday Writing in the Graeco-Roman East* (Vol. 69). Univ of California Press.

Banducci, L.M., 2017. 'Tastes of Roman Italy: Early Roman expansion and taste articulation'. In *Taste and the Ancient Senses* (pp. 134–151). Routledge. London

Barkai. O., Lernau, O., Ahanovy, K., 2013. 'Analysis of Fish Bones from the Tantura F Shipwreck, Israel,' *Archaeofauna* 22. 189–199. Madrid.

Barney, S.A., Lewis, W.J., Beach, J.A. and Berghof, O., 2006. *The Etymologies of Isidore of Seville*. Cambridge University Press, Cambridge.

Barnes, D.J., Munks, B., Kaucher, D., 1944. 'The effect of vitamin D from cod-liver oil and a tuna-liver oil upon serum phosphatase concentrations in rachitic infants,' *The Journal of Pediatrics* vol.24, issue 2, 159–166. Elsevier, Amsterdam.

Bateman, N., and Locker, A. 1982. 'The sauce of the Thames,' *The London Archaeologist* 4 (8), 204–7. London.

Beddows, C.G., Ardeshir, A.G. and Daud, W.J.B., 1979. 'Biochemical changes occurring during the manufacture of Budu,' *Journal of the Science of Food and Agriculture*30 (11), 1097–1103. Wiley, Hoboken, NJ.

Beddows, C.G., 1998. 'Fermented fish and fish products' in Wood, J.B.B. (ed.), *Microbiology of fermented foods*, vol., pp. 416–434. Blackie, London.

Benoît, F. 1958 Nouvelles èpaves de Provence, *Gallia* 16, pp. 5–39.

Benoit, F. 1962, 'Nouvelles épaves de Provence (III)', *Gallia* 20 (1), 147–176. CNRS, Paris.

Bernal-Casasola, D., Arévalo, A., Lorenzo, L. and Aguilera, L., 2003. 'Imitations of italic amphorae for fish sauce in Baetica. New evidence from the salt-fish factory of Baelo Claudia (Hispania)' *Rei Cretariae Romanae Fautorum Acta* 38, 305–313. Dr Rudolf Habelt GmbH · Verlag, Bonn.

Bernal-Casasola, D., and Romero, A.M.S., 2006. 'Infundibula gaditana" acerca de los vasos troncocónicos perforados para filtrar" garum" y otros usos industriales en la bahía de Cádiz, *Romula* 5, 167–218. Universidad Pablo de Olavide de Sevilla. Seville.

Bernal-Casasola, D., Arévalo, A., Morales, A. y Roselló, E., 2007. 'Un ejemplo de conservas de pescado Baelonenses en el siglo ii a.c.,' in *LAS CETARIAE DE BAELO CLAUDIA: Avance de las investigaciones arqueológicas en el barrio industrial meridional* (2000-2004), 355–374. *Monografías de Arqueología, Junta de Andalucía*, Seville.

Bernal-Casasola D. and Sáez Romero, A., 2008. 'Fish-Salting Plants and Amphorae Production in the Bay of Cadiz (Baetica, Hispania). Patterns of Settlement from the Punic Era to Late Antiquity', in Vanhaverbeke, H., Poblome, J., Waelkens, M., Vermeulen, F., Brulet, R. (eds.), *Thinking about Space. The potential of surface survey and contextual archaeology in the definition of space in Roman times. Studies in Eastern Mediterranean Archaeology* 8. 45–113. Brepols, Turnhout.

Bernal Casasola, D., and Gonzáles, A., 2008a. 'Baelo Claudia y sus industrias haliéuticas. Síntesis de las últimas actuaciones arqueológicas (2000–2004)' in J. Napoli (ed.), *Ressources et activités maritimes des peuples de l'Antiquité*. Les Cahiers du Littoral, 2, pp.9–30. Boulogne-sur-Mer.

Bernal-Casasola, D., and Domingues-Bella, S., 2012. 'Colorantes y pigmentos en las pesquerías hispanorromanas,' in *Cuadernos de Prehistoria y Arqueología de la Universidad Autónoma de Madrid*, 37–38 (2011-2012), 671–685. Madrid.

Bernal-Casasola, D., García Vargas, E. and Sáez Romero, A., 2013. 'Ánforas itálicas en la Hispania meridional,' in Olcese, G. (ed.) *Immensa Aequora. Ricerche archeologiche, archeometriche e informatiche per la ricostruzione dell'economia e dei commerci nel bacino occidentale del Mediterraneo* (metà IV sec. aC-I sec. aC): atti del convegno: Roma 24-26 gennaio 2011, 351–372. Quasar, Rome.

Bernal-Casasola, D., Cottica, D., García-Vargas, E., Toniolo, L., Rodríguez-Santana, C.G., Acqua, C., Marlasca, R., Sáez, A.M., Vargas, J.M., Scremin, F. and Landi, S., 2014. 'Un Contexto Excepcional en Pompeya: la pila de ánforas de la bottega del garum (i, 12, 8).'. *Rei Cretariae Romanae Fautorum, acta* 43, 219–232. Dr Rudolf Habelt GmbH · Verlag, Bonn.

Bernal-Casasola, D., 2015. 'What contents do we characterise in Roman amphorae? methodological and archaeological thoughts on a 'trending topic,' in Oliveira, C., Morais, R. and Morillo, A. (eds.), *Archaeo-Analytics -Chromatography and DNA analysis in archaeology*. University of Oporto, pp. 61–83.

Bernal Casasola, D., Marlasca, R., Rodrigues Santana, C.G., Ruiz Zapata, B., Gil Garcia, M.J. and Alba, M., 2016. 'Garum de sardinas en Augusta Emerita. Caracterización arqueológica, epigráfica, ictiológica y palinológica del contenido de un ánfora Beltrán IIB.' *Rei Cretariae Romanae Fautorum Acta* 44. Xanten pp. 737–49. Dr Rudolf Habelt GmbH · Verlag, Bonn.

Bernal-Casasola, D., 2016. 'Lusitanian amphorae in the straits of Gibraltar,' in Pinto, I.V., de Almeida, R. and Martin, A. (eds.), *Lusitanian Amphorae: Production and Distribution*, pp 299–309. Archaeopress, Oxford.

Bernal-Casasola, D., Arévalo, A., Díaz, J.J., Expósito, J.A., 2016a. 'Baelo Claudia y sus actividades haliéuticas. Una nueva cetaria y una posible domus en el barrio meridional (2005–2009),' *IIas Jornadas Internacionales Baelo Claudia* (Cádiz y Baelo Claudia, 2010), 147–176. Junta de Andalucía, Seville.

Bernal-Casasola, D., Cottica, D., 2017. 'Produzione e vendita di pesce sotto sale e suoi derivati a Pompei nel 79 d.C.: le evidenze dalla cosidetta "Bottega del garum" (I, 12, 8),' in González-Villaescusa, R., Schörle, K., Gayet, F., Rechin, F. (eds.). *L'exploitation des ressources maritimes de l'Antiquité. Activités productives et organisation des territoires. Actes des XXXVIe rencontres internationales d'archéologie et d'histoire ibes & Xlle colloque de l'association AGER*, pp. 235–252. Éditions APDCA, Antibes.

Bernal-Casasola, D., Díaz, J.J., Expósito, J. A., Palacios, V., Vargas, J.M., Lara, M., Pascual, M.A., Retamosa, J.A., Eid, A., Blanco, E., Portillo, J. L., 2018. 'Atunes y Garum en Baelo Claudia: nuevas investigaciones 2017,' *AL QANTIR* 21, 73–86. Cadiz.

Bernal-Casasola, D., 2018a. 'The Baelo Claudia Paradigm: The Exploitation of Marine Resources in Roman Cetariae,' *Journal of Maritime Archaeology* 13, 329–351. Springer, Berlin/Heidelberg.

Bernardes, J. P., 2015. "Consumo e transformação de peixe entre o mundo romano e o mundo islâmico", in Gómez Martinez, Susana (coord.) *Memória dos Sabores do Mediterrâneo*, (CAM/CEAACP) Campo Arqueológco de Mértola/Centro de Estudos em Arqueologia, Artes e Ciências do Patrimóni. pp. 55–68

Berdowski, P., 2003. 'Tituli picti und die antike Werbesprache für Fischprodukte,' *Münstersche Beiträge z. antiken Handelsgeschichte*, Bd. 22, H. 2, 18–54. University of Marburg; Marburg.

Berdowski, P., 2006. 'Garum of Herod the great (a Latin-Greek inscription on the amphora from Masada,' *Analecta Archaeologica Ressoviensia* 1. 239–257. Rzeszow.

Berdowski, P., 2008. Roman Businesswomen: the case of the producers and distributors of garum in Pompeii. *Analecta Archaeologica Ressoviensia* 3. 251–272. Rzeszow

Berdowski, P., 2013. In search of the lexical meaning of the Latin terms *cetarius* and *cetaria*. *Glotta*, 89(1-4), pp.47–61.

Biddulph, E., 2008. 'Form and function: the experimental use of Roman Samian ware cups,' *Oxford Journal of Archaeology*, 27 (1), 91–100.

Boethius, A., 2016. 'Something rotten in Scandinavia: the world's earliest evidence of fermentation,' *Journal of Archaeological Science* 66. 169–180. Elsevier, Amsterdam.

Bonifay, M., 2004. *Etudes sur la céramique romaine tardive d'Afrique*. BAR-IS 1301; Archaeopress, Oxford.

Botte, E., 2009. *Salaisons et sauces de poissons en Italie du Sud et en Sicile durant l'Antiquité* Coll. Centre Jean Bérard 31, Naples.

Botte, E., 2017. 'L'exploitation de la mer en Italie centrale tyrrhénienne (Étrurie et Latium): production et commerce durant l'Antiquité,' *Melanges de I'Ecole Francaise de Rome - Antiquité* 129.2.

Bowman, A.K., 1998. *Life and letters on the Roman frontier: Vindolanda and its people*. Psychology Press.

Bowman, A. and Wilson, A. (eds.), 2009. *Quantifying the Roman economy: methods and problems* (Vol. 1). Oxford University Press, Oxford.

Broekaert, W. (2012). Vertical integration in the Roman economy: a response to Morris Silver. *Ancient Society*, 42, 109–125. Retrieved April 21, 2020, from www.jstor.org/stable/44079962

Broekaert, W., 2016. 'The soldiers' kitchen' in Broekaert, R. Nadeau, J. Wilkins (edd.) *Food, Identity and Cross-Cultural Exchange in the Ancient World. Collection Latomus* 354. Brussels. pp. 64–87.

Bruschi T. and Wilkens B., 'Conserves de poisson à partir de quatre amphores romaines', *Archaeofauna: International Journal of Archaeozoology* 5, 166.

Buonopane, A., 2009. 'La produzione olearia e la lavorazione del pesce lungo il medio e l'alto Adriatico: le fonti letterarie,' in: Pesavento Mattioli, S. and Carre, M.B. (eds.), *Olio e pesce in epoca romana. Produzione e commercio nelle regioni dell'alto adriatico. Atti del Convegno, Padova, 16 febbraio 2007*, pp.25–36. Editrice Diana, Cassino.

Butler C., 1948. *The fish liver oil industry. Fishery leaflet 233, Fish and wildlife service*, United States Department of the Interior. Washington DC.

Callender, M.H., 1965. *Roman amphorae: with index of stamps*. Oxford, UK: Oxford University Press.

Cappelletto, E., Bernal-Casasola, D., Cottica, D., Álvarez M. B., Medina, M. L., and Sáez Romero, A. M., 2013. 'Urcei per salse di pesce da Pompei-Ercolano: una prima analisi,' in *Hornos, talleres y focos de producción alfarera en Hispania: I Congreso Internacional de SECAH, Ex Officina Hispana, Cádiz, 3-4 de marzo de 2011*, pp. 271–280. University of Cádiz.

Carannante, A., Chilardi, S., Della Vecchia Resti, M., 2006. 'Archeozoologici dalla casa pompeiana di Marco Fabio Rufo: risultati preliminari,' *Atti del V Convegno Nazionale di Archeozoologia, Rovereto 10-12 dicembre 2006*, pp. 139–142. Edizioni Osiride, Rovereto.

Carannante, A., 2008. 'L'ultimo garum di Pompei. Analisi archeozoologiche sui resti di pesce dalla cosiddetta officina del garum,' *Automata*, 3(3), pp.43–53. L'Erma di Bretschneider, Rome.

Carannante, A., Giardino, C., Y Saverese, U., 2011. 'In Search of Garum. The "Colatura d'alici" from the Amalfitan Coast (Campania, Italy): an Heir of the Ancient Mediterranean Fermented Fish Sauces,' in Lugli, F., Stoppiello, A. A., Biagetti, S. (eds.), *Atti del 4.° Convegno Nazionale di Etnoarcheologia* (Roma, 17-19 Maggio 2006), British Archaeological Report S2235, pp. 69–79. Archaeopress, Oxford.

Carannante, A., 2019. 'The last Garum of Pompeii: archaeozoological analyses on fish remains from the "Garum Shop" and related ecological inferences,' *International Journal of Osteoarchaeology*, 29(3), 377–386.

Cesteros, H.G., de Almeida, R.R., Costello, J.C., 2016. 'Special Fish Products for the Jewish Community? A Painted Inscription on a Beltrán 72 Amphora from Augusta Emerita (Mérida, Spain),' *HEROM*, 5(2), pp.197–237. *Institute* for Archaeological and Monumental Heritage of the National Research Council *(IBAM-CNR), Catania.*

Cesteros, H.G., 2012. 'Scomber gaditano en Efeso. Una Dressel 12 con titulus encontrada en la "Casa Aterrazada 2" de Efeso,' *Dialogues d'histoire ancienne*, 38/1, 111–124. Presses Universitaires de Franche-Comté, Besançon.

Clavel, B., Lepetz, S., 2014 'La consommation des poissons en France du nord à la période romaine'. Consommer dans les campagnes de la Gaule Romaine, *Actes du Xe congrès de l'Association AGER, Revue du Nord*, Hors série. Collection Art et Archéologie N° 21. p 93–108

Comis L. RE. C., 2009. 'The archaeology of taste: Gargilius Martialis's Garum,' *Journal of Reconstruction and Experiment in Archaeology*, 33–38. EXARC, Leiden.

Corcoran, T.H., 1962. 'Roman fish sauces,' *Classical Journal* 58, 204–210.

Cotton, H., Lernau, O., Goren., Y., 1996. 'Fish sauce from Herodian Masada,' *Journal of Roman Archaeology* 9, 223–238.

Crisan, E.V. and Sands, A., 1975. 'Microflora of four fermented fish sauces,' *Applied Microbiology*, 29(1), 106.

Cribiore, R., 2019, 'The Dissemination of Texts in the High Empire,' *American Journal of Philology*, 140(2), 255–290.

Curtis R., 1979 The Garum Shop of Pompeii (I.12.8), *Cronache Pompeiane 5*: 5–23.

Curtis, R., 1984. 'Product identification and advertising on Roman commercial amphorae,' *Ancient Society*, 15/17, 209–228. Retrieved from http://www.jstor. org/stable/44080242

Curtis, R., 1984. "Negotiatores Allecarii" and the Herring. *Phoenix, 38*(2), 147–158. doi:10.2307/1088898

Curtis, R.I., 1988. 'A. Umbricius Scaurus of Pompeii,' in Curtis, R.I., (ed.), *Studia pompeiana et classica in honor of Wilhelmina F. Jashemski, 1*, pp.19–50. A.D. Caratzas, New Rochelle, N.Y.

Curtis, R.I., 1991. *Garum and salsamenta: production and commerce in materia medica*. Brill, Leiden.

Curtis R., 2009. 'Umami and the foods of classical antiquity,' *American Journal of Clinical Nutrition* 90 (suppl.), 712–718. American Society for Nutrition, Rockville, Maryland.

Curtis R., 2016 Ancient processed fish products in Bekker-Nielsen, T., Gertwagen, R. (eds.), *The Inland Seas: Towards an Ecohistory of the Mediterranean*, pp. 23–56. Franz Steiner Publishers, Stuttgart.

Dalby, A., 1996. *Siren feasts: a history of food and gastronomy in Greece*. Routledge, London.

Dalby, A. and Dalby, A.K., 2000. *Empire of pleasures: luxury and indulgence in the Roman world*. Psychology Press, London.

Dalby, A., 2003. *Flavours of Byzantium*. Prospect Books, Totnes.

Dalby A., 2011 *Geoponika: Farm Work. A Modern Translation of the Roman and Byzantine Farming Handbook*. Prospect Books, Totnes.

Dalby, A., 2013. *Food in the Ancient World from A to Z*. Routledge, London.

Darmon, J. P., 1964. 'Notes sur le tarif de Zaraï,' *Les cahiers de Tunisie*, XII (1964), 6–23.

Davidson, J.N., 1998. *Courtesans & fishcakes: the consuming passions of classical Athens*. Macmillan, London.

Davidson, A., 1981 reprint. *Mediterranean Seafood*. Penguin Books, Harmondsworth.

Davies, R.W., 1971. 'The Roman military diet,' *Britannia* 2, 122–142.

Daveau, I., Delaval, E., Pellegrino, E., Sternberg, M., Sabatier, P. and Poignant, E., 2007. *Garum et pissalat: de la pêche à la table, mémoires d'une tradition. Catalogue d'exposition*. Musée d'histoire et d'archéologie d'Antibes/Snoeck, Louvain.

De Lima, R.A., 2018. As Cetariae na Província da Hispânia Bética: O Garum enquanto Continuação de uma Prática Econômica e Cultural Fenício-Púnico em Gades e sua Ressignificação em Contexto Romano. *Mare Nostrum (São Paulo)*, 9(1), pp.115–134.

De Melo, W., 2011. Plautus. Amphitryon. The Comedy of Asses. The Pot of Gold. The Two Bacchises. The Captives. Loeb Classical Library *Cambridge, MA*.

Delaval, E., Poignant, E., 2007. 'Une tranche de vie originale d'Antibes à Menton: la pêche à la poutine,' in Daveau, I., Delaval, E., Pellegrino, E., Sternberg, M., Sabatier, P. and Poignant, E., 2007 (eds.), *Garum et pissalat: de la pêche à la table.*

Delussi F., and Wilkens B., 2000. 'La conserve di pesce. Alcuni dati da contesti Italiani,' *Mélanges de l'École Française de Rome Antiquité* 112. 53–65.

Demir, M. (2007). 'The trade in salt-pickled hamsi and other fish from the Black Sea to Athens during the Archaic and Classical periods,' in Erkut. G. and Mitchell, S. (Eds.), *The Black Sea: Past, Present and Future*, pp. 57–64. London: British Institute at Ankara. Retrieved from http://www.jstor.org/stable/10.18866/j.ctt1n7qjwz.9

Desse-Berset N., and Desse J., 2000. 'Salsamenta, garum, et autres preparations de poison. Ce qu'en dissent les os,' *Mélanges de l'École Française de Rome Antiquité* 112. 73–97.

Desse-Berset, N. 1993. 'Contenus d'amphores et surpêche: l'exemple de San Perduto,' in J. Des and F. Audoin-Rouzeau (eds.), *Exploitation des animaux sauvages à travers le temps*. Colloque int. de l'homme et l'animal (Juan-les-Pins), pp. 341–46.

Dickey, E. (ed.), 2012. *The Colloquia of the Hermeneumata Pseudodositheana (Cambridge Classical Texts and Commentaries)*. Cambridge University Press, Cambridge.

Djaoui, D., Piquès, G., Botte. E., 2014. 'Nouvelles données sur les pots dits à garum" du Latium, d'après les découvertes subaquatiques du Rhône (Arles),' in Botte, E., Leitch, V. (eds.), *Fish & Ships. Production et commerce des salsamenta durant l'Antiquité. Actes de l'atelier doctoral, Rome 18-22 juin*, pp.175–197, Bibliothèque d'Archéologie Méditerranéenne et Africaine 17. Errance, Paris.

Djaoui, D., 2016. The Myth of '*Laccatum*': a study starting from a new *titulus* on a Lusitanian Dressel 14. In *Lusitanian Amphorae: Production and Distribution* (p. 117–127) Archaeopress, Oxford.

Dobney, K.M., Jaques, S.D. and Irving, B.G., 1996. Of Butchers and Breeds: Report on Vertebrate Remains from Various Sites in the City of Lincoln (Lincoln Archaeological Studies 5). *Lincoln: Lincoln Archaeology*.

Dorrego, F., Carrera, F. and Luxán, M.P., 2004. 'Investigations on Roman amphorae sealing systems,' *Materials and Structures* 37.5, 369–374. https://doi.org/10.1007/BF02481686.

Drexhage, H. J., 1993. 'Garum und Garumhandel im römischen und spätantiken Ägypten. *Münstersche Beiträge zur antieken Handelsgeschichte* 12, 27–55.

Dressel, H., 1879. 'Di un grande deposito di anphore rinvenuto nel nuovo quartiere del Castro Pretorio,' *BCAR* 7, pp 36–112.

Driard, C., 2012. 'Les sauces de poisson dans l'ouest de la province romaine de Lyonnaise: réflexions sur l'élaboration et la nature des produits,' in Botte, E., Leitch, V. (eds.), *Fish & Ships. Production et commerce des salsamenta durant l'Antiquité. Actes de l'atelier doctoral, Rome 18-22 juin*, pp.47–59. Bibliothèque d'Archéologie Méditerranéenne et Africaine 17. Errance, Paris.

Dueck, D., 2000. *Strabo of Amasia: A Greek Man of Letters in Augustan Rome*. Routledge, London.

Dunbabin, K.M., 2003. *The Roman Banquet: Images of Conviviality*. Cambridge University Press, Cambridge.

Dütting M., Laarman F., Wouters W. & Van Neer W., 2020. Spanish mackerels and other faunal remains from two Augustan latrines at the Kops Plateau (Nijmegen, the Netherlands). *Nederlandse Archeologische Rapporten* 70: 73–86. Amersfoort, Rijksdienst voor het Cultureel Erfgoed.

Ehmig, U., 1996. 'Garum för den Statthalter: Eine saucenamphore mit Besitzeraufschrift aus Mainz,' *Mainzer Archäologische Zeitschrift* 3. 25–56. Philipp von Zabern, Mainz.

Ehmig, U., 2001. 'Hispanische Fischsaucen in Amphoren aus dem mittleren Rhônetal,' *Münstersche Beiträge zur Antiken Handelsgeschichte* 20.2, 62–71.

Ehmig, U., 2003. *Die römischen Amphoren aus Mainz*. Bibliopolis.

Ehmig, U. 2007: Die römische Amphoren im Umland von Mainz, Wiesbaden

Ehmig, U., 2009. 'Tituli picti auf Amphoren in Köln 11,' *Kölner Jahrbuch* 42. 393–445.

Ellis, S.J. and Devore, G., 2010. 'The Fifth season of Excavations at VIII. 7.1-15 and the Porta Stabia at Pompeii: Preliminary Report,' *FastiOnLine documents & research*, 202.1–21. Estreicher, S. K., 2006. *Wine from Neolithic Times to the 21st Century*. Algora Publishing, New York.

Essuman, K. M., 1992. *Fermented fish in Africa: a study on processing, marketing and consumption*, Fish Utilization and Marketing Service Fishery Industries Division, FAO fisheries technical paper 329, pp. 80. Food and Agriculture Organization of the United Nations, Rome, (http://www.fao.org/DOCREP/T0685E/T0685E00.HTM12/08/2010 12:48)

Ettienne, R., 1970. 'À propos du *garum sociorum*,' *Latomus* 29. 297–313

Fairclough, H. R., 1929. Horace, Satires. Epistles. The Art of Poetry, Loeb Classical Library.

Feldman, C., 2005. Roman Taste, *Food, Culture & Society*, 8:1, 7–30. Taylor & Francis, London.

Feldman, C., 2014 Roman taste in *Food History Critical and Primary Sources* Ed Jeffrey M. Pilcher, Bloomsbury, London 404–417.

Flohr, M. and Wilson A., 2017. *The Economy of Pompeii*. Oxford University Press.

Funari, P.P.A., 2001. 'Monte Testaccio and the Roman economy' in Blàzquez Martínez, J-M. and Remesal Rodríguez, J. (eds.), *Estudios sobre el Monte Testaccio (Roma)* 1. 558. University of Barcelona; *Journal of Roman Archaeology* 14, 585–588.

Gabriel, S., Fabiao, C. and Filipe, I., 2009. Fish remains from the Casa do Governador–A Roman fish processing factory in Lusitania. In *Fishes, culture, environment: Through archaeoichthyology, ethnography & history. The 15th meeting of the ICAZ Fish Remains Working Group (FRWG), 3–9 September 2009, Poznań & Toruń, Poland* (pp. 117–119).

Gabriel, S. and Tavares da Silva, C., 2016. 'Fish Bones and Amphorae: New Evidence for the Production and Trade of Fish Products in Setúbal (Portugal),' *Lusitanian Amphorae: Production and Distribution* 10. 111–16.

Garcia Vargas, E., 2006. 'Pesca y Salazones en la Bética Altoimperial,' in *Historia de la pesca en el ámbito del estrecho. I Conferencia internacional*, (Puerto de Santa María Cádiz, Junho de 2004), pp. 533–576. Instituto de Investigación y Formación Agraria y Pesquera, Córdoba.

Garcia Vargas, E. and Bernal Casasola, D., 2009. 'Roma y la producción de garum y salsamentaen la costa meridional de Hispania. Estado actual de la investigación,' in Bernal Casasola, D. (ed.), *Arqueología de la pesca en el Estrecho de Gibraltar. De la Prehistoria al fin del Mundo Antiguo,* pp. 133–81. Monografías del Proyecto Sagea 1, Cádiz.

García Vargas, E.A., Bernal Casasola, D., Palacios Macías, V., Roldán Gómez, A.M., Rodríguez Alcántara, Á. and Sánchez García, J., 2014. 'Confectio Gari Pompeiani. Procedimiento experimental para la elaboración de salsas de pescado romanas.' *SPAL*, 23. 65–82.

Garnier, N., Bernal-Casasola, D., Driard, C., and Pinto, I. V., 2018. 'Looking for Ancient Fish Products Through Invisible Biomolecular Residues in the Roman Production Vats from the Atlantic Coast,' *Journal of Maritime Archaeology*, 13(3), 285–328.

Goold, G.P., 1977. (ed. & trans.) *Manilius: Astronómica*. Loeb Classical Library, Harvard, Heinemann.

Gowers, E. 1993. *The Loaded Table: Representations of Food in Roman Literature*. Clarendon Press, Oxford.

Grainger S., 2010 *Roman fish sauce: an experiment in Archaeology,* unpublished MA dissertation, Archaeology department, Reading University.

Grainger, S., 2012. 'Roman fish sauce: an experiment in Archaeology', *EuroREA. Journal for (Re)construction and Experiment in Archaeology,* http://journal.exarc. net/issue-2012-1.

Grainger, S., 2013. 'Roman Fish Sauce: Fish Bone Residues and the Practicalities of Supply,' in *Fish and Fishing: Archaeological, Anthropological, Taphonomical and Ecological perspectives. Proceedings from the I.C.A.Z. Fish Remains Working Group, Jerusalem, October 22nd- 30th, 2011,* pp. 13–28. *Archaeofauna: International Journal of Zooarchaeology* 22, Madrid.

Grainger, S., 2018. 'Garum and Liquamen, What's in a Name?' *Journal of Maritime Archaeology*, 13(3), 247–261.

Grainger, S., Biddulph, E., In press Ancient fish sauce at the Roman table: Identifying archaeological traces of ancient recipes via experimental research into the use and deposition of cooking wares in *Understanding Roman Archaeology by Experiment: A Handbook*, (Ed.) Lee Graña, Dr. Tatiana Ivleva, Bill Griffiths, Bloomsbury London.

Grant, M, 1996. *Anthimus: On the Observance of Foods*. Prospect Books, Totnes.

Grocock, C.W. and Grainger, S. eds., 2020. *Apicius*. Prospect Books.

Grimaldi, M., et al. 2011. 'La casa di Marco Fabio Rufo. Lo scavo del diardino e i materiali.' *The Journal of Fasti on line* (http://www.fastionline.org/docs/FOLDER-it-2011-217.pdf) accessed 08.11.2019.

Hall, G.M., 2012. *Fish processing technology*. Springer Science & Business Media, Berlin.

Hamilton Dyer. S., 2008. Fish bones from selected contexts, Supplementary material to Trevarthen M. (ed.), *Suburban life in Roman Durnovaria*. http://www.wessexarch.co.uk/projects/dorset/dorchester_hospital/specialist_reports/index.html accessed 08.11.2019.

Harcum, C.G., 1914. *Roman Cooks*. J.H. Furst Company, Baltimore.

Höster Plogmann, H., 2006. Der Mensch lebt nicht von Brot allein. Gesellschaftliche Normen und Fischkonsum, in Höster Plogmann, H. (ed.), *Fisch und Fischer aus zwei Jahrtausenden: Eine fischereiwirtschaftliche Zeitreise durch die Nordwestschweiz*, 187–99. Römerstadt Augusta Raurica, Augst.

Irby, G. L., 2016. *A Companion to Science, Technology, and Medicine in Ancient Greece and Rome*, John Wiley & Sons, Hoboken, N.J.

Jones, A.K., 1986. 'Fish bone survival in the digestive systems of the pig, dog and man: some experiments,' in *Fish and archaeology: Studies in osteometry, taphonomy, seasonality and fishing methods* (Vol. 294, pp. 53–61). British Archaeological Reports, Oxford.

Jones, R. and Robinson, D., 2007. Intensification, heterogeneity and power in the development of insula VI. 1. In ed. Foss. P., Dobbin J.J., *The World of Pompeii*, London.

Jusup, M., Klanjscek, T., Matsuda, H., Kooijman Salm, 2011. A Full Lifecycle Bioenergetic Model for Bluefin Tuna. *PLoS ONE* 6(7), ed. Robert Planque, Vrije Universiteit, Netherlands. https://doi.org/10.1371/journal.pone.0021903, accessed 08.11.2019.

Klomklao, S., Benjakul, S., Visessanguan, W., Kishimura, H. and Simpson, B.K., 2006. Effects of the addition of spleen of skipjack tuna (Katsuwonus pelamis) on the liquefaction and characteristics of fish sauce made from sardine (Sardinella gibbosa), *Food Chemistry*, 98(3), 440–452.

Klomklao, S., Benjakul, S., Visessanguan, W., Kishimura, H. and Simpson, B.K., 2006. Proteolytic degradation of sardine (Sardinella gibbosa) proteins by trypsin from skipjack tuna (Katsuwonus pelamis) spleen, *Food Chemistry* 98(1), 14–22.

Koder J., 2013. Liutprand of Cremona. A critical guest at the Byzantine emperor's table, in Anagnostakis A., 2013 (ed.), *Flavours and Delights. Tastes and Pleasures of Ancient and Byzantine Cuisine*, pp. 105–109. Armos Publications, Athens.

Lagóstena, L., Bernal Casasola, D., Arévalo, A. (eds.), 2005. *Cetariae: Salsas y salazones de pescado en Occidente durante la Antigüedad, Actas del Congreso Internacional* (Cádiz, 7-9 de noviembre de 2005). BAR International Series 1686, Oxford.

Lagóstena B., Lázaro, G., 2007. 'Sobre la elaboración del garum y otros productos piscícolas en las costas béticas,'*Mainake* XXIX, pp. 273–289 http://hdl.handle.net/10498/14385, accessed 08.11.2019.

Lagóstena, L., Bernal, D. and Arévalo, A., 2007. *Cetariae 2005. Salsas y salazones de pescado en Occidente durante la Antigüedad. BAR International Series, 1686.*

Langkavel B. (ed.), 1868. *Simeon Seth: On the properties of foods.* Teubner, Leipzig.

Latifa, G.A., Chakraborty, S.C., Begüm, M., Nahid, M.N. and Farid, F.B., 2014. Nutritional quality analysis of Bangladeshi fish species, M. Tengra (Hamilton-Buchanan, 1822) preserved with different salt curing methods in laboratory condition, *American Journal of Food and Nurition*, 2(6), 100–107. http://pubs.sciepub.com/ajfn/2/6/2 30/09/2019.

Lauffer, S., 1971. *Diokletians Prisedikt.* De Gruyter, Berlin.

Leigh, M., 2015. *Food in Latin Literature. A Companion to Food in the Ancient World*, John Wiley & Sons, Hoboken, N.J.

Lauwerier, R.C.G.M., 1993. Twenty-eight bird briskets in a pot; Roman preserved food from Nijmegen, *Archaeofauna* 2, 15–19

Lentacker, A., A. Ervynck and W. Van Neer 2004. The symbolic meaning of the cock. The animal remains from the Mithraeum at Tienen (Belgium), in M. Martens and G. De Boe (edd.), *Roman mithraism: the evidence of the small finds* (Zellik-Tienen) 57–80.

Lee, Soon-Chun, Woo, Kang-Lyung, 1992. A Study on Development of Effective Utilization Method of Skipjack Tuna Viscera, *Korean Journal of Food Science and Technology* 24.1, 86–91. Korean Society of Food Science and Technology.

Leon, J.M., 2001. A propósito de la marca Soc y en torno al Garum Sociorum, *HABIS* 32, 171–184.

Lepsiksaar, J., 1986, 'Tierreste in einer Römischen amphora aus Salstburg (Mozartsplatz 4),' *Bayerische Vorgeschichtsblatter 51.* 163-85.

Liou, B. and E. Rodríguez-Almeida 2000. "Les inscriptions peintes des amphores du Pecio Gandolfo (Almería)," *MEFRA* 112, 7–25.

Locker, A., 2007. *In piscibus diversis*; the bone evidence for fish consumption in Roman Britain. *Britannia, 38*, pp.141–180.

Lopetcharat, K., Choi, Y.J., Park, J.W. and Daeschel, M.A., 2001. 'Fish sauce products and manufacturing: a review,' *Food Reviews International* 17(1), 65–88.

Low, B., 2016. 'The consumption of salted fish in the Roman empire,'in de Souza, P., Arnaud, P., Buche, C. (eds.) *The Sea in History - The Ancient World*, Boydell and Brewer, Martlesham.

Lowe B. J.,2017 The consumption of salted fish in the Roman World. In: de Souza P(ed) *The sea in history.* Boydell & Brewer, Martlesham, pp 307–318

Lyding Will, E., 1987 The Roman amphoras in McCann, A.M., 1987 (reissued 2017). *The Roman port and fishery of Cosa: a centre of ancient trade* (Vol. 4898). Princeton University Press.

Lytle, E., 2018. 'The Economics of Saltfish Production in the Aegean During the Classical and Hellenistic Periods,' *Journal of Maritime Archaeology*, 13(3), 407–418.

Maganto Martinez. J., 2005. 'Una inscripción inédita de pecio gandolfo. el complejo análisis de los tituli picti en ánforas salarias y el comercio de salsamenta,' in Logostena, L., Bernal, D., and Arevalo, A. (eds.), *Cetariae: Salsas y Salazones de pescado en Occidente durante la Antiguedad, Actas del congreso International (Cadiz 7-9 de Noviembre de 2005).* BAR International Series 1686, 2007.

Maire, B. (trs), 2001. *Gargilius Martialis. Medicinae ex holeribus et pomis*. Les Belles Lettres, Paris.

Malcovati H., 1955, *Oratorum Romanorum Fragmenta Liberae Rei Publicae iteratis curis recensuit collegit (Corpus Scriptorum Latinorum Paravianum). Aug. Taurinorum: Paravia.*

Malfitana, D., 2008. 'Roman Sicily project: Ceramic and Trade. A multidisciplinary approach to the study of material culture assemblages. First overview. The transport amphorae evidence,' *Facta: a journal of Roman material culture studies* 2. 127–192. Rome.

Maravela-Solbakk, A., 2009. 'Byzantine Inventory Lists of Food Provisions and Utensils on an Ashmolean Papyrus,' *Zeitschrift für Papyrologie und Epigraphik*, 170. 127–46.

Maritan, L., Iacumin, P., Zerboni, A., Venturelli, G., Dal Sasso, G., Linseele, V., Talamo, S., Salvatori, S. and Usai, D., 2018. 'Fish and salt: the successful recipe of White Nile Mesolithic hunter-gatherer-fishers,' *Journal of Archaeological Science*, 92. 48–62.

Martin-Kilcher, S., 1994. 'Die römischen Amphoren aus Augst und Kaiseraugst. Ein Beitrag zur römischen Handels- und Kulturgeschichte II: die Amphoren für Wein, Fischsauce, Südfrüchte (Gruppen 2-24) und Gesamtauswertung,' *Forschungen in Augst* 7.2, 313–612. Augst.

Martin-Kilcher, S., 2002. 'Lucius Urittius Verecundus, négociant à la fin du Ier siècle, et sa marchandise découverte à Mayence,' in: Rivet, L. and Sciallano, M. (eds.), *Vivre, produire et échanger: Reflets méditerranéens. Mélanges offerts à Bernard Liou*. Archéologie et Histoire Romaine 8, 343–53. Montagnac.

Martin-Kilcher, S., 2003. 'Fish sauce amphorae from the Iberian peninsula: the forms and observations on trade with the north-west provinces,' *Amphorae in Britain and the western Empire, Journal of Roman Pottery Studies* 10. 69–84.

Marzano, A., 2013. *Harvesting the sea: the exploitation of marine resources in the Roman Mediterranean*. Oxford University Press.

Marzano, A., 2018. 'Fish and seafood,' in Erdkamp, P. and Holleran, C. (eds.), *The Routledge Handbook of Diet and Nutrition in the Roman World*, pp. 163–175. Routledge, London and New York.

Mathieu, J. R., 2002. 'Introduction,' in J. R. Mathieu (ed.), *Experimental Archaeology: Replicating Past Objects, Behaviours and Processes. BAR International Series 1035*, pp. 1–4. Archaeopress, Oxford.

Mathews, J., 2006. *The Journey of Theophanes: Travel, Business, and Daily Life in the Roman East*. Yale University Press, New Haven. Ct. and London.

Mateo Corredor, D. and Molina Vidal, J., 2016. Archaeological quantification of pottery: the rims count adjusted using the modulus of rupture (MR). *Archaeometry*, 58(2), pp.333–346.

Mazzocchin, S. and Wilkens, B., 2013. 'Fish and Crustaceans from a Roman Amphora in Northern Italy,' *Archaeofauna* 22.1, 105-111.

McCann, A.M., 1987 (reissued 2017). *The Roman port and fishery of Cosa: a centre of ancient trade* (Vol. 4898). Princeton University Press.

McIver, R.C., Brooks, R.I. and Reineccius, G.A., 1982. Flavor of fermented fish sauce. *Journal of Agricultural and Food Chemistry*, 30(6), pp.1017–1020.

Miller, J.I., 1969. *The Spice Trade of the Roman Empire: 29 BC to AD 641*. Clarendon Press, Oxford.

Molina Vidal, J., and Mateo Corredor, D., 2018. 'The Roman Amphorae Average Capacity (AC),' *Oxford Journal of Archaeology*, 37: 299–311. https://doi.org/10.1111/ojoa.12143, accessed 9.11.2019.

Morais, R., Oliveira, C., Araujo, A., 2016. 'Lusitanian amphorae of the Augustan era and their content: Organic residue analysis,' in Pinto, I.V., de Almeida, R. and Martin, A. (eds.), *Lusitanian Amphorae: Production and Distribution*, pp. 105–7. Archaeopress, Oxford.

Morales Muñiz, A., 1993. 'Where are the tunas? Ancient Iberian fishing industries from an archaeozoological perspective,' in Clason, A., Payne, S., Uerpmann, H-P. (eds.), *Skeletons in her Cupboard. Festschrift to Juliet Clutton-Brock*, Oxbow Monograph 34, pp. 135–41.

Morales Muñiz, A and Roselló Izquierdo, E., 2008. 'Iruña-Veleia (Álava, Spain): an overview of the fish remains from the domus at Pompeia Valentina' in Béarez, P.; Grouard, S. y Clavel, B. (eds.), *Archéologie du poisson: 30 ans d'archéo-ichtyologie au CNRS. Hommage aux travaux de Jean Desse et Nathalie Desse-Berset, (actes des XXVIIIe Rencontres internationales d'Archéologie et d'Histoire d'Antibes et XIVe conf. ICAZ)*. Actes des Rencontres Internationales d'Archéologie et d'Histoire et d'Antibes.

Morales-Muñiz, A. and Roselló-Izquierdo, E., 2016. 'Fishing in Mediterranean prehistory: an archaeo-ichthyological overview,' in Bekker-Nielsen, T., Gertwagen, R. (eds.), *The Inland Seas: towards an ecohistory of the Mediterranean*, pp. 23–56. Franz Steiner Publishers, Stuttgart.

Morales-Muniz, A., Ephrem, B., de Agüero, E.G., Rodriguez, C.F., López-Arias, B., Rodríguez, L.L., Rey, F.S. and Izquierdo, E.R., 2017. Fishes as indicators of seasonality in Roman non-industrial fisheries: an overview from the southern NE Atlantic. In *L' EXPLOITATION DES RESSOURCES MARITIMES DE L'ANTIQUITÉ Activités productives et organisation des territoires*. Association pour la promotion et la diffusion des connaissances archéologiques, Antibes.

Morris, S., 2018. *The Early Eastern Orthodox Church: A History, AD 60-1453.* McFarland

Mylona, D. 2008. *Fish-eating in Greece from the Fifth Century B.C. to the Seventh Century A.D.* BAR International Series 1754. Archaeopress, Oxford.

Mylona, D., 2014. 'Aquatic animal resources in Prehistoric Aegean Greece,' *Journal of Biological Research-Thessaloniki* 21:2; https://jbiolres.biomedcentral.com/articles/10.1186/2241-5793-21-2, accessed 09.11.2019.

Mylona, D., 2016. 'Fish and seafood consumption in the Aegean: variations on a theme,' in Bekker-Nielsen, T., Gertwagen, R. (eds.), *The Inland Seas: towards an ecohistory of the Mediterranean*. Franz Steiner Publishers, Stuttgart; also at https://www.academia.edu/30436208/The_Inland_Seas_-_fishing_variations_on_a_theme, accessed 09.11.2019.

Mužinić, R., 1977. 'On the shoaling behaviour of sardines (Sardina pilchardus) in aquaria, *ICES Journal of Marine Science* 37.2, 147–55, https://doi.org/10.1093/icesjms/37.2.147, accessed 09.11.2019.

Nadeau, R., 2015. 'Cookery Books,' in Wilkins, J. and Nadeau, R. (eds.), *A Companion to Food in the Ancient World*, pp. 53–8. Wiley-Blackwell, Malden/Oxford.

Noall, C., 1972. *Cornish Seines and Seiners: a History of the Pilchard Fishing Industry*. Bradford Barton, Truro.

Nghia, N. D., Trung, T. S., Dat, P. V., 2017. '"Nuoc Mam" Fish Sauce in Vietnam: A Long History from Popular Seasoning to Health Benefit Bioactive Peptides,' *Ann. Food Process Preservation* 2(2). 1017.

Nicholson R., Robinson, J., Robinson, M., & Rowen E., 2018. 'From the Waters to the Plate to the Latrine: Fish and Seafood from the Cardo V Sewer, Herculaneum,' in *The Bountiful Sea: Fish Processing and Consumption in Mediterranean Antiquity, Journal of Maritime Archaeology (2018)* 13:263–284, https://doi.org/10.1007/s11457-018-9218-y, accessed 09.11.2019.

Nolla, J. M., and Nieto, X. 1980. 'La villa romana baix-imperial de la Ciutadella de Roses,' *Revista de Girona*, 93. 267–274.

Noret J., 1982. *Vitae duae antiquae sancti Athanasii Athonitae. Corpus Christianorum Graeca 9.* Brepols, Turnhout.

Olson, S.W., 2006. *Athenaeus, The Learned Banqueters. 8 vols. Loeb Classical Library*; Harvard University Press, Cambridge, MA & London.

Olson, S.D. and Sens, A. (ed. and trans.), 2000. *Archestratos of Gela: Greek Culture and Cuisine in the Fourth Century BCE.* Oxford University Press.

Osbaldeston, T.A. and Wood, R.P., 2000. *Dioscórides, De materia medica: Being an herbal with many other medicinal materials: written in Greek in the first century of the common era: a new indexed version in modern English.* Ibidis Press, Johannesburg.

Opait, A., 2007. 'A weighty matter: Pontic fish amphorae,' in Gabrielson, V. and Lund, J. (eds.), *The Black Sea in Antiquity: Regional and Interregional Economic Exchanges*, pp. 101–21. Aarhus University Press, Aarhus.

Ørsted, P., 1998. 'Salt, fish and the sea in the Roman Empire,' in Nielsen, I., Nielsen, H.S. (eds.), *Meals in a Social Context. Aspects of the Communal Meal in the Hellenistic and Roman World.* (Aarhus Studies in Mediterranean Antiquity; 2nd ed. revised), pp.13–35. Aarhus University Press, Aarhus.

Outram, A.K., 2008. 'Introduction to experimental archaeology,' *World Archaeology* Vol. 40(1): 1–6.

Owens, J.D., and Mendoza, L.S., 1985. 'Enzymically hydrolysed and bacterially fermented fishery products, *International Journal of Food Science & Technology*, 20(3), 273–293.

Palmer. R., 1969. *Hermeneutics: Interpretation Theory in Schleiermacher, Dilthey, Heidegger, and Gadamer.* Northwestern University Press, Evanston.

Palacios, V., Garcia, E., Bernal Casasola, D., Roldan, A., Rodrigues, Á. and Sanchez, J., 2016. 'Conservas antiguas y gastronomía contemporánea,' in Bernal Casasola, D., Espósitó Álvarez, J. Á., Medina Grande, L. and Vicente-Franqueira Garca, J.S. (eds.), *Un Estrecho de Conservas. Del Garum de Baelo Claudia a la melva de Tarifa*, pp. 89–105. University of Cádiz.

Paludan-Müller, C., 1998. *The Microbiology of low-salt fermented fish products, Fish Utilisation in Asia and the Pacific.* Proceedings of the APFIC Symposium Beijing, RAP publications, Bangkok.

Parker, A.J., 1992. *Ancient shipwrecks of the Mediterranean & the Roman provinces* (Vol. 580). *Tempus Reparatum*, Oxford.

Pearson, A.C., 1917. *The Fragments of Sophocles.* Cambridge University Press, Cambridge.

Peckel Möller F., 1895. *Cod-Liver Oil and Chemistry.* P. Möller, London and Christiana.

Pecci, A., Domínguez-Bella, S., Buonincontri, M.P., Miriello, D., De Luca, R., Di Pasquale, G., Cottica, D. and Bernal-Casasola, D., 2018. 'Combining residue analysis of floors and ceramics for the study of activity areas at the Garum Shop at Pompeii,' *Archaeological and Anthropological Sciences* 10(2), 485–502.

Peña, J.T., 2007. *Roman pottery in the archaeological record.* Cambridge University Press.

Peterson, T.S., 1994. *Acquired Taste: The French Origins of Modern Cooking.* Cornell University Press, Ithaca, NY.

Pinkster, H., 2005. 'The language of Pliny the Elder,' in Reinhardt, T., Adams, J.N., Lapidge, M., Adams, J.N. and Winterbottom, M. (eds.), *Aspects of the Language of Latin Prose.* Oxford University Press.

Pinto, I.V., Magalhães, A.P. and Brum, P., 2012. 'An overview of the fish-salting production centre at Tróia (Portugal),' in Botte, E., Leitch, V. (eds.), *Fish & Ships. Production et commerce des salsamenta durant l'Antiquité. Actes de l'atelier doctoral, Rome 18-22 juin,* pp. 18–22. Bibliothèque d'Archéologie Méditerranéenne et Africaine 17. Errance, Paris.

Piqués, G., 2005. 'Les déchets d'une fabrication de sauce de poisson dans le comblement d'un puit gallo-romain et la question du sel à Lattes,' *Lattara 18.* 293–306.

Piqués, G., Chibechinni, F., De Juan, C., Djaoui, D., Rovira, N., Sanchez, C., Tillier, M., 2015 New studies about sauces and salted fishes in the analysis of the palaeocontent of Roman jars and amphorae, *Poster presentation at Roman Amphora Contents International Interactive Conference (RACIIC) Reflecting on Maritime Trade in foodstuffs in Antiquity,* In tribute to Miguel Beltrán Lloris. 2015 Cadiz

Ponsich, M., & Tarradell, M., 1965. *Garum et industries antiques de salaison dans la Méditerranée occidentale.* Presses Universitaires de France, Paris.

Prete S. (ed.), 1978. *Ausonius. Opuscula.* Teubner, Leipzig.

Regia, E., 2018. 'The Circulation, Distribution and Consumption of Marine Products in Byzantium: Some Considerations,' *Journal of Maritime Archaeology,* 13.449–466.

Rice, C., 2016. Shipwreck cargoes in the western Mediterranean and the organization of Roman maritime trade. *Journal of Roman Archaeology,* 29, pp.165–192.

Richardson-Hay, C., 2009. 'Dinner at Seneca's Table: the philosophy of food,' *Greece & Rome* 56(1), 71–96.

Roberts, Rev. Can. (ed.), 1905. *Titus Livy The History of Rome.* Everyman's Library; J. M. Dent & Sons Ltd., London.

Rose, V. 1874. 'Aringus und der Hering,' *Hermes* 8.224–7.

Rodríguez-Alcántara, Á., Roldán-Gómez, A.M., Bernal-Casasola, D., García-Vargas, E. and Palacios-Macías, V.M., 2018. 'New technological contributions to Roman *garum* elaboration from chemical analysis of archaeological fish remains from the 'garum shop 'at Pompeii (I. 12.8),' *Zephyrus 82.* 149–63.

Romero, A. S. 2019. 'Indicios del aprovechamiento de restos de pescado como combustible de los hornos cerámicos de la Gadir púnica,' *Boletín De La SECAH,* 10.

Rowan, E., 2014. 'The fish remains from the Cardo V sewer: New insights into consumption and the fishing economy of Herculaneum,' in Botte, E., Leitch, V. (eds.), *Fish & Ships. Production et commerce des salsamenta durant l'Antiquité. Actes de l'atelier doctoral, Rome 18-22 juin,* pp. 61–73. Bibliothèque d'Archéologie Méditerranéenne et Africaine 17. Errance, Paris.

Sáez Romero A.M, Theodoropoulou T., Belizón Aragón R. 2020 'Atunes púnicos y vinos egeos en una taberna de la Grecia clásica. resultados iniciales del Corinth Punic Amphora Building Project in *A Journey between East and West in the Mediterranean Proceedings IX International Congress of Phoenician and Punic Studies MYTRA 5*, 2020: 817–836

Sanquer, R. and Galliou, P., 1972. 'Garum, sel et salaisons en Armorique gallo-romain,' *Gallia* 30(1), 199–223.

Scarborough, J., 1981. 'The Galenic question,' *Sudhoffs Archive* 65(1), 1–31.

Scheidel, W. ed., 2012. *The Cambridge companion to the Roman economy.* Cambridge University Press. Cambridge.

Sikorski, Z.E., 1990. *Seafood: resources, nutritional composition, and preservation.* CRC Press, Boca Raton, Florida.

Slim, L., Bonifay, M., Piton, J. and Sternberg, M., 2007. 'An example of fish salteries in Africa Proconsularis: the officinae of Neapolis (Nabeul, Tunisia),' *Actas del Congreso Internacional CETARIAE. Salsas y salazones de pescado en Occidente durante la Antigüedad, Universidad de Cádiz, Noviembre de 2005*, pp. 21–44. BAR International Series 1686, Oxford.

Smigra, M., Mizukoshi, T., Iwahata, D., Eto, S., Miyaho H., et al. 2010. 'Amino acids and minerals in ancient remnants of fish sauce (garum) sampled in the "Garum Shop" of Pompeii, Italy,' *Journal of Food Composition Analysis* 23:442–446.

Smith, A. F., 1996. *Pure Ketchup: A History of America's National Condiment, with Recipes.* University of South Carolina Press, Columbia, SC.

Soliman, W. S., Shaapan, R.M., Mohamed, L.A., Younes, A.M., Elgendy, M. and Salah El din, A., 2017. 'Laboratory Screening of Biogenic Amines Producing Bacteria Potentially Threatens Human Health in Some Egyptian Fish and Fish Products,' *Journal of Fisheries and Aquatic Science* 12: 134–140.

Speth, J.D., 2017. 'Putrid Meat and Fish in the Eurasian Middle and Upper Paleolithic: Are We Missing a Key Part of Neanderthal and Modern Human Diet?' *PaleoAnthropology* 44–72, and https://www.greenpasture.org/pub/media/imported/500.pdf, accessed 09.11.2019.

Spieler, M., 2018. *A Taste of Naples: Neapolitan Culture, Cuisine, and Cooking.* Rowman & Littlefield, Lanham, Maryland.

Sternberg, M., 2000. 'Données sur les produits fabriqués dans une officine de Neapolis (Nabeul, Tunisie),' *Mélanges de l'École Française de Rome - Antiquité* 112.135–153.

Sternberg M., 2006 'Les poissons', in: M. Bats (dir.), *Olbia de Provence à l'époque romaine*, Études massaliètes, 9, Centre Camille-Jullian, Éd. Édisud, p. 432–449.

Sternberg, M., 2008 'La pêche dans l'économie des sociétés du Bronze final au iiie siècle après J.-C. de la Méditerranée occidentale: apports de l'archéo-ichtyologie Archéologie. In *Archeologie du poisson 30 Ans d'archéo ichtyologie au CNRS*, XXVIII e rencontres internationales d'archéologie et d'histoire d'Antibes, 14th ICAZ Fish remains working group meeting Sous la direction de P. Béarez, S. Grouard et B. Clavel Éditions APDCA, Antibes, 369–377

Studer, J., 1994. 'Roman fish sauce in Petra, Jordan,' in Van Neer, W. (ed.), *Fish Exploitation in the Past. Proceedings of the 7th meeting of the ICAZ fish remains working group. Annales du Musée Royal de L'Afrique Centrale, Sciences Zoologiques* 274, Tervuren.

Swanson., D., Block, R., Mousa, S.A., 2012. 'Omega-3 Fatty Acids EPA and DHA: Health Benefits Throughout Life,' *Advances in Nutrition, 3.1*, 1–7.

Thee, F.C., 1984. *Julius Africanus and the Early Christian View of Magic.* Hermeneutische Untersuchungen Zur Theologie (Vol. 19). Mohr Siebeck Verlag, Heidelberg, Tübingen.

Theodoropoulou, T., 2018. 'To Salt or Not to Salt: A Review of Evidence for Processed Marine Products and Local Traditions in the Aegean Through Time,' *Journal of Maritime Archaeology*, 13(3). 389–406.

Teodor, E.D., Badea, G.I., Alecu, A. et al., 2014. 'Interdisciplinary study on pottery experimentally impregnated with wine,' *Chemical Papers* 68: 1022. https://doi.org/10.2478/s11696-014-0559-1 accessed 09.11.2019. De Gruyter, Berlin, Boston.

Totolin, L., 2017. 'The Third Way: Galen, Pseudo-Galen, Metrodora, Cleopatra and the Gynaecological Pharmacology of Byzantium,' in Lehmhaus, L. and Martelli, M. (eds.), *Collecting Recipes: Byzantine and Jewish Pharmacology in Dialogue. Science, Technology and Medicine in Ancient Cultures* 4, p. 103. De Gruyter, Berlin, Boston.

Trakadas, A., 2005. 'The Archaeological Evidence for Fish Processing in the Western Mediterranean,' in Bekker-Nielsen, T. (ed.), *Ancient Fishing and Fish Processing in the Black Sea Region*, pp. 47–82. Aarhus: Aarhus University Press.

Trakadas, A., 2006. '"Exhausted by Fishermen's Nets;" Roman Sea Fisheries and their Management,' *Journal of Mediterranean Studies* 16(1), 259–72.

Valdes, M.G., *Dioscórides: Plantas y Remedios Medicinales: De materia medica Libros 1-111.* Editorial Gredos, Madrid.

Vanderhoeven, A. et al. 2001. Interdisziplinäre Untersuchungen in dem römischen Vicus von Tienen (Belgien). Die Integration von ökologischen und archäologischen Daten, in M. Frey and N. Hanel (edd.) *Archäologie, Naturwissenschaften, Umwelt* (BAR S929; Oxford) 13–31.

Van Neer, W. and A. Lentacker 1994. New archaeozoological evidence for the consumption of locally-produced fish sauce in the northern provinces of the Roman empire, *Archaeofauna* 3, 53–62.

Van Neer, W., and Ervynck. A., 2002. 'Remains of traded fish in archaeological sites: indicators of status or bulk food?' in Jones O'Day, S., Van Neer, W., Ervynck, A. (eds.), *Behaviour Behind Bones. Proceedings of the 9*th *ICAZ Conference, Durham 2002*, pp. 203–214. Oxbow, Oxford.

Van Neer, W., Lernau, O., Friedman, R., Mumford, G., Poblóme, J., Waelken, M., 2004. 'Fish remains from archaeological sites as indicators of former trade connections in the Eastern Mediterranean,' *Paléorient*, 30.1, 101–147.

Van Neer, W., Wouters, W., Ervynck, A. and Maes, J., 2005. 'New evidence from a Roman context in Belgium for fish sauce locally produced in northern Gaul,' *Archaeofauna* 14, 171–82.

Van Neer, W. and Parker, S.T., 2008. 'First archaeozoological evidence for haimation, the 'invisible' garum,' *Journal of Archaeological Science* 35(7). 1821–27.

Van Neer, W., Ervynck, A. and Monsieur, P., 2010. 'Fish bones and amphorae: evidence for the production and consumption of salted fish products outside the Mediterranean region,' *Journal of Roman Archaeology* 23. 161–95.

Van Neer, W., Wouters, W., Dolors Codina, R., Fournet, J.L., Preiss S., 2015 'Découverte de deux salaisons de poissons à Oxyrhynchus, el-Bahnasa, Égypte' In: Castellano N., Mascort M., Piedrafita C., Vivó J. (eds.) *Ex Aegypto lux et sapientia. Nova Studia Aegyptiaca* 9: 567–578.

Vargas, E.G. and Casasola, D.B., 2009. Roma y la producción de "garvum" y "sal-samenta" en la costa meridional de "Hispania": estado actual de la investigación. In *Arqueología de la pesca en el Estrecho de Gibraltar: de la Prehistoria al fin del Mundo Antiguo* (pp. 133–182). Servicio de Publicaciones

Vargas, E.G., Casasola, D.B., Macías, V.P., Gómez, A.M.R., Alcántara, A.R. and García, J.S., 2014. 'Confectio gari Pompeiani. Experimental procedure for the preparation of Roman fish sauces,' *SPAL* (Revista de Prehistoria y Arqueología de la Universidad de Sevilla) 23. 65–82. Seville.

Vassiliou, A., 2016. 'Middle Byzantine chafing dishes from Argolis,' *Deltion of the Christian Archaeological Society* 37. 251–276.

Vedat, Onar (forthcoming) Animals in food consumption during the Byzantine period in light of the Yenikapı metro and Marmaray excavations, Istanbul, *in* S.Y. Waksman (ed.), *Multidisciplinary approaches to food and foodways in the medieval Eastern Mediterranean*, Maison de l'Orient et de la Méditerranée, Lyon.

Vicente, M., Reyes, A.F. and De, A., 2009. 'La pesca y las conservas en la Bahía de Cádiz en época fenicio-púnica,'in Bernal Casasola, D. (ed.) *Arqueología de la pesca en el Estrecho de Gibraltar. De la Prehistoria al fin del Mundo Antiguo*, pp. 81–132. Monografías del Proyecto Sagea 1, Cádiz.

Von Den Driesch, A., 1980. 'Osteoarchäologische Auswertung von Garum-Resten des Cerro del Mar,' *Madrider Mitteilungun* 21. 151–4.

Waines, D., 1991. 'Murri. The tale of a condiment,' in *Al-Qantara. Revista de Estudios Árabes* 12. 371–388.

Wang, X. Q., Terry, P. D., & Yan, H., 2009. 'Review of salt consumption and stomach cancer risk: epidemiological and biological evidence,' *World Journal of Gastroenterology* 15(18). 2204–13.

Weingarten, S., 2005. 'Mouldy Bread and Rotten Fish: Delicacies in the Ancient World,' *Food and History* 3 (2005) 61–72.

Weingarten, S., 2018. 'Fish and Fish Products in Late Antique Palestine and Babylonia in Their Social and Geographical Contexts: Archaeology and the Talmudic Literature,' *Journal of Maritime Archaeology* 13(3). 235–45.

Wilkins, J., 2000. *The Boastful Chef. The Discourse of Food in Ancient Greek Comedy*. Oxford University Press.

Wilkins, J., 2005. 'Fish as a source of food in antiquity,' in Bekker-Nielsen, T. (ed.), *Ancient Fishing and Fish Processing in the Black Sea Region*. Aarhus University Press, Aarhus.

Wilkins, J., 2018. 'Cooking and Processing Fish in Antiquity: Questions of Taste and Texture,' *Journal of Maritime Archaeology* 13(3). 225–34.

Wilson, 2007. 'Quantification of fish-salting infrastructure capacity in the Roman world,' http://oxrep.classics.ox.ac.uk/working%20papers/quantification_fishsalting_infrastructure_capacity_roman_world, accessed 09.11.2019.

Wheeler, A and Jones, A.K.G., 1989 *Fishes, Cambridges*

Woolf, G., 2015. Ancient illiteracy? *Bulletin of the Institute of Classical Studies*, 58(2), pp.31–42.

Woodman, J. E., 1964. *The Expositio totius mundi et gentium: Its Geography and Its Language*. MA Thesis, The Ohio State University.

Woys-Weaver W., *forthcoming Food and Drink in Medieval Cyprus: The French Court, The Greek Gentry, and the Village Serfs*, (being considered by) Ludwig Reichert Verlag, Wiesbaden.

Yizhou Fang, Saiqi Gu, Shulai Liu, Jianyou Zhang, Yuting Dinga and Jianhua Liu, 2018. 'Extraction of oil from high-moisture tuna liver by subcritical dimethyl ether: feasibility and optimization by the response surface method,' *Royal Society Chemistry Advances*, 8(5). 2723–32.

Yule, B., 2005. *A prestigious Roman building complex on the Southwark waterfront: excavations at Winchester Palace, London, 1983-90*. Museum of London Archaeology Service, Monograph 23.

Zimmerman Munn, M.-L., 2003. 'Corinthian trade with the Punic West in the Classical period,' in Bookidis, N. (ed.), *Corinth. Results of Excavations Conducted by the American School of Classical Studies at Athens*, pp. 195–217. Princeton University Press, Princeton, NJ.

Index